UCLA Symposia on Molecular and Cellular Biology, New Series

Series Editor, C. Fred Fox

RECENT TITLES

Volume 74
Growth Regulation of Cancer, Marc E. Lippman, *Editor*

Volume 75
Steroid Hormone Action, Gordon Ringold, *Editor*

Volume 76
Molecular Biology of Intracellular Protein Sorting and Organelle Assembly, Ralph A. Bradshaw, Lee McAlister-Henn, and Michael G. Douglas, *Editors*

Volume 77
Signal Transduction in Cytoplasmic Organization and Cell Motility, Peter Satir, John S. Condeelis, and Elias Lazarides, *Editors*

Volume 78
Tumor Progression and Metastasis, Garth L. Nicolson and Isaiah J. Fidler, *Editors*

Volume 79
Altered Glycosylation in Tumor Cells, Christopher L. Reading, Senitiroh Hakomori, and Donald M. Marcus, *Editors*

Volume 80
Protein Recognition of Immobilized Ligands, T.W. Hutchens, *Editor*

Volume 81
Biological and Molecular Aspects of Atrial Factors, Philip Needleman, *Editor*

Volume 82
Oxy-Radicals in Molecular Biology and Pathology, Peter A. Cerutti, Irwin Fridovich, and Joe M. McCord, *Editors*

Volume 83
Mechanisms and Consequences of DNA Damage Processing, Errol C. Friedberg and Philip C. Hanawalt, *Editors*

Volume 84
Technological Advances in Vaccine Development, Laurence Lasky, *Editor*

Volume 85
B Cell Development, Owen N. Witte, Norman R. Klinman, and Maureen C. Howard, *Editors*

Volume 86
Synthetic Peptides: Approaches to Biological Problems, Emil Thomas Kaiser and James P. Tam, *Editors*

Volume 87
Gene Transfer and Gene Therapy, I. Verma, R. Mulligan, and A. Beaudet, *Editors*

Volume 88
Molecular Biology of the Eye: Genes, Vision, and Ocular Disease, Joram Piatigorsky, Toshimichi Shinohara, and Peggy S. Zelenka, *Editors*

Volume 89
Liposomes in the Therapy of Infectious Diseases and Cancer, Gabriel Lopez-Berestein and Isaiah J. Fidler, *Editors*

Volume 90
Cell Biology of Virus Entry, Replication, and Pathogenesis, Richard W. Compans, Ari Helenius, and Michael B.A. Oldstone, *Editors*

Volume 91
Bone Marrow Transplantation: Current Controversies, Richard E. Champlin and Robert Peter Gale, *Editors*

Volume 92
The Molecular Basis of Plant Development, R. Goldberg, *Editor*

Volume 93
Cellular and Molecular Biology of Muscle Development, Laurence H. Kedes and Frank E. Stockdale, *Editors*

Volume 94
Molecular Biology of RNA, Thomas R. Cech, *Editor*

Volume 95
DNA-Protein Interactions in Transcription, Jay D. Gralla, *Editor*

Volume 96
Stress-Induced Proteins, Mary Lou Pardue, James R. Feramisco, and Susan Lindquist, *Editors*

Volume 97
Molecular Biology of Stress, Shlomo Breznitz and Oren Zinder, *Editors*

Volume 98
Metal Ion Homeostasis: Molecular Biology and Chemistry, Dean H. Hamer and Dennis R. Winge, *Editors*

Volume 99
Human Tumor Antigens and Specific Tumor Therapy, Richard S. Metzgar and Malcolm S. Mitchell, *Editors*

Please contact the publisher for information about previous titles in this series.

UCLA Symposia Board

Metal Ion Homeostasis

Molecular Biology and Chemistry

Metal Ion Homeostasis

Molecular Biology and Chemistry

Proceedings of a UCLA Colloquium
Held at Frisco, Colorado
April 10–16, 1988

Editors

Dean H. Hamer
Laboratory of Biochemistry
National Cancer Institute
National Institutes of Health
Bethesda, Maryland

Dennis R. Winge
Biochemistry & Medicine
University of Utah Medical Center
Salt Lake City, Utah

Alan R. Liss, Inc. • New York

Address all Inquiries to the Publisher
Alan R. Liss, Inc., 41 East 11th Street, New York, NY 10003

Printed in the United States of America

Library of Congress Cataloging-in-Publication Data

Metal ion homeostasis : molecular biology and chemistry : proceedings of a UCLA Symposium held at Frisco, Colorado, April 10–16, 1988 / editors, Dean H. Hamer, Dennis R. Winge.

p. cm.—(UCLA Symposia on Molecular and Cellular Biology ; new sero., v. 98)

Includes index.

1. Metal ions—Metabolism—Congresses. 2. Metal ions—Metabolic detoxication—Congresses. 3. Metalloproteins—Congresses. 4. Molecular genetics—Congresses. 5. Homeostasis—Congresses. I. Hamer, Dean H. II. Winge, Dennis R. III. Series.

QP532.M46 1989

574.19′214—dc19 88-29560

ISBN 0-8451-2697-0 CIP

Contents

IV. MECHANISMS OF METAL DETOXIFICATION

V. PHYSIOLOGY AND DISEASES OF UNBALANCED METAL ION HOMEOSTASIS

Contributors

L.G. Albrigo, Citrus Research and Education Center, University of Florida, Lake Alfred, FL 33850 **[385]**

H. Appel, Institut für Experimentelle Kernphysik, Universität Karlsruhe, D-7500 Karlsruhe 1, Federal Republic of Germany **[179]**

Koji Arizono, Department of Biochemistry, School of Medicine, University of South Dakota, Vermillion, SD 57069; present address: Faculty of Pharmaceutical Sciences, Nagasaki University, Nagasaki 852, Japan **[81]**

Georg Auling, Institute of Microbiology, University of Hannover, D-3000 Hannover 1, Schneiderberg 50, Federal Republic of Germany **[459]**

Ruud Bank, Institute of Human Genetics, Medical Faculty, Free University, 1007 MC Amsterdam, The Netherlands **[347]**

Mariano Beltramini, Biochemisches Institut der Universität Zürich, CH-8057 Zürich, Switzerland; present address: Department of Biology, University of Padova, Italy **[227]**

Lawrence W. Bergman, Department of Chemistry, Clippinger Laboratories, Ohio University, Athens, OH 45701 **[101]**

D. Bielser, Department of Biology, Washington University, St. Louis, MO 63130 **[117]**

G. Boulay, Department of Microbiology, Faculty of Medicine, Centre Hospitalier Universitaire, Sherbrooke, Quebec, Canada J1H 5N4 **[31]**

Frank O. Brady, Department of Biochemistry, School of Medicine, University of South Dakota, Vermillion, SD 57069 **[81]**

J.F. Briat, Laboratoire de Biologie Moléculaire Végétale, CNRS UA 1178, Université Joseph Fourier, F-38041 Grenoble Cédex, France **[137]**

Marius Brouwer, Duke University Marine Laboratory, Beaufort, NC 28516 **[247]**

Thea Brouwer-Hoexum, Duke University Marine Laboratory, Beaufort, NC 28516 **[247]**

P.H. Brown, Department of Biology, Washington University, St. Louis, MO 63130 **[117]**

Stephen A. Brown, Department of Biochemistry, University of Vermont College of Medicine, Burlington, VT 05405 **[169]**

The numbers in brackets are the opening page numbers of the contributors' articles.

Angès Cadic-Jacquier, Laboratoire d'Embryologie Moléculaire et Expérimentale, Université Paris XI, 91405 Orsay Cedex, France **[91]**

J. Camakaris, Department of Genetics, University of Melbourne, Parkville, Victoria 3052, Australia **[439,469]**

O. Capasso, Instituto Nazionale della Nutrizione, Rome, Italy 00179 **[71]**

John L. Casey, Branch of Cell Biology and Metabolism, National Institute of Child Health, Rockville, MD 20892 **[109]**

S. Wright Caughman, Branch of Cell Biology and Metabolism, National Institute of Child Health, Rockville, MD 20892 **[109]**

C.D. Chase, Department of Vegetable Crops, University of Florida, Gainesville, FL 32611 **[385]**

N.D. Chasteen, Department of Chemistry, University of New Hampshire, Durham, NH 03824 **[161]**

Shiow-meei Chen, Department of Chemistry, University of Wisconsin-Milwaukee, Milwaukee, WI 53201 **[197]**

Mark Chernaik, Department of Biochemistry, The Johns Hopkins University, Baltimore, MD 21205 **[237]**

Lien Chu, Department of Microbiology and Immunology, University of Illinois, Chicago, IL 60680 **[417]**

Mary Cismowski, Department of Biochemistry, The Johns Hopkins University, Baltimore, MD 21205 **[237]**

Joseph E. Coleman, Department of Molecular Biophysics and Biochemistry, Yale University School of Medicine, New Haven, CT 06510 **[3]**

R.N. Cooley, Department of Genetics, University of Liverpool, Liverpool L69 3BX, England; present address: Department of Genetics, University of Birmingham, Birmingham B15 2TT, England **[375]**

J.B. Cornelius, Biotechnology Resource in Pulsed EPR Spectroscopy, Albert Einstein College of Medicine, Bronx, NY 10461 **[161]**

C.T. Dameron, Departments of Medicine and Biochemistry, University of Utah Medical Center, Salt Lake City, UT 84132 **[301]**

S. Daniels-McQueen, Department of Biology, Washington University, St. Louis, MO 63130 **[117]**

Emmanuel Delhaize, Genetics Group, Life Sciences Division, Los Alamos National Laboratory, Los Alamos, NM 87545 **[337]**

Peter B. Dervan, Division of Chemistry and Chemical Engineering, California Institute of Technology, Pasadena, CA 91125 **[21]**

M. Ebadi, Department Pharmacology, University of Nebraska College of Medicine, Omaha, NE 68105 **[257]**

Wilfried H.O. Ernst, Department of Ecology and Ecotoxicology, Faculty of Biology, Free University, 1007 MC Amsterdam, The Netherlands **[347]**

Nour-Eddine Erraïss, Laboratoire d'Embryologie Moléculaire et Expérimentale Université Paris XI, 91405 Orsay Cedex, France **[91]**

M. Ettinger, Department of Biochemistry, SUNY/Buffalo, Buffalo, NY 14214 **[449]**

James A. Fee, Isotope and Structural Chemistry Group, INC-4 C345, Los Alamos National Laboratory, Los Alamos, NM 87545 **[149]**

Betsy Frantz, Department of Chemistry, Northwestern University, Evanston, IL 60208 **[407]**

Jonathan H. Freedman, Institute for Structural and Functional Studies, University City Science Center, Philadelphia, PA; and Department of Molecular Pharmacology, Albert Einstein College of Medicine, Bronx, NY 10461 **[269]**

L. Gaffield, Department of Microbiology and Immunology, College of Medicine, The University of Illinois at Chicago, Chicago, IL 60612 **[117]**

Scott H. Garrett, Department of Biochemistry, School of Medicine, University of South Dakota, Vermillion, SD 57069 **[81]**

Lashitew Gedamu, Department of Biological Sciences, University of Calgary, Calgary, Alberta, Canada T2N 1N4 **[49]**

Ursula A. Germann, Biochemisches Institut der Universität Zürich, CH-8057 Zürich, Switzerland; present address: Laboratory of Tumor Virus Biology, NCI, NIH, Bethesda, MD 20892 **[227]**

David P. Giedroc, Department of Molecular Biophysics and Biochemistry, Yale University School of Medicine, New Haven, CT 06510 **[3]**

David J. Gingrich, The Department of Chemistry and The Aquatic Biomedical Research Center, The University of Wisconsin-Milwaukee, Milwaukee, WI 53201 **[313]**

John H. Griffin, Division of Chemistry and Chemical Engineering, California Institute of Technology, Pasadena, CA 91125 **[21]**

Erwin Grill, Institut für Pharmazeutische Biologie, Universität München, Federal Republic of Germany **[283]**

Dean H. Hamer, Laboratory of Biochemistry, National Cancer Institute, National Institutes of Health, Bethesda, MD 20892 **[xix,13]**

Joanna L. Hansen, Department of Medicine, University of Utah, Salt Lake City, UT 84132 **[479]**

Joe B. Harford, Branch of Cell Biology and Metabolism, National Institute of Child Health, Rockville, MD 20892 **[109]**

M. Hassan, Department of Microbiology, Faculty of Medicine Centre Hospitalier Universitaire, Sherbrooke, Quebec, Canada J1H 5N4 **[31]**

John D. Helmann, Department of Biological Chemistry and Molecular Pharmacology, Harvard Medical School, Boston, MA 02115 **[397]**

Andreas Heltzel, Department of Microbiology, University of Georgia, Athens, Georgia 30602 **[427]**

Matthias W. Hentze, Branch of Cell Biology and Metabolism, National Institute of Child Health, Rockville, MD 20892 **[109]**

R.C. Hider, Department of Pharmacy, King's College, University of London, London, England SW3 6LX **[189]**

K.E. Hodges, Department of Genetics, University of Liverpool, Liverpool L69 3BX, England **[375]**

Stella T. Hu, Laboratory of Biochemistry, National Cancer Institute, National Institutes of Health, Bethesda, MD 20892 **[13]**

P.C. Huang, Department of Bochemistry, The Johns Hopkins University, Baltimore, MD 21205 **[237]**

A.K. Hughes, Departments of Medicine and Biochemistry, University of Utah Medical Center, Salt Lake City, UT 84132 **[301]**

P.L. Iversen, Department of Pharmacology, University of Nebraska College of Medicine, Omaha, NE 68105 **[257]**

Paul J. Jackson, Genetics Group, Life Sciences Division, Los Alamos National Laboratory, Los Alamos, NM 87545 **[337]**

Nadia Jahroudi, Department of Biological Sciences, University of Calgary, Calgary, Alberta, Canada T2N 1N4 **[49]**

Richard D. Klausner, Branch of Cell Biology and Metabolism, National Institute of Child Health, Rockville, MD 20892 **[109]**

David M. Koeller, Branch of Cell Biology and Metabolism, National Institute of Child Health, Rockville, MD 20892 **[109]**

Paul Koevoets, Department of Ecology and Ecotoxicology, Faculty of Biology, Free University, 1007 MC Amsterdam, The Netherlands **[347]**

Rachel M. Krotz, Plant Physiology/Biochemistry/Molecular Biology Program, Department of Agronomy, University of Kentucky, Lexington, KY 40546-0091 **[325]**

Urs Kupper, Biochemisches Institut der Universität Zürich, CH-8057 Zürich, Switzerland **[227]**

James P. Kushner, Department of Medicine, University of Utah, Salt Lake City, UT 84132 **[479]**

A.M. Labouré, Laboratoire de Biologie Moléculaire Végétale, CNRS UA 1178, Université Joseph Fourier, F-38041 Grenoble Cédex, France **[137]**

J.P. Laulhère, Laboratoire de Biologie Moléculaire Végétale, CNRS UA 1178, Université Joseph Fourier, F-38041 Grenoble Cédex, France **[137]**

B.T.O. Lee, Department of Genetics, University of Melbourne, Parkville, Victoria 3052, Australia **[439,469]**

A. Leone, Instituto di Biochimica Cellulare e Molecolare, University of Naples, Naples, Italy 80131 **[59,71]**

Konrad Lerch, Biochemisches Institut der Universität Zürich, CH-8057 Zürich, Switzerland **[227]**

A.M. Lescure, Laboratoire de Biologie Moléculaire Végétale, CNRS UA 1178, Université Joseph Fourier, F-38041 Grenoble Cédex, France **[137]**

Xiucai Liu, Department of Chemistry, University of Wisconsin-Milwaukee, Milwaukee, WI 53201 **[197]**

Anne B. Mason, Department of Biochemistry, University of Vermont College of Medicine, Burlington, VT 05405 **[169]**

J. McCracken, Biotechnology Resource in Pulsed EPR Spectroscopy, Albert Einstein College of Medicine, Bronx, NY 10461 **[161]**

R.K. Mehra, Departments of Medicine and Biochemistry, University of Utah Medical Center, Salt Lake City, UT 84132 **[301]**

M.C. Mellone, Instituto Scienze Biochimiche, University of Naples, Naples, Italy 80131 **[71]**

Tapan K. Misra, Department of Microbiology and Immunology, University of Illinois, Chicago, IL 60680 **[417]**

Raja Mokdad-Gargouri, Laboratoire d'Embryologie Moléculaire et Expérimentale, Université Paris XI, 91405 Orsay Cedex, France **[91]**

G. Morelli, Instituto Nazionale della Nutrizione, Rome, Italy 00179 **[71]**

Karl Münger, Biochemisches Institut der Universität Zürich, CH-8057 Zürich, Switzerland; present address: Laboratory of Molecular Biology, NCI, NIH, Bethesda, MD 20892 **[227]**

Cleo M. Naranjo, Isotope and Structural Chemistry Group, INC-4 C345, Los Alamos National Laboratory, Los Alamos, NM 87545 **[149]**

J.B. Neilands, Biochemistry Department, University of California, Berkeley, CA 94720 **[189]**

M. Neu-Müller, Institut für Experimentelle Kernphysik, Universität Karlsruhe, D-7500 Karlsruhe 1, Federal Republic of Germany **[179]**

Eric C. Niederhoffer, Isotope and Structural Chemistry Group, INC-4 C345, Los Alamos National Laboratory, Los Alamos, NM 87545 **[149]**

Giuseppina Nucifora, Department of Microbiology and Immunology, University of Illinois, Chicago, IL 60680 **[417]**

Thomas V. O'Halloran, Department of Chemistry, Northwestern University, Evanston, IL 60208 **[407]**

James Otvos, Department of Chemistry, University of Wisconsin-Milwaukee, Milwaukee, WI 53201 **[197]**

F. Palida, Department of Biochemistry, SUNY/Buffalo, Buffalo, NY 14214 **[449]**

V.K. Paliwal, Department of Pharmacology, University of Nebraska College of Medicine, Omaha, NE 68105 **[257]**

Soon-Jung Park, Department of Microbiology, University of Georgia, Athens, GA 30602 **[397]**

M.C. Pascale, Instituto di Biochimica Cellulare e Molecolare, University of Naples, Naples, Italy 80131 **[59]**

J. Peisach, Departments of Molecular Pharmacology, Physiology, and Biophysics, Albert Einstein College of Medicine, Bronx, NY 10461 **[269]**

David H. Petering, The Department of Chemistry and The Aquatic Biomedical Research Center, The University of Wisconsin-Milwaukee, Milwaukee, WI 53201 **[313]**

L. Powers, AT&T Bell Laboratories, Murray Hill, NJ 07974-2070 **[269]**

D. Proudhon, Laboratoire de Biologie Moléculaire Végétale, CNRS UA 1178, Université Joseph Fourier, F-38041 Grenoble Cédex, France **[137]**

Diana M. Ralston, Department of Biochemistry, Molecular and Cellular Biology, Northwestern University, Evanston, IL 60208 **[407]**

R.N. Reese, Departments of Medicine and Biochemistry, University of Utah Medical Center, Salt Lake City, UT 84132 **[301]**

P. Remondelli, Instituto di Biochimica Cellulare e Molecolare, University of Naples, Naples, Italy 80131 **[59]**

In-Koo Rhee, Department of Biochemistry, The Johns Hopkins University, Baltimore, MD 21205; present address: Department of Agricultural Chemistry, Kyungpook National University, Taegu, Korea **[237]**

Nigel J. Robinson, Genetics Group, Life Sciences Division, Los Alamos National Laboratory, Los Alamos, NM 87545; present address: Department of Botany, University of Durham Science Laboratory, Durham, DH1 3LE, England **[337]**

Wilma Ross, Department of Microbiology, University of Georgia, Athens, GA 30602 **[397]**

Tracey A. Rouault, Branch of Cell Biology and Metabolism, National Institute of Child Health, Rockville, MD 20892 **[109]**

D. Rouch, Department of Genetics, University of Melbourne, Parkville, Victoria 3052, Australia **[439,469]**

David E. Salt, Department of Botany, University of Liverpool, Liverpool, L69 3BX, England **[367]**

M.L. Scarino, Instituto Nazionale della Nutrizione, Rome, Italy 00179 **[71]**

F. Schwab, Institut für Experimentelle Kernphysik, Universität Karlsruhe, D-7500 Karlsruhe 1, Federal Republic of Germany **[179]**

Steven R. Sczekan, Department of Biochemistry, North Carolina State University, Raleigh, NC 27695-7622 **[127]**

C. Frank Shaw III, The Department of Chemistry and The Aquatic Biomedical Research Center, The University of Wisconsin-Milwaukee, Milwaukee, WI 53201 **[313]**

Lisa M. Shewchuk, Department of Biological Chemistry and Molecular Pharmacology, Harvard Medical School, Boston, MA 02115 **[397]**

Myung Shin, Department of Chemistry, Northwestern University, Evanston, IL 60208 **[407]**

Philippe Silar, Laboratoire d'Embryologie Moléculaire et Expérimentale, Université Paris XI, 91405 Orsay Cedex, France **[91]**

Simon Silver, Department of Microbiology and Immunology, University of Illinois, Chicago, IL 60680 **[417]**

P.A. Snetsinger, Department of Chemistry, University of New Hampshire, Durham, NH 03824 **[161]**

John C. Steffens, Department of Plant Breeding, Cornell University, Ithaca, NY 14853 **[359]**

Anne O. Summers, Department of Microbiology, University of Georgia, Athens, GA 30602 **[397,427]**

S.K. Swope, Department of Chemistry, University of New Hampshire, Durham, NH 03824 **[161]**

T. Takahashi, Department of Pharmacology, University of Nebraska College of Medicine, Omaha, NE 68105 **[257]**

Y.-C. Tam, Department of Microbiology, Faculty of Medicine Centre Hospitalier Universitaire, Sherbrooke, Quebec, Canada J1H 5N4 **[31]**

E.B. Tarbet, Departments of Medicine and Biochemistry, University of Utah Medical Center, Salt Lake City, UT 84132 **[301]**

D.M. Taylor, Institut für Genetik und für Toxikologie von Spaltstoffen, Kernforschungszentrum Karlsruhe, D-7500 Karlsruhe 1, Federal Republic of Germany **[179]**

K.C. Taylor, Department of Vegetable Crops, University of Florida, Gainesville, FL 32611 **[385]**

R.E. Thach, Department of Biology, Washington University, St. Louis, MO 63130 **[117]**

Elizabeth C. Theil, Department of Biochemistry, North Carolina State University, Raleigh, NC 27695-7622 **[127]**

W.-G. Thies, Institut für Genetik und für Toxikologie von Spaltstoffen, Kernforschungszentrum Karlsruhe, D-7500 Karlsruhe 1, Federal Republic of Germany **[179]**

J.-P. Thirion, Department of Microbiology, Faculty of Medicine, Centre Hospitalier Universitaire, Sherbrooke, Quebec, Canada J1H 5N4 **[31]**

David A. Thurman, Department of Botany, University of Liverpool, Liverpool, L69 3BX, England **[367,375]**

Cynthia R. Timblin, Department of Chemistry, Clippinger Laboratories, Ohio University, Athens, Ohio 45701; present address: NCI-Naval Medicine Oncology, National Medical Center, Bethesda, MD 20814 **[101]**

A. Brian Tomsett, Department of Genetics, University of Liverpool, Liverpool, L69 3BX, England **[367,375]**

Paul Totis, Department of Microbiology, University of Georgia, Athens, Georgia 30602 **[427]**

Jan van 't Riet, Biochemical Laboratory, Faculty of Chemistry, Free University, 1007 MC Amsterdam, The Netherlands **[347]**

Mary C. van Rossenberg, Biochemical Laboratory, Faculty of Chemistry, Free University, 1007 MC Amsterdam, The Netherlands **[347]**

H. van Willigen, Department of Chemistry, University of Massachusetts at Boston, Boston, MA 02125 **[161]**

Milan Vašák, Biochemisches Institut der Universität Zürich, Zürich, Switzerland **[207]**

Jos A.C. Verkleij, Department of Ecology and Ecotoxicology, Faculty of Biology, Free University, 1007 MC Amsterdam, The Netherlands **[347]**

George J. Wagner, Plant Physiology/Biochemistry/Molecular Biology Program Department of Agronomy, University of Kentucky, Lexington, KY 40546-0091 **[325]**

W.E. Walden, Department of Microbiology and Immunology, College of Medicine, The University of Illinois at Chicago, Chicago, IL 60612 **[117]**

G. Waldrop, Department of Biochemistry, SUNY/Buffalo, Buffalo, NY 14214 **[449]**

Christopher T. Walsh, Department of Biological Chemistry and Molecular Pharmacology, Harvard Medical School, Boston, MA 02115 **[397]**

Yuh-Hwa Wang, Department of Biochemistry, North Carolina State University, Raleigh, NC 27695-7622 **[127]**

Daniel N. Weber, The Department of Chemistry and The Aquatic Biomedical Research Center, The University of Wisconsin-Milwaukee, Milwaukee, WI 53201 **[313]**

H.W. Weber, Institut für Experimentelle Kernphysik, Universität Karlsruhe, D-7500 Karlsruhe 1, Federal Republic of Germany **[179]**

Maurice Wegnez, Laboratoire d'Embryologie Moléculaire et Expérimentale, Université Paris XI, 91405 Orsay Cedex, France **[91]**

Bill G. Williams, BioVision, Inc., San Jose, CA **[359]**

R.J.P. Williams, Inorganic Chemistry Laboratory, University of Oxford, Oxford OX1 3OR **[189]**

Dennis R. Winge, Departments of Medicine and Biochemistry, University of Utah Medical Center, Salt Lake City, UT 84132 **[xix,301]**

M.R. Wormald, Inorganic Chemistry Laboratory, University of Oxford, Oxford OX1 3OR **[189]**

Jeffrey G. Wright, Department of Chemistry, Northwestern University, Evanston, IL 60208 **[407]**

Preface

This is an exciting time for studies of the biology and chemistry of metal ions. In the last few years, it has been shown that zinc is an essential component of a new class of DNA binding "finger" proteins, which have been implicated in processes as diverse as the synthesis and storage of 5S RNA, mRNA transcription, the control of early development, and sex determination. Metals also have been found in a regulatory protein of the AIDS virus, HIV. Studies of iron, mercury, and cadmium metabolism have revealed a variety of strategies by which metals can activate the expression of specific genes. Thus, we are beginning to understand the role of metals in perhaps the most fundamental of all biological processes—the transmission of genetic information.

At the same time, research in the more traditional areas of bioinorganic chemistry has proceeded apace. Analytical and spectroscopic methods have become more and more powerful. The increasing library of known metalloprotein structures has made it possible to suggest plausible models for proteins that have not yet been crystallized. Processes such as metal transport and detoxification are finally being studied at the molecular level rather than at the purely phenomenological level.

Scientists who are studying metals in biology range from cloners to spectroscopists, from physiologists to chemists. The rationale behind the meeting that produced this volume was to bring together the diverse groups of scientists involved in metal ion research. The 1988 UCLA Colloquium on **Metal Ion Homeostasis: Molecular Biology and Chemistry**, held at Frisco, Colorado, April 10–16, 1988, succeeded in providing a forum for exchange. The papers in this volume derive from the many excellent talks and posters that were presented during the course of the five-day meeting. We hope that the meeting generated enough excitement to inspire similar meetings in the future.

We wish to thank the UCLA Symposia staff for their invaluable help in organizing the colloquium. We also gratefully acknowledge support from E.I. DuPont Medical Products, Smith Kline & French Laboratories, and Atom Sciences, Inc. Additional support was provided by contributions to the Director's Fund by Pfizer, Inc.; Amoco Technology Co., Biotechnology Division; Bio-Rad Laboratories; Bristol-Myers Company, Pharmaceutical Research and Development Division; Codon; DNAX Research Institute; Eastman Kodak Company; Immunex Corporation; Life Technologies, Inc.; Minnesota Mining & Manufacturing Co.; New England Biolabs; Sandoz Research Institute; Ciba-Geigy Corporation, Pharmaceuticals Division; Wyeth Laboratories; and Warner Lambert/Parke-Davis. The generous support of these groups helped make this scientific meeting possible.

Dean H. Hamer
Dennis R. Winge

I. METALS IN GENE REGULATION

Metal Ion Homeostasis: Molecular Biology and Chemistry, pages 3–12

DH Hamer & D R Winge, eds.

ZINC DOMAINS IN NONSPECIFIC NUCLEIC ACID BINDING PROTEINS[1]

David P. Giedroc and Joseph E. Coleman

Department of Molecular Biophysics & Biochemistry,
Yale University School of Medicine,
New Haven, CT 06510

ABSTRACT Gene 32 protein (g32P) the single-stranded (ss) DNA binding protein from bacteriophage T4 contains 1 g at Zn(II)/mol which can be exchanged with Cd(II). NMR of ^{113}Cd(II)-substituted g32P shows a single resonance 637 ppm downfield from aqueous ^{113}Cd(II) with no evidence for anion (solvent) access to the inner metal sphere; all data are consistent with tetrahedral NS_3 coordination with Cys^{77}, His^{81}, Cys^{87} and Cys^{90} as providing metal ligand donor atoms. Noncooperatively binding DNA ligand $[d(pA)_6]$ causes a small two ppm downfield change in the ^{113}Cd(II) g32P chemical shift ruling out metal site expansion, ligand exchange or large conformational changes transmitted through the metal ion chelate as a result of DNA binding. A model of g32P bound to ssDNA is presented which emphasizes the role of the metal ion in DNA binding and synthesizes all physicochemical data accumulated to date. 14-amino acid Cys_3/His domains quite similar to the metal binding site in g32P are found in the C-terminal processed product of retroviral gag-gene polyproteins as well as the gag-like gene from a subset of retrotransposable elements. This protein is thought to function in nonspecific RNA binding in packing the virion RNA into the mature nucleocapsid. To address whether such domains specifiy zinc binding

[1]This work was supported by NIH Grant (DK09070) to J.E.C., instrumentation support from NIH (RR03475) and NSF (DMB-8610557) and an NIH postdoctoral fellowship (F32-GM10972) to D.P.G.

sites, we have constructed a plasmid which directs the expression of the p15(p7/p9)^{gag} from HIV-1 under control of the lambda phage P_L promoter in E. coli; preliminary purification and characterization experiments are discussed.

INTRODUCTION

Gene 32 protein (g32P) from bacteriophage T4 plays an essential role in T4 replication, recombination and repair by binding stoichiometrically and cooperatively to single-stranded (ss) DNA transiently formed during any one of these processes (1). G32P binds to ssDNA relatively nonspecifically (i.e. without sequence specificity) (1). The protein contains an intrinsic Zn(II) ion within the DNA binding core fragment bound in a rather regular tetrahedral coordination geometry to three Cys side chains (2). The bound metal ion stabilizes the folded form of the protein and directly enhances cooperative binding to an "infinite" length ssDNA lattice with little effect on noncooperative binding to a single binding site oligodeoxyribonucleotide (3-4). In this report, we present ongoing experiments which further define the structural and functional characteristics of the metal ion site through the use of NMR spectroscopy of the ^{113}Cd(II)-substituted g32P.

Another member of this class of proteins which appear to bind relatively nonspecifically to ss nucleic acids includes the low molecular weight, basic nucleic acid binding protein (NABP) processed from retroviral gag gene polyprotein (5). These proteins contain one or two sequence repeats which conform to ..Cys-X_2-Cys-X_4-His-X_4-Cys.., a nearly perfect inversion of the g32P metal ligand donor set (6). Herein, we present experiments which will permit elucidation of the in vitro ss nucleic acid binding properties and specifically the role of the evolutionarily conserved Cys-rich domain in the NABP derived from HIV-1.

METHODS

Homogeneous g32P and the proteolytically-prepared DNA-binding core fragment [g32P core, g32P-(A+B)] were obtained as previously described (2). ^{113}Cd(II) substitution of the intrinsic Zn(II) in g32P was carried out either by addition of slightly greater than stoichiometric $^{113}CdCl_2$ (96% isotopically enriched) to the metal-free apo protein or via direct metal ion exchange of the Zn(II) protein. ^{113}Cd NMR spectra were collected on g32P samples (2 mL, 0.5-0.7 mM, buffered in 50 mM NaP_i, pH8, 30 mM NaCl) at 298 K on a Bruker AM500 spectrometer equipped with a broad-band tunable probe operating at an applied magnetic field strength of 11.5 Tesla (110.9 MHz for ^{113}Cd). Routine acquisition parameters were a pulse angle of 45°, a sweep width of 10-15,000 Hz and a recycle delay time of 1-1.5 sec.

RESULTS

^{113}Cd NMR of Gene 32 Protein.

The intrinsic Zn(II) of g32P can be exchanged with other divalent metal ions, including Co(II) and Cd(II), with retention of high affinity, cooperative binding to "infinite" length polynucleotides, stabilization against thermal denaturation especially when bound to polynucleotides, and protection from proteolysis of the DNA-binding core fragment (2-4). Thus, we are justified in investigating the spectroscopic properties of the 113Cd(II)-substituted g32P to probe the g32P Zn(II) site structure and function.

^{113}Cd chemical shifts derived from ^{113}Cd substitution of a number of Zn(II), Cu(II) and Ca(II) metalloproteins have proven to be qualitatively dependent on the number and type of ligand donor atoms be they oxygen, nitrogen, neutral sulfur or cysteine sulfur (8). Chemical shifts extend from -100 ppm [upfield from aqueous ^{113}Cd(II)] characterized by oxygen donor sets of Ca(II)-binding proteins to some +800 ppm [downfield from aqueous ^{113}Cd(II)] exhibited by tetrathiolate donor sets (8).

Fig. 1 shows the 110.9 MHz NMR spectrum of ^{113}Cd(II)-g32P core fragment at pH 8, 298 K. A single resonance of rather broad linewidth (110 ± 20 Hz) occurs positioned at 637 ppm downfield from aqueous ^{113}Cd(II).

Table I presents the chemical shifts for a number of sulfur-containing metalloenzyme sites substituted with ^{113}Cd(II). In non-bridged tetrahedral metal sites, a downfield change in the chemical shift accompanies increased cysteine sulfur ligation. As can be seen, the chemical shift of ^{113}Cd g32P is consistent with three

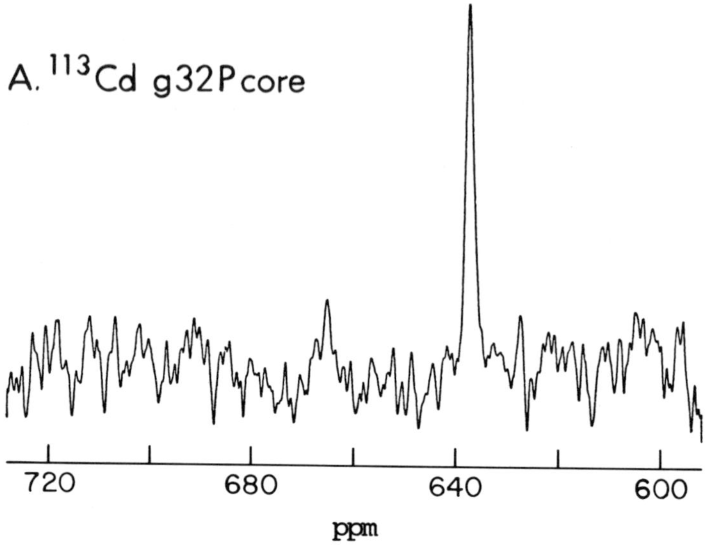

FIGURE 1. NMR spectrum of ^{113}Cd(II) g32P core.

cysteine sulfur atoms as ligands in addition to a fourth atom of lower deshielding potential, probably a histidine nitrogen atom but less likely, a side chain or peptide carbonyl oxygen atom. ^{35}Cl NMR and a Cl^- titration of the

TABLE 1
^{113}CD CHEMICAL SHIFTS OF SULFUR-CONTAINING METALLOPROTEIN SITES[a]

Protein	Ligand Donor Set	Chemical Shift (ppm)
Plastocyanin	SS^*N_2	432
Alcohol Dehydrogenase (ADH), catalytic sites	S_2NO	483
ADH, catalytic sites + imidazole	S_2N_2	519
T4 Gene 32 Protein	S_3N	637
ADH, structural sites	S_4	751

[a]S=cysteine sulfur, S*=methionine, N=imidazole, O=oxygen.

^{113}Cd g32P give no evidence for an open or solvent (anion) accessible coordination site; thus, the tetrahedral geometry is created by four protein ligands. As proposed, Cys^{77}-X_3-His^{81}-X_5-Cys^{87}-X_2-Cys^{90} remains a strong candidate for the zinc domain in g32P and is consistent with all the physicochemical data accumulated (2,7).

The ^{113}Cd chemical shift of g32P can be used to probe for any changes in coordination geometry or number as well as any global conformational changes transmitted through the metal domain as a result of ligand (DNA) binding. Fig. 2 compares the NMR spectra of unliganded g32P (resonance A) with that of a 1:1 complex between oligodeoxyribonucleotide d(pA)$_6$ and g32P core protein (resonance B). Note only a 2.5 ppm change in ^{113}Cd chemical shift upon complex formation. The intermediate resonance line C results from mixing 0.5 equivalents of oligonucleotide with 1 equivalent of protein. Given that the linewidths of all three species are approximately the same, the free and complexed forms of ^{113}Cd(II) g32P must be in fast exchange on the NMR time scale, in this case, the frequency difference expected for a minimum K_{off} of 220 sec^{-1} for d(pA)$_6$. The rather small change in chemical shift upon noncooperative DNA binding to d(pA)$_6$ effectively rules out expansion of the coordination sphere

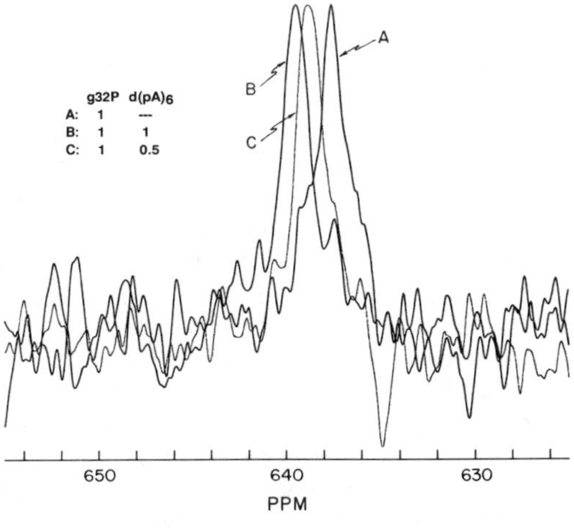

FIGURE 2. Effect of p(dA)$_6$ binding on the chemical shift of ^{113}Cd(II) g32P core protein at the indicated protein:nucleotide stoichiometry.

or ligand exchange as a result of DNA binding. Similarly, the electronic spectrum of Co(II) g32P shows little if any perturbation in the "d-d" transitions upon noncooperative binding to d(pT)$_8$ or metal-dependent cooperative binding to poly(dT) (D. Giedroc and J. Coleman, unpublished).

Are the NABPs Derived from the Retroviral gag Gene Zinc Metalloproteins?

The gag-gene of retroviruses and the gag-like genes of various retrotransposable elements are known or thought to encode the structural proteins of the nucleocapsid. Near the C-terminus of the retroviral gag-polyprotein is situated the NABP which by a variety of *in vivo* and *in vitro* criteria makes direct contact, in a relatively nonspecific fashion, with the daughter RNA.

Fig. 3 presents the primary structures of the NABPs from representative isolates of known retroviruses in the region characterized by one or two repeats of a 14-amino acid sequence element conforming to ..Cys-X$_2$-Cys-X$_4$-His-X$_4$-Cys... These are compared to similar sequences found in a select number of eukaryotic transposable elements which are thought to replicate via an RNA intermediate and form recognizable intracellular nucleocapsid structures (9). Finally, the proposed metal binding sequence from T4 g32P is shown. When the sequences in Fig. 3 are taken as a group, there is minimal absolute sequence conservation outside of the 3 Cys and 1 His residues. What is common, however, is a propensity for turn residues especially between the second Cys and unique His residues with deviation, in some instances, to β-sheet or α-helix prediction toward the N- and C-terminal ends of the elements. Such a secondary structure prediction would seem well suited to accommodate a "loop" structure about a central metal ion ligated to the conserved Cys and His side chains as proposed by Berg (10). In the case of g32P, the zinc binding domain has been modeled without violation of acceptable boundary conditions (i.e. bond angle, bond distance, van der Waals radii) by considering the three-dimensional constraints implicite in the primary structure of this sequence combined only with the additional constraints described by the proposed tetrahedral ligation of the Zn(II) ion (D. Giedroc and B. Johnson, unpublished).

To address whether the retroviral and like sequences

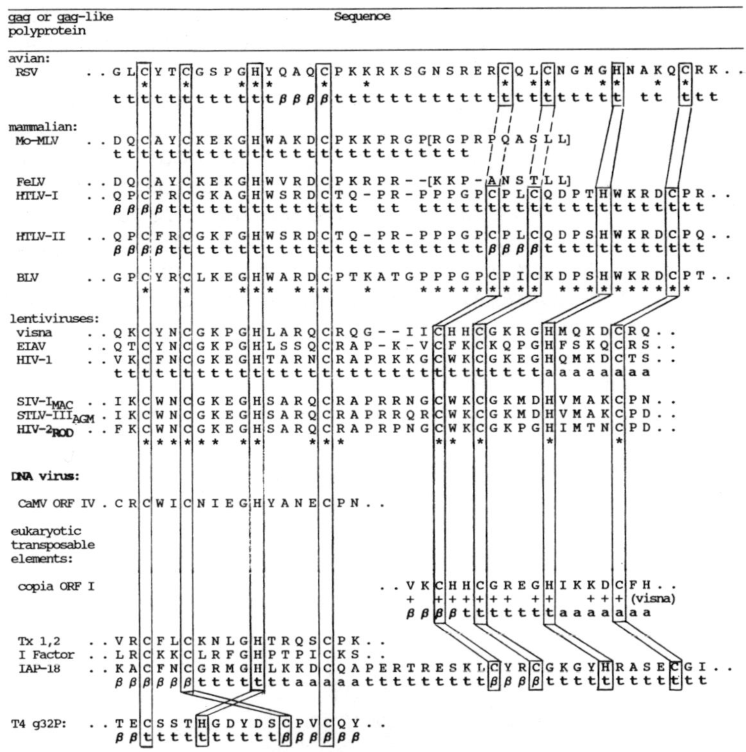

FIGURE 3. Primary and predicted secondary structures of ..Cys-X_2-Cys-X_4-His-X_4-Cys.. 14-amino acid (aa) sequence elements found in the gag-genes of retroviruses and the gag-like ORFs of various eukaryotic transposable or other putative RNA-based reverse-transcribing elements as compared to the Zn(II) binding domain of phage T4 g32P. Proceeding down the figure, the * compares the avian and mammalian aa sequences with one another, the mammalian sequences as a group, and finally the lentivirus sequences as a group and denotes identities and conservative substitutions in each comparison. The symbol + compares the copia ORF I and the visna sequences. a, β, and t indicate 2° structure prediction of α-helix, β-sheet and β-turn according to Garnier-Osguthorpe-Robson (window=5).

specify zinc binding domains, we have decided to clone, express, purify and characterize one or more of these proteins. Fig. 4 shows inducible expression of HIV-1 p15(p7/p9)^{gag} (11) in E. coli driven by λP_L promoter. Purification protocols have utilized large pore gel exclusion chromatography under denaturing conditions to free the protein preparation of contaminating nucleic acid, followed by refolding with or without zinc ions and chromatography on ssDNA-cellulose. Such manipulations, which are presently being optimized with regard to yield, etc., result in a p15 fraction of ~70% purity. In vitro oligo- and polynucleotide binding assays are underway to determine the quantitative features of ss nucleic acid binding by p15.

DISCUSSION

The role or potential role that closed, sulfur-rich zinc chelates play in mediating the association of

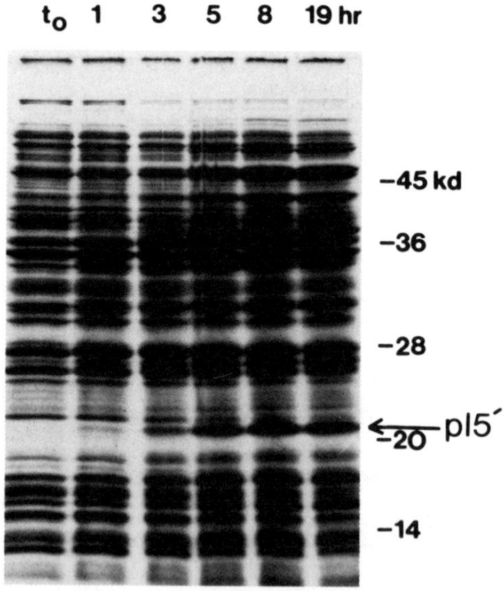

FIGURE 4. Protein profile of the time-course of overexpression of HIV-1 p15^{gag} (p15') in E. coli AR120 following induction (at t_o) of cell culture harboring pDG71407 (to be described) with naladixic acid. p15' subjected to automated N-terminal sequencing gives the expected sequence to 38 cycles.

proteins which bind relatively nonspecifically to ss nucleic acids (DNA or RNA) has been discussed. In the case of g32P, the ^{113}Cd NMR data has provided additional evidence for NS_3 coordination perhaps involving Cys^{77}, His^{81}, Cys^{87} and Cys^{90} just N-terminal to several known determinants of the ssDNA binding site including the side chains of one or more of the regularly-disposed tyrosyl residues between amino acids 72 and 116 (7). However, the binding of any DNA ligand perturbs the structure of the metal site quite minimally.

These and other data allow us to conclude that the metal site appears only indirectly coupled to major DNA binding surface as evidenced by the fact that metal-free apo g32P binds oligonucleotide nearly or as well as Zn(II) g32P (3). The major role of Zn(II) in g32P is in stabilizing the interactions between contiguous g32P monomers, perhaps in conjunction with the N-terminal "B" domain, which gives rise to high affinity, cooperative binding to polynucleotides. A model synthesizing these points is presented in Fig. 5.

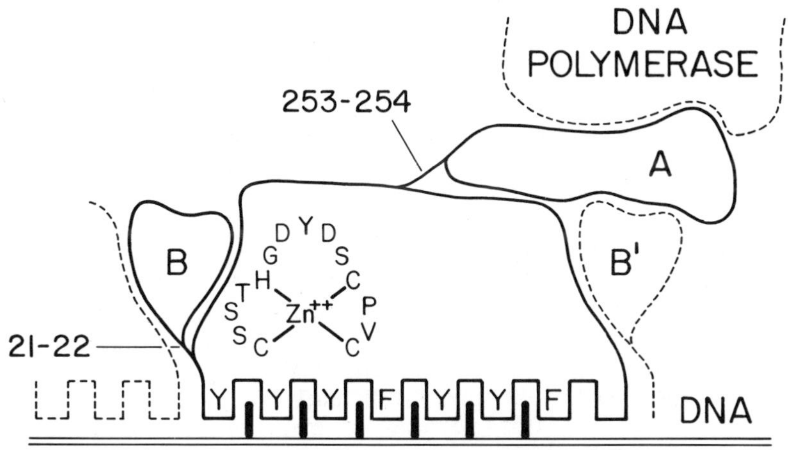

FIGURE 5. Schematic representation of T4 g32P bound cooperatively to an "infinite"-length DNA lattice.

At present, the role that the Cys-rich sequence elements play in retroviral *gag*-gene polyprotein or processed protein binding to virion RNA is still unclear (5). The availability of reasonable quantities of overexpressed NABP from HIV-1 will permit detailed *in vitro* binding experiments as well as structural studies.

REFERENCES

1. Chase JW, Williams KR (1986). Single-stranded DNA binding proteins required for DNA replication. Ann Review Biochem 55:103.
2. Giedroc DP, Keating KM, Williams KR, Konigsberg WH, Coleman JE (1986) Gene 32 protein, the single-stranded DNA binding protein from T4, is a zinc metalloprotein. Proc Natl Acad Sci USA 83:8452.
3. Giedroc DP, Keating KM, Williams KR, Coleman JE (1987). The function of zinc in gene 32 protein from T4. Biochemistry 26:5251.
4. Keating KM, Ghosaini LR, Giedroc DP, Williams KR, Coleman JE, and Sturtevant JM (1988). Thermal denaturation of T4 gene 32 protein: effects of zinc removal and substitution. Biochemistry (in press).
5. Karpel RL, Henderson LE, and Oroszlan S (1987). Interactions of retroviral structural proteins with single-stranded nucleic acids. J Biol Chem 262:4961.
6. Henderson LE, Copeland TD, Sowder RC, Smythers GW, and Oroszlan S (1981). Primary structure of the low molecular weight nucleic acid-binding proteins of murine leukemia viruses. J Biol Chem 256:8400.
7. Coleman JE, and Giedroc DP (1988). Zinc binding proteins involved in nucleic acid replication. In Sigel H (ed.): "Metals Ions in Biological Systems," New York: Marcell Dekker, Vol 25 (in press).
8. Armitage IM, and Otvos JD (1982). Principles and applications of ^{113}Cd NMR to biological systems. In Berliner and Reuben (eds.): "Biological Magnetic Resonance," New York: Plenum Vol 4, p. 79.
9. Fink GR, Boeke JD, and Garfinkel DJ (1986). The mechanism and consequences of retrotransposition. Trends Genet 2:118.
10. Berg JM (1986). Potential metal binding domains in nucleic acid binding proteins. Science 232:485.
11. Henderson LE, Copeland TD, Sowder RC, Schultz AM and Oroszlan S (1988). Analysis of proteins and peptides purified from sucrose banded HTLV-III. In Bolognesi D (ed.): UCLA Symp Mol Cell Biol, New Series. New York: Alan R Liss Vol 71, p. 135.

Metal Ion Homeostasis: Molecular Biology and Chemistry, pages 13–20

SACCHAROMYCES CEREVISIAE DNA SEQUENCES GRATUITOUSLY HOMOLOGOUS TO THE CONSERVED DOMAIN OF *DROSOPHILA KRÜPPEL*

Stella T. Hu and Dean H. Hamer

Laboratory of Biochemistry, National Cancer Institute
National Institutes of Health, Bethesda, MD 20892

ABSTRACT *Krüppel*, a segmentation gene of the fruit fly *Drosophila melanogaster*, encodes multiple potential zinc finger domains. The H/C link sequences connecting the fingers are highly conserved within the *Krüppel* repeats and in related genes in *Drosophila* and mice. While sequences homologous to this region can readily be detected in the genome of the yeast *Saccharomyces cerevisiae*, cloning and analysis of four such sequences revealed that none of them could encode multiple zinc finger domains.

INTRODUCTION

Krüppel (*Kr*) is a member of the gap class of Drosophila segmentation genes. In strong *Kr* mutant embryos, the first five anterior abdominal segments and all thoracic segments are absent (1). Cloning and sequence analysis of *Kr* revealed that it encodes a series of potential zinc and DNA binding finger domains homologous to those of the archetypal zinc finger protein TFIIIA (2). The zinc liganding sequences CXXC and HXXXH are conserved between *Kr*, TFIIIA and several other known and putative DNA binding regulatory proteins (reviewed in 3,4). In addition, the repeated sequences in *Kr* are strongly internally conserved within the H/C link or "web" sequence HTGE$^{R}_{K}$P$^{F}_{Y}$XC that connects the last histidine of one repeat with the first cysteine of the next.

Recently it was shown, by Southern blotting experiments, that the Kr gene is a member of a multigene family in Drosophila and is also homologous to DNA sequences within the genomes of a variety of eukaryotes including yeast, invertebrates, rodents and humans (5,6). The cloning of Kr-like genes from Drosophila (5) and mouse (6,7) has shown that the homologous sequences in these organisms do encode multiple finger proteins in which the H/C link is the predominent conserved region responsible for their detection by nucleic acid hybridization. Despite the homology of these genes to Kr, their function is not clear because mutants are not yet available.

To further study the evolution and function of Kr-like genes, we searched for homologous sequences in the genome of the unicellular, genetically manipulatable eukaryote Sacharomyces cerevisiae. Although sequences homologous to the H/C link region were readily detected and cloned, none of them have the capacity to encode multiple finger proteins.

RESULTS AND DISCUSSION

In order to detect yeast sequences homologous to Kr, we hybridized Southern blots of yeast genomic DNA to a 544 bp fragment containing the Drosophila Kr finger region (5) and to a series of synthetic oligonucleotides encoding the H/C link, the H/C link plus 7 N-terminal amino acids, or the H/C link plus 7 C-terminal amino acids. After hybridizing and washing the blots at various criteria, the strongest and most consistent signal was observed with the following oligonucleotide which encodes just the H/C link:

HisThrGlyGluArgProPheGluCys
3' GTGTGACCACTTGCCGGTAAGCTCACG 5'

This probe was used to screen a library of approximately 10^5 λgt11-yeast genomic DNA recombinants representing roughly 20 genome equivalents. Restriction and Southern blotting analysis of 10 recombinants showed that they fell into 4 classes. One subcloned representative of each class was digested with a frequently-cutting restriction endonuclease and a single, small fragment homologous to the Kr oligonucleotide was inserted into an M13 vector and sequenced by the dideoxy method. The sequencing results are shown on the following 3 pages and are summarized below:

```
                                        CLONE 3

                                             27                                          54
GGT AGA TTC TTT ACA ACT TCA TTA CTT CCC TGA AGA CCA TTT TAT GTC TTG TTT
Gly Arg Phe Phe Thr Thr Ser Leu Leu Pro  .  Arg Pro Phe Tyr Val Leu Phe
 Val Asp Ser Leu Gln Leu His Tyr Phe Pro Glu Asp His Phe MET Ser Cys Phe
   .  Ile Leu Tyr Asn Phe Ile Thr Ser Leu Lys Thr Ile Leu Cys Leu Val Leu


                                             81                                         108
TAG AAA AAG TAA ACC AGC AAT AAC GGT AAA ATC CAA TAA GAA GTT TTA TTT AAA
 .  Lys Lys  .  Thr Ser Asn Asn Gly Lys Ile Gln  .  Glu Val Leu Phe Lys
 Arg Lys Ser Lys Pro Ala Ile Thr Val Lys Ser Asn Lys Lys Phe Tyr Leu Asn
  Glu Lys Val Asn Gln Gln  .  Arg  .  Asn Pro Ile Arg Ser Phe Ile  .  MET


                                            135                                         162
TGG GAA GGT TTT CAA CAA GGA CAA AGA ATA TTT CAA TGA AAC TAG GAT ATC AAA
Trp Glu Gly Phe Gln Gln Gly Gln Arg Ile Phe Gln  .  Asn  .  Asp Ile Lys
 Gly Lys Val Phe Asn Lys Asp Lys Glu Tyr Phe Asn Glu Thr Arg Ile Ser Lys
  Gly Arg Phe Ser Thr Arg Thr Lys Asn Ile Ser MET Lys Leu Gly Tyr Gln Lys


                                            189                                         216
AGC TAA TGA AGT CGA ACT AAG CAA AAT TCA AAC AGC GAT GAC ACG GCT TAC AAA
Ser  .   .  Ser Arg Thr Lys Gln Asn Ser Asn Ser Asp Asp Thr Ala Tyr Lys
 Ala Asn Glu Val Glu Leu Ser Lys Ile Gln Thr Ala MET Thr Arg Leu Thr Asn
  Leu MET Lys Ser Asn  .  Ala Lys Phe Lys Gln Arg  .  His Gly Leu Gln Ile


                                            243                                         270
TAG ACA TAG AAA CAG TAT TCC CTC TGA ATT TCG CTA CTT GAG ACG AGA TTT AAA
 .  Thr  .  Lys Gln Tyr Ser Leu  .  Ile Ser Leu Leu Glu Thr Arg Phe Lys
 Arg His Arg Asn Ser Ile Pro Ser Glu Phe Arg Tyr Leu Arg Arg Asp Leu Lys
  Asp Ile Glu Thr Val Phe Pro Leu Asn Phe Ala Thr  .  Asp Glu Ile  .  Asn


                                                            His Thr Gly Glu Ar
                                                            Ca cAc TGG TGA ac-
ATT AAA GGT GAA AAC AAC GTT TAT CAA GGA GTG GTG TAA ACT AAA TGG TGA TAA
Ile Lys Gly Glu Asn Asn Val Tyr Gln Gly Val Val  .  Thr Lys Trp  .   .
 Leu Lys Val Lys Thr Thr Phe Ile Lys Glu Trp Cys Lys Leu Asn Gly Asp Lys
   .  Arg  .  Lys Gln Arg Leu Ser Arg Ser Gly Val Asn  .  MET Val Ile Arg

g Pro Phe Glu Cys
GGC CAT TCG AGt Gc
GGC CAT TCG AGA GTA CGT TAA TCT TAA CAG ATC CTC TAA GAG A
Gly His Ser Arg Val Arg  .  Ser  .  Gln Ile Leu  .  Glu
 Ala Ile Arg Glu Tyr Val Asn Leu Asn Arg Ser Ser Lys Arg
  Pro Phe Glu Ser Thr Leu Ile Leu Thr Asp Pro Leu Arg
```

CLONE 19

```
                     His Thr Gly  Glu Arg   Pro Phe Glu Cys
                     cA cAC TGG T-G AAC -GG CCA TTc gag tgc
AAA ATT GAG GTT GAA GAA AAC TGG TAG AAC GGG CCA TTG CTT CAA CTT CTC CAA
Lys Ile Glu Val Glu Glu Asn Trp  .  Asn Gly Pro Leu Leu Gln Leu Leu Gln
 Lys Leu Arg Leu Lys Lys Thr Gly Arg Thr Gly His Cys Phe Asn Phe Ser Lys
  Asn  .  Gly  .  Arg Lys Leu Val Glu Arg Ala Ile Ala Ser Thr Ser Pro Arg

                                              81                      108
GAA ATC TCC ACA TGG AAA TTT TTC AGC CCA TTG GAC TGC CAA ATC CTT TGG ATT
Glu Ile Ser Thr Trp Lys Phe Phe Ser Pro Leu Asp Cys Gln Ile Leu Trp Ile
 Lys Ser Pro His Gly Asn Phe Ser Ala His Trp Thr Ala Lys Ser Phe Gly Leu
  Asn Leu His MET Glu Ile Phe Gln Pro Ile Gly Leu Pro Asn Pro Leu Asp Trp

                                             135                      162
GGC ACC TTT GAT TCT CAA TCT TGG GAT AGG AAT CAA CAA ATC GCC TCT CTC CAT
Gly Thr Phe Asp Ser Gln Ser Trp Asp Arg Asn Gln Gln Ile Ala Ser Leu His
 Ala Pro Leu Ile Leu Asn Leu Gly Ile Gly Ile Asn Lys Ser Pro Leu Ser Ile
  His Leu  .  Phe Ser Ile Leu Gly  .  Glu Ser Thr Asn Arg Leu Ser Pro  .

                                             189
AGT ATT TGT CCA TTC TAA TGC TGG AAA AAT AAG AGA CGA GTC TAC GCC G
Ser Ile Cys Pro Phe  .  Cys Trp Lys Asn Lys Arg Arg Val Tyr Ala
 Val Phe Val His Ser Asn Ala Gly Lys Ile Arg Asp Glu Ser Thr Pro
  Tyr Leu Ser Ile Leu MET Leu Glu Lys  .  Glu Thr Ser Leu Arg
```

CLONE 10

```
                                             27                       54
TGC AGG TCG ACT CTA GAG GAC TGA AGT ATT CTG ATA ATG AAC CAA ACT CAT TTC
Cys Arg Ser Thr Leu Glu Asp  .  Ser Ile Leu Ile MET Asn Gln Thr His Phe
 Ala Gly Arg Leu  .  Arg Thr Glu Val Phe  .   .   .  Thr Lys Leu Ile Ser
  Gln Val Asp Ser Arg Gly Leu Lys Tyr Ser Asp Asn Glu Pro Asn Ser Phe Leu

                                              81                      108
TTT CTT CTT TGC CAT CTT CTT TAT TGC TAT TAT TAT CTT TAT TAC CAT TGT TCT
Phe Leu Leu Cys His Leu Leu Tyr Cys Tyr Tyr Tyr Leu Tyr Tyr His Cys Ser
 Phe Phe Phe Ala Ile Phe Phe Ile Ala Ile Ile Ile Phe Ile Thr Ile Val Leu
  Ser Ser Leu Pro Ser Ser Leu Leu Leu Leu Leu Ser Leu Leu Pro Leu Phe Leu

                                             135                      162
TGC CCT CAC CCT CTT TGT TTG GTC CTA TTT TCA TTT AAC CTT GTT CCA TGA GAT
Cys Pro His Pro Leu Cys Leu Val Leu Phe Ser Phe Asn Leu Val Pro  .  Asp
 Ala Leu Thr Leu Phe Val Trp Ser Tyr Phe His Leu Thr Leu Phe His Glu Ile
  Pro Ser Pro Ser Leu Phe Gly Pro Ile Phe Ile  .  Pro Cys Ser MET Arg Tyr

                His Thr Gly Glu Arg Pro Phe Glu
                cAc act GGt GAA CGG ccA TTc gag
ATC GGA AGG AAC GAT TTG GGA GAA CGG TGA TTG TGT
Ile Gly Arg Asn Asp Leu Gly Glu Arg  .  Leu Cys
 Ser Glu Gly Thr Ile Trp Glu Asn Gly Asp Cys
  Arg Lys Glu Arg Phe Gly Arg Thr Val Ile Val
```

```
                         CLONE 12

                           27                         54
TCG AAT CTT TTA ATT GGG TAA AAT GTA AAG CAT CAT CCT CAC AGC GAC ACA TGC
Ser Asn Leu Leu Ile Gly  .  Asn Val Lys His His Pro His Ser Asp Thr Cys
 Arg Ile Phe  .  Leu Gly Lys MET  .  Ser Ile Ile Leu Thr Ala Thr His Ala
  Glu Ser Phe Asn Trp Val Lys Cys Lys Ala Ser Ser Ser Gln Arg His MET His


                           81                         108
ACA CCC ATT TAT CAT AGT ACA TAG ATT TGT TAA GCT CTT TAT ATG GGA ATG TTT
Thr Pro Ile Tyr His Ser Thr  .  Ile Cys  .  Ala Leu Tyr MET Gly MET Phe
 His Pro Phe Ile Ile Val His Arg Phe Val Lys Leu Phe Ile Trp Glu Cys Phe
  Thr His Leu Ser  .  Tyr Ile Asp Leu Leu Ser Ser Leu Tyr Gly Asn Val Leu


                                            His Thr Gly Glu Arg
                                            ca cac Tgg tgA aCG
TGA TGT TAC GGG AGT AAT TTT GTG CCT GTA ATA GCT CCA AAT TTT TTA AAA TCG
 .  Cys Tyr Gly Ser Asn Phe Val Pro Val Ile Ala Pro Asn Phe Leu Lys Ser
 Asp Val Thr Gly Val Ile Leu Cys Leu  .   .  Leu Gln Ile Phe  .  Asn Arg
  MET Leu Arg Glu  .  Phe Cys Ala Cys Asn Ser Ser Lys Phe Phe Lys Ile Gly

 Pro Phe Glu Cys
GCC ATT cGA GTG C
GCC ATT TGA GTG CTT TGA GCT TGC CCC TCG CTA CAA AGT CTC TAG AAG ATT GTA
Ala Ile  .  Val Leu  .  Ala Cys Pro Ser Leu Gln Ser Leu  .  Lys Ile Val
 Pro Phe Glu Cys Phe Glu Leu Ala Pro Arg Tyr Lys Val Ser Arg Arg Leu  .
  His Leu Ser Ala Leu Ser Leu Pro Leu Ala Thr Lys Ser Leu Glu Asp Cys Asn


                           243                        270
ACA GCC CTT GTA AAA ATT CTT CTC TCG TAA TAC GTA CCG CTT GCT GTA AGG ATT
Thr Ala Leu Val Lys Ile Leu Leu Ser  .  Tyr Val Pro Leu Ala Val Arg Ile
 Gln Pro Leu  .  Lys Phe Phe Ser Arg Asn Thr Tyr Arg Leu Leu  .  Gly Phe
  Ser Pro Cys Lys Asn Ser Ser Leu Val Ile Arg Thr Ala Cys Cys Lys Asp Leu


                           297
TGG GTA AAT CCT
Trp Val Asn Pro
 Gly  .  Ile
  Gly Lys Ser
```

<u>Clone 3</u>. This sequence, which showed the strongest hybridization to the oligonucleotide probe, was isolated in 6 independent, overlapping phage. The best alignment to the oligonucleotide gives 20/27 matches. However, because it's necessary to introduce a one nucleotide gap to obtain this alignment, there is essentially no conservation of the H/C link protein sequence. This clone does have an uninterrupted open reading frame, but it shows no obvious homology to <u>Kr</u> or other zinc finger proteins.

<u>Clone 19</u>. This intermediate hybridizing clone showed 18/27 matches but only with the introduction of 2 gaps. The open reading frame showed no obvious homology to known zinc finger proteins.

Clone 12. This intermediate hybridizing clone showed 16/27 matches with no gaps but there was no long open reading frame nor potential zinc binding sites.

Clone 10. This very weakly hybridizing clone had only 12/24 matches to the oligonucleotide within the sequenced region and no long open reading frame.

In order to determine the relationship between the cloned DNA sequences and the *Kr*-homologous sequences in the genome that had been detected with the oligonucleotide probe, each clone was nick translated and used as a probe against genomic Southerns. As shown in Fig. 1, clone 3 was homologous to the most prominent EcoRI and Hind3 fragments hybridizing to the oligonucleotide probe. Subcloning experiments showed that the additional bands displayed by this probe were due to homology to a region not containing *Kr*-like sequences. Clones 10 and 12 hybridized to single bands coincident with oligonucleotide bands, while clone 19 displayed several bands which also overlapped the set of oligonucleotide-homologous bands. Thus of the approximately 6 major bands hybridizing to the oligonucleotide probe, sequences corresponding to four have been cloned. Since clone 10 showed only 50% homology to the probe, it did not appear encouraging to continue searching for homologues with this probe.

Although several yeast regulatory proteins containing zine fingers are now known (3,4,8), none of them contain the conserved H/C link motif characteristic of the *Kr* class. The present results suggest that yeast genes containing this motif are, if present at all, not common. One possible explanation is that the *Kr* proteins play a role in developmental processes that do not occur in a unicellular eukaryote such as yeast. Further understanding of the evolution and function of *Kr*-like proteins may best be approached by further characterization of these genes from other eukaryotic species.

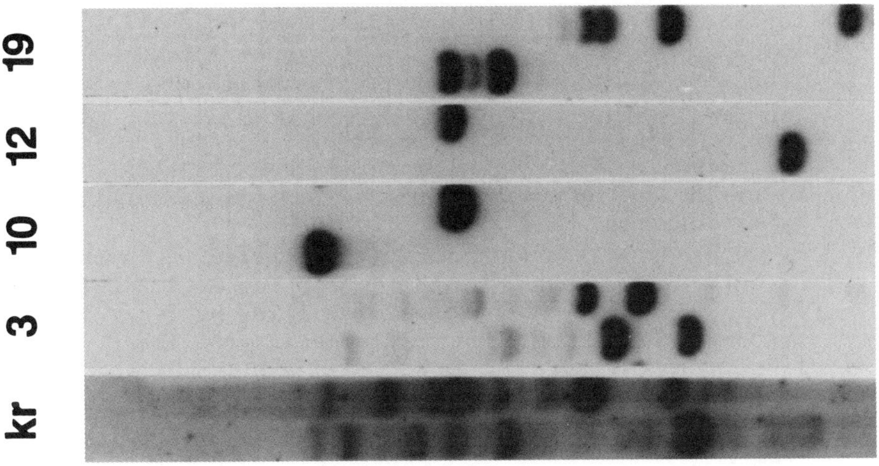

Figure 1. Hybridization of the Kr oligonucleotide and four cloned yeast DNA fragments to yeast genomic DNA. Yeast DNA was digested with EcoRI (left lane in each panel) or Hind3 (right lane in each panel), electrophoresed through a 1% agarose gel, denatured and transferred to nitrocellulose paper. Strips were hybridized to 10^7 cpm of Kr oligonucleotide (labelled with T4 kinase) or of purified EcoRI fragments obtained from subclones (labelled by nick translation). Hybridization was in 10 ml of 6XSSC, 05% SDS, 5X Denhardt's solution, 100 µg/ml denatured salmon sperm DNA at 50° for the oligonucleotide or at 65° for the clones. Washing was in 2XSSC, 0.1% SDS at 35° for the oligonucleotide or in 0.1XSSC, 0.1% SDS at 52° for the clones.

REFERENCES

1. Preiss A, Rosenberg UB, Kienlin A, Seifert E, Jackle H (1985). Molecular genetics of Kruppel, a gene required for segmentation of the Drosophila embryo. Nature 313:27.

2. Rosenberg UB, Schroder C, Preiss A, Kienlin A, Cote S, Riede E, Jackle M (1986). Structural homology of the product of the Drosophila Kruppel gene with Xenopus transcription factor IIIA. Nature 319:336.
3. Berg J (1986). Potential metal-binding domains in nucleic acid binding proteins. Science 232:485.
4. Evans RM, Hollenberg SM (1988). Zinc fingers: gilt by association. Cell 52:1.
5. Schuh R, Aicher W, Gaul U, Cote S, Preiss A, Maier D, Seifert E, Nouber U, Schroder C, Kemler R, Jackle H (1986). A conserved family of nuclear proteins containing structural elements of the finger protein encoded by Kruppel, a Drosophila segmentation gene. Cell 47:1025.
6. Chowdhury K, Deutsch U, Gruss P (1987). A multigene family encoding several "finger" structures is present and differentially active in mammalian genomes. Cell 48:771.
7. Chavrier P, Lemaire P, Revelant O, Bravo R, Charnay P (1988). Characterization of a mouse multigene family that encodes zinc finger structures. Mol. Cell Biol. 8:1319.
8. Blumberg H, Eisen A, Sledziewski A, Bader D, Young ET (1987). Two zinc fingers of a yeast regulatory protein shown by genetic evidence to be essential for its function. Nature 328:443.

Metal Ion Homeostasis: Molecular Biology and Chemistry, pages 21–30

METALLOREGULATION IN THE SEQUENCE SPECIFIC BINDING OF SYNTHETIC MOLECULES TO DNA[1]

John H. Griffin and Peter B. Dervan

Division of Chemistry and Chemical Engineering, California Institute of Technology, Pasadena, CA 91125

ABSTRACT The design, synthesis, and DNA binding studies of a series of bis(netropsin)s linked by homologous polyether tethers has been undertaken. DNA affinity cleaving experiments show that one compound, bis(netropsin)-3,6,9,12,15-pentaoxaheptadecanediamide-EDTA:Fe(II) (**1b:Fe(II)**) is positively activated for specific DNA binding in the presence of Sr^{2+} or Ba^{2+} ions. The effects of linker structure, metal ion concentration, and DNA sequence on the metalloregulated binding are reported.

INTRODUCTION

Gene expression at the transcriptional level may be influenced by environmental conditions such as temperature, the presence of specific metal ions, or the presence of specific organic molecules. Although the mechanisms involved in such transcriptional control are still not well understood, our interest in the sequence specific recognition of double helical DNA challenged us to design a synthetic DNA binding molecule which could be regulated by specific metal cations (1).

---

[1]This work was supported by NIH and by NSF and Merck Sharp & Dohme/ACS fellowships to JHG.

Our approach was to combine reasonably well understood DNA binding subunits with a known metal binding subunit. The result was a homologous series of netropsin dimers (Figure 1), which bind in the minor groove of DNA at locations rich in adenine-thymine (AT) base pairs (2,3), connected by polyether tethers, which are capable of coordinating specific metal cations (4).

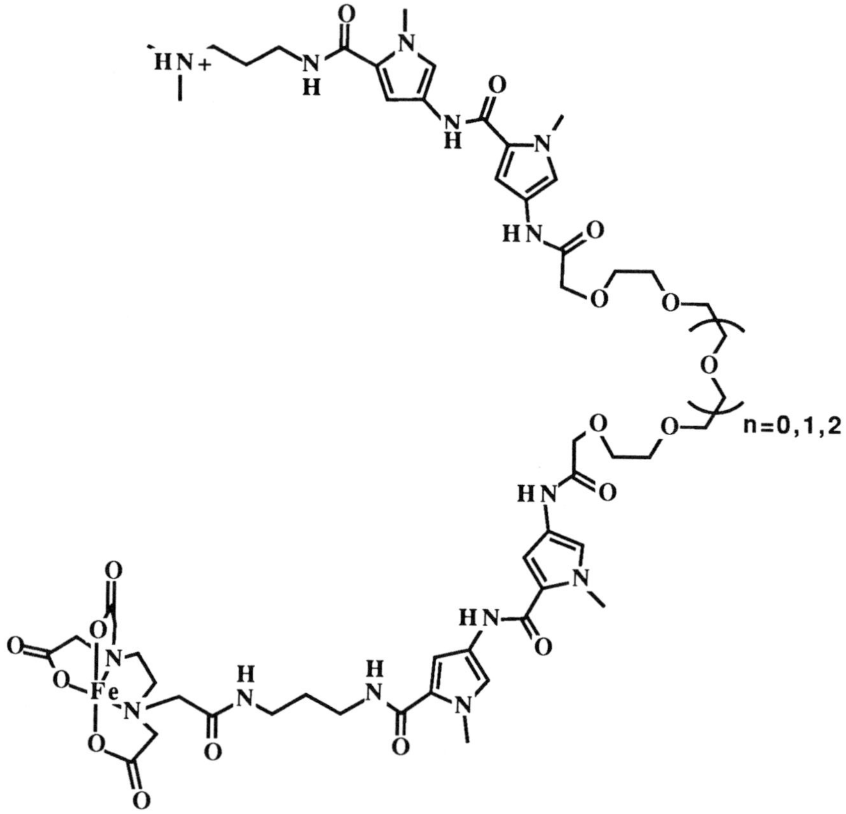

FIGURE 1. **1a:Fe, n=0; 1b:Fe, n=1; 1c:Fe, n=2.**

Covalent attachment of EDTA to one terminus of **1a-c** allows use of the DNA affinity cleaving method (5) to determine binding properties of **1a-c**, in the presence and absence of cations, on ^{32}P end labeled DNA restriction fragments.

RESULTS

Synthetic procedures to **1a-c** will be reported elsewhere. DNA substrate for the first stage of the affinity cleaving analysis was Sty 1 linearized pBR322 plasmid DNA (4362 base pairs) which could be labeled on either end using routine enzymatic procedures. Double strand cleavage of this plasmid DNA (made up to a total DNA concentration of 100 μM in base pairs using calf thymus DNA) by **1a-c:Fe(II)** (5-50 μM) in the absence of added cations followed by agarose gel electrophoresis and autoradiography afforded the cleavage patterns shown in Figure 2.

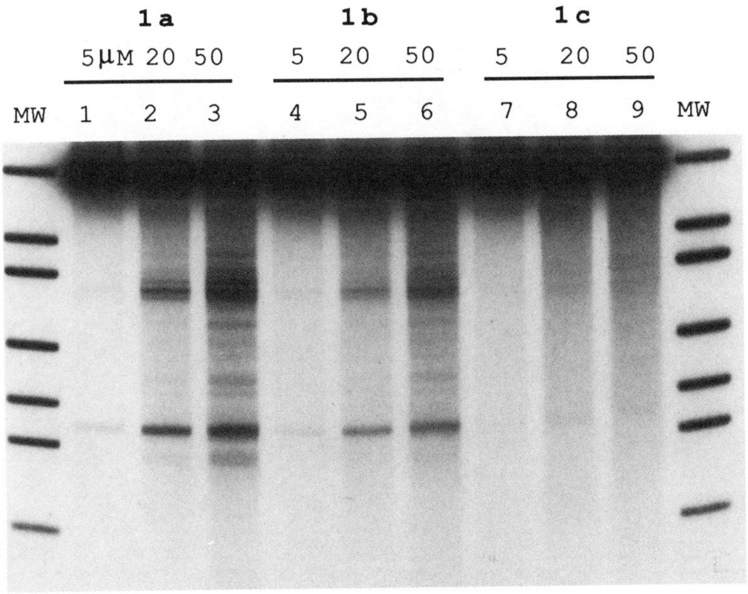

FIGURE 2.

In the absence of added cations, relatively weak binding by **1a-c** is observed, with binding affinity decreasing in the order **1a>1b>1c.** For **1a** and **1c,** similar experiments carried out in the presence of various cations (at 1 mM concentration) afforded no significant changes in apparent binding affinity or specificity. Moreover, reactions performed with **1b:Fe(II)** (5 μM) in the presence of 1 mM Li^+, Na^+, K^+, Rb^+, Cs^+, Mg^{2+}, Ca^{2+}, NH_4^+, Ni^{2+}, Cd^{2+}, Hg^{2+}, Ag^+, Tl^+, Tl^{3+}, Co^{2+}, Cu^{2+}, Zn^{2+}, Fe^{2+}, Fe^{3+}, Pb^{2+}, In^{3+}, Au^{3+}, Lu^{3+}, Ho^{3+}, Eu^{3+}, Bi^{3+}, La^{3+}, Ce^{3+}, or Ce^{4+} showed no apparent change in binding affinity or specificity. However, addition of Sr^{2+} or Ba^{2+} ions (1 mM) produced enhanced DNA cleavage by **1b:Fe(II)** (5 μM, Figure 3). Two regions of major DNA cleavage were observed.

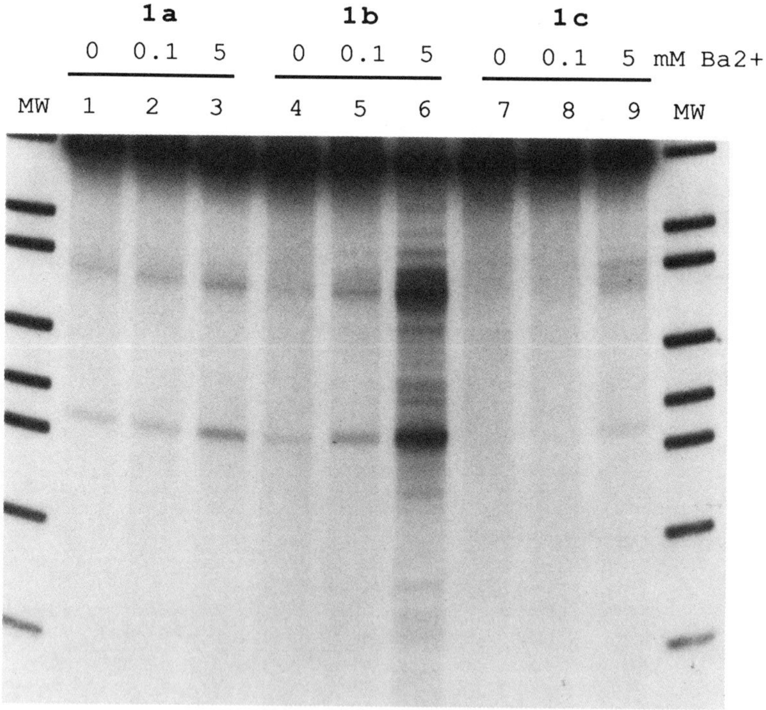

FIGURE 3.

For a more precise analysis of the binding locations, cleavage of ^{32}P end-labeled DNA restriction fragments by **1a-c:Fe(II)** with and without added Ba^{2+} was visualized by denaturing high resolution polyacrylamide gel electrophoresis followed by autoradiography. A 517 base pair ^{32}P end labeled DNA restriction fragment (EcoRI/RsaI) from plasmid pBR322, which contained the higher

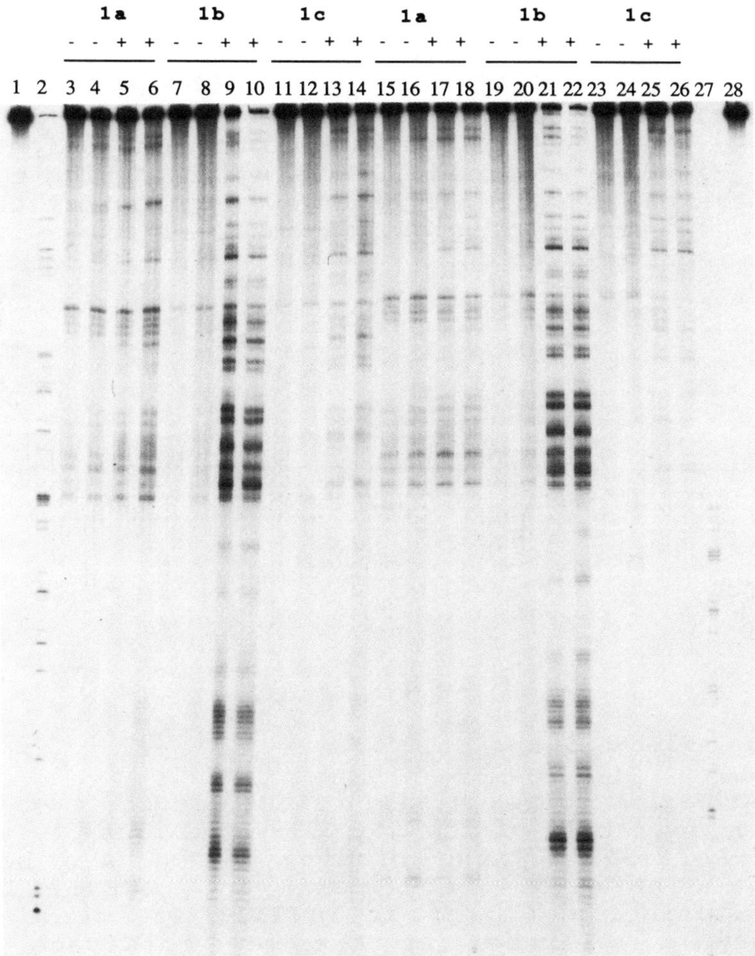

FIGURE 4.

cleavage locus observed in Figure 3, was prepared and cleaved by **1a-c:Fe(II)**. Lanes 1-14 are with DNA labeled on one 3' terminus, and lanes 15-28 are with DNA labeled on one 5' terminus. Cleavage of this fragment by **1a-c:Fe(II)** in the absence of added cations (lanes marked "-" in Figure 4) showed weak binding, most notably to the sequence 5'-ATTTTTAT-3' in the lower part of the gel. Addition of 5 mM Ba^{2+} (lanes marked "+" in Figure 4) afforded a strong and specific pattern of DNA cleavage by **1b:Fe(II)** (lanes 9,10,21,22). This pattern is not only stronger, but also different from the pattern observed in the absence of added Ba^{2+}.

Densitometric analysis of the autoradiogram in Figure 4 afforded the histogram in Figure 5, where the lengths of the arrows are proportional to the amount of cleavage observed at a given nucleotide position.

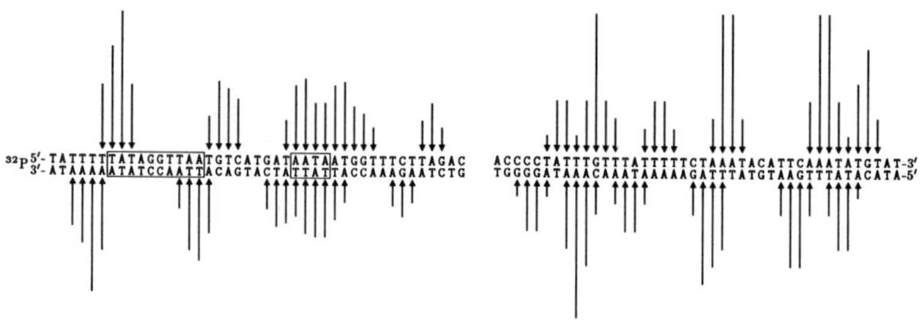

FIGURE 5.

Two well-resolved binding sites are observed in the lower half of the gel, 5'-TATAGGTTAA-3', and 5'-AATA-3'. These binding sites have been independently confirmed by MPE:Fe(II) footprinting (1) of **1b:In(III)** (results not shown). Further up the gel, cleavage is observed flanking several sites of four to five AT base pairs adjacent to one GC base pair. A plot of relative cleavage by **1b:Fe(II)** versus log [Ba^{2+}] indicates binding saturation above 10 mM Ba^{2+} (Figure 6).

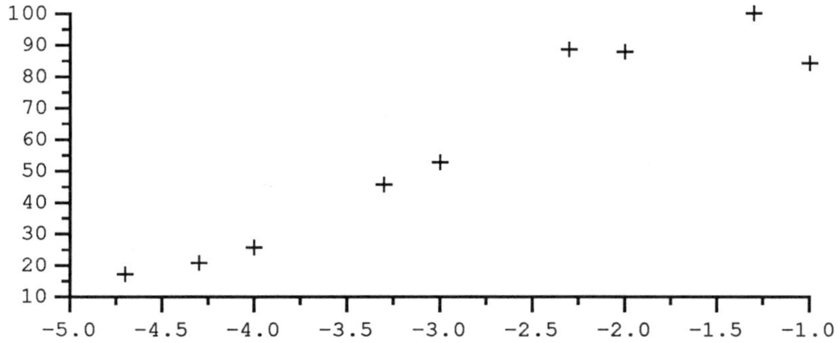

FIGURE 6.

DISCUSSION

Addition of Sr^{2+} or Ba^{2+} ions converts **1b:Fe(II)** from a species which produces little DNA cleavage (at μM concentrations) to a specific DNA binding/cleaving molecule. We have thus demonstrated positive metalloregulation in the sequence specific binding of a small molecule to DNA. The effect is metal ion specific, occurring with the heavier alkaline earth cations, which are known to bind more strongly to 18-crown-6 in water than most other cations (6). That Ba^{2+} induces stronger binding by **1b** than Sr^{2+} is consistent with their relative binding affinities to 18-crown-6 in water (log K 3.9 versus 2.8). Why certain metals (such as Hg^{2+} and Pb^{2+}) that exhibit strong binding to 18-crown-6 in water but do not activate **1b** is not fully understood, but may be related to the "harder" alkaline earth cations having high affinity for the phosphate oxygen anions of the DNA backbone while the "softer" post-transition metals should have high affinity for heterocyclic nitrogen coordination sites found in the major groove of B DNA (7). The

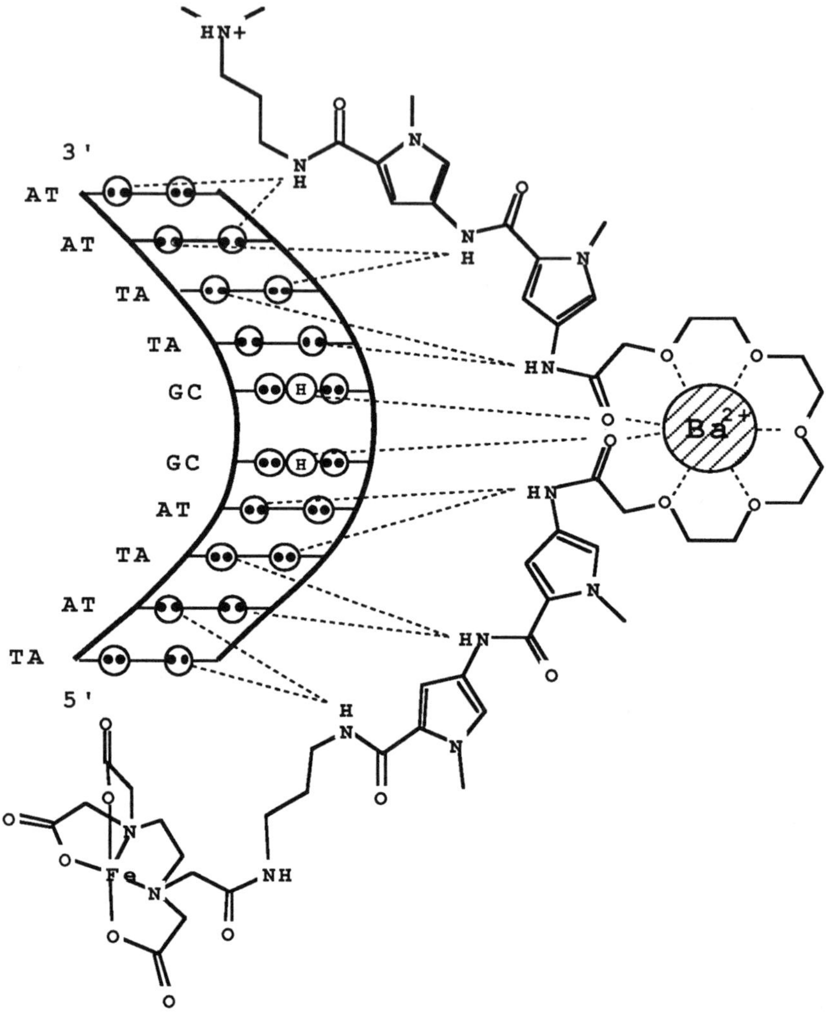

FIGURE 7.

metalloregulatory binding is dependent on the structure of the acyclic polyether, being significant only with **1b**. CPK models indicate that **1b** may heptacoordinate a cation (using five ether oxygens and two carbonyl oxygens from the tether's terminal glycolamide groups), while maintaining an overall crescent shape complementary to right-handed double helical DNA. Models indicate that **1a** and **1c** should be less successful in meeting both criteria.

A single netropsin subunit, having three amide NHs, should bind to four contiguous AT base pairs (8). The observation of four base pair (5'-AATA-3') and ten base pair (5'-TATAGGTTAA-3') binding sites indicates that both monomeric and dimeric binding by **1b** is metalloregulated. We propose a hypothetical model for dimeric binding (9), wherein bound netropsin subunits flank a Ba^{2+}:ionophore subunit which forms specific contacts with guanine N2 amino groups protruding from the floor of the DNA minor groove (Figure 7). Positive metalloregulation may be considered to arise by Ba^{2+} filling a specific cation binding cavity consisting of a neutral polyoxa macrocycle "capped" by phosphate oxygen anions from the nearby DNA backbone. For metalloregulated monomeric binding, one or two GC base pairs are often found neighboring the AT rich binding sites. Thus, it is plausible that Ba^{2+}:ionophore interactions are similar for both monomeric and dimeric binding modes.

REFERENCES

1a. Dervan, PB (1986). Design of Sequence-Specific DNA-Binding Molecules. Science 232:464.

1b. Sluka, JP, Horvath, SJ, Bruist, MF, Simon, MI, Dervan, PB (1987). Synthesis of a Sequence-Specific DNA-Cleaving Peptide. Science 238:1129.

1c. Moser, H, Dervan, PB (1987). Sequence Specific Cleavage of DNA by Triple Helix Formation. Science 238:645.

2. Zimmer, C, Wahnert, U (1986). Nonintercalating DNA Binding Ligands: Specificity of the Interaction and Their Use in Biophysical, Biochemical, and Biological Investigations of the Genetic Material. Prog Biophys Mol Biol 47:31.

3. Kopka, ML, Yoon, C, Goodsell, D, Pjura, P, Dickerson, RE (1985). Binding of an Antitumor Drug to DNA: Netropsin and C-G-C-G-A-A-T-T-^{Br}C-G-C-G J Mol Biol 183:553.

4. Vogtle, F, Weber, E (1979). Multidentate Acyclic Neutral Ligands and Their Complexation. Angew Chem Int Ed Engl 18:753.

5. Taylor, JS, Schultz, PG, Dervan, PB (1984). DNA Affinity Cleaving: Sequence Specific Cleavage of DNA by Distamycin-EDTA:Fe(II) and EDTA-Distamycin:Fe(II). Tetrahedron 40:457.

6. Izatt, RM, Bradshaw, JS, Nielsen, SA, Lamb, JD, Christensen, JJ, Debabrata, S (1985). Thermodynamic and Kinetic Data for Cation Macrocycle Interaction. Chem Rev 85:271.

7. Saenger, W (1984). "Principles of Nucleic Acid Structure." New York: Springer Verlag, pp 201-219.

8. Youngquist, RS, Dervan, PB (1985). Sequence-Specific Recognition of B-DNA by Oligo(*N*-Methylpyrrolecarboxamide)s. Proc Natl Acad Sci USA 82:2565.

9. Griffin, JH, Dervan, PB (1987). Metalloregulation in the Sequence Specific Binding of Synthetic Molecules to DNA. J Am Chem Soc 109:6840.

Metal Ion Homeostasis: Molecular Biology and Chemistry, pages 31–47

GENETIC EVIDENCE FOR A METALLOTHIONEIN GENE FAMILY IN NEW ZEALAND RABBIT

J.-P. Thirion, Y.-C. Tam, M. Hassan and G. Boulay

Department of Microbiology, Faculty of Medicine
Centre Hospitalier Universitaire, Sherbrooke,
Quebec, Canada J1H 5N4

ABSTRACT Screening of a partial lambda gt10 rabbit genomic library with mouse metallothionein MT-I cDNA yielded two phages with a 4.4 Kb and a 4.8 Kb inserts containing an MT-I metallothionein gene and an MT-II processed pseudogene. The MT-I gene has exon/intron structures similar to other mammalian MT genes. Comparison of the promotor sequence of this gene with the promoters of other MT genes identified a number of oligodeoxynucleotide sequences which are most probably recognized by _trans_-acting proteins involved in the regulation of this gene by heavy metals, glucocorticoids and alpha-interferon. The MT-II pseudogene shows features of a processed retrogene. Comparison of the amino acid sequences of the MT-II protein and the amino acid sequence coded by the pseudogene shows that the two sequences diverged approximately 5.7 million years ago. Besides these two genes described, Southern blot analysis using a probe containing the MT-I second and third exon revealed a family of MT genes in the rabbit.

INTRODUCTION

MTs are small, 6,000 dalton, cysteine-rich proteins that bind the heavy metal ions. Their synthesis is induced by Cd, Zn, Cu, glucocorticoids, alpha interferons, carbon tetrachloride, iodoacetate, stress, etc.(1) These proteins are thought to be involved in homeostasis(1) and metal detoxification(2). MTs are present in a wide variety or organisms. They are usually isolated as two major isoforms (MT-I and MT-II). At the DNA level, Southern blot analysis reveals numerous MT-related genes (for ex. there are 12 such bands in human). As a first step to the study of their synthesis, their promoter sequences, their trans-acting factors and their evolution in the rabbit, we have cloned and characterized one MT-I and one MT-II pseudogene.

METHODS

New Zealand rabbit DNA was purified from 8 grams of a female rabbit liver as described(3) and digested with EcoRI. The 4-6 Kb EcoRI-digested DNA fragments were fractionated by sucrose gradient and inserted by ligation at the EcoRI site of lambda gt10(4). The DNA was then packaged(5) and the phages were grown in E. coli C600 Hfl (F^-, lacY1, leu, B6, tonA21, thi-1, thr-1, HC1, lambda$^-$). The phages were screened as described by Benton & Davis(6). About $2x10^6$ phages DNA of the genomic library were probed with a nick-translated mouse MT-I cDNA(7). Hybridization was carried out in 50% formamide at 37°C for 16-60 hours. The filters were washed 30 min at 25°C in 0.5xSSC, 0.1% SDS. Plaques that hybridized with the probe were picked and purified through 3 further rounds of plaque screening. Two clones MT44 and MT48 (containing a 4.4 Kb DNA insert and a 4.8 Kb insert) that hybridized with the mouse MT-I cDNA were isolated. The insert fragments were subcloned in pUC13(8) and amplified. The restriction map of the two fragments were determined and the appropriate fragments were subcloned in M13(9) vectors mp18 and mp19 for sequence analysis by the chain termination procedure(10). To identify other MT-related genes in the rabbit, Southern genomic blots were carried out. Rabbit liver DNA was digested with different restriction enzymes separately. Digested DNA was separated on a 1% agarose gel, transferred to nitrocellulose, hybridized *à la* Southern (11) and probe with a 1.4 Kb fragment spanning the SmaI-StuI-PstI-SphI-PstI-EcoRI sites from the MT-1 (MT44, figure 1) rabbit 4.4 Kb insert. The DNA probe used contains the second and third exon of MT-I. Hybridization was carried out at 42°C and washing of the nitrocellulose paper at 58°C in 0.1xSSC, 0.1% SDS.

RESULTS

Figure 1 shows the restriction map of the 4.4 Kb insert of clone MT44 and the 4.8 Kb insert of clone MT48. A fragment of 2.7 Kb (from the 4.4 Kb insert) and a fragment of 1.8 Kb (from the 4.8 Kb insert) that hybridize with the mouse MT-cDNA were sequenced. Figure 2 shows the sequence of the 2.7 Kb fragment that contains the rabbit MT-I coding gene. Figure 3 shows the sequence of the 1.8

Kb fragment that contains the rabbit MT-II pseudo gene. That the first gene belongs to the MT-I family and the second gene to the MT-II family are indicated by the absence of an aspartate at positions 10 or 11 of the peptide for the first gene while the presence of an aspartate is characteristic of the MT-II family (19). Southern genomic blots (fig. 4) using the second and third exons of the cloned MT-I gene as a probe (see Materials and Methods) show many bands of different intensities. As expected, digestion with BamHI yields a 1.8 Kb band while digestion with EcoRI yields a 4.4 and a 4.8 Kb band showing that our two clones were not rearranged. In addition, one finds in EcoRI digests, 2, 2.8, 3 and 8 Kb and in BamHI digest, 2.2, 2.6, 2.9, 4.3, 4.8, 6 and 8 Kb MT-I related DNA fragments that are most probably other rabbit metallothionein genes not yet isolated.

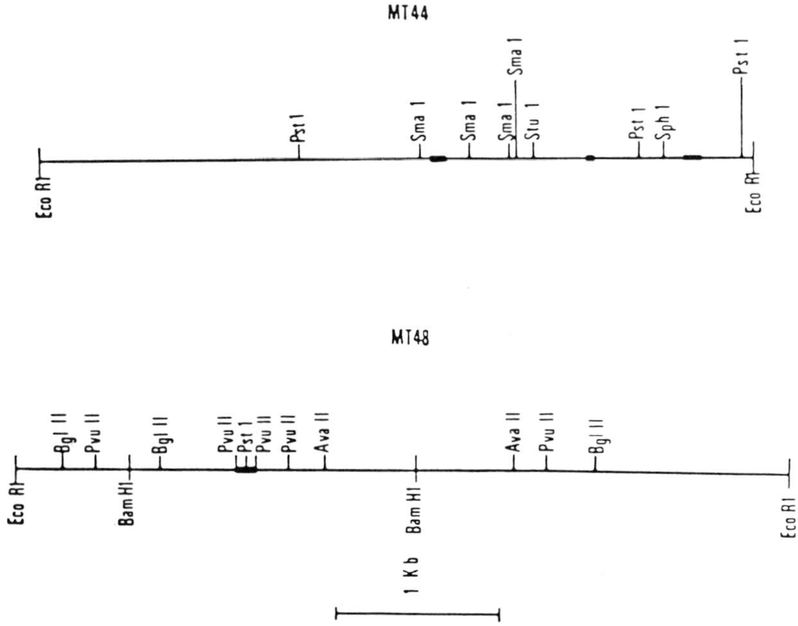

FIGURE 1. Restriction enzyme maps of the cloned insert of lambda MT44 and MT48 (transcription proceeds from left to right). The exons are shown with thickened lines.

```
              CTGCAGTGCCAGCATCCCATATGGGCGCCAGTTCAAGTCCTGGCTGCTCCACTTCCAAT   -1050
               PstI
CCAGCTCTCTGCTGTGGCCTGGGAAAGCAGTAGAAGATGGCCCAGGTCCTTGGGCCCCTGCACCCACGTGGGAGA   -975
CCCTGAAGAAGCTCCTGGCTTCGGATCAGTGCAGCTCCAGCCGTTGCAGCCAGTTGGGGAATGGACCAACAGATA   -900
GAAGACCTCTCTCTCTCTCTCTCTCCCTCTCTCTGCCTTTCCTTCTCTCTCCGTGTAACTCTGACTTTCAAATAA   -825
ATAAATCTTTAAAAAAAAATGAAAGAAAGAAAAGGATGATAAACAGAGCTTGGCACGGTCTACAGGCTATATAAT   -750
AGCTAAAGATCCTGAAAGAGAAACCATCACTTTTCTGCTTTTTAAGGTAACACAGGCTCTGTATACTGCCTAAGG   -675
                    Alfa Interferon                    MRE11   API
ACAAGACTTTAAAATGAGCTTTTCCCCTGGGGCATCCTGGAGTAGGTGGGGTGGCTAGGCTGAGGGGAGCCCC     -600
TGGAGAGTCTTCCAGATAGGAGGACCCAGACAAATGTGCCCAGCCACCCCTCAGGCTCAGGGAATTGGGGGAGG    -525
                            GRE                  AP2               MRE10
                                      Z
GAGGCAGAGCCCCGGATCTGCTTTCCTGCTGTGTACGAGAGAAACGGGGCCAAGGACTGGCATGGAACAAGGAC    -450
                                                       Z
AGGCACAGCGGGGGACGCAGGGGTCATGTTCATGGCCAGACCTCTGCACACAACTCCCTCGCTGCGCTCCACGGC   -375
                       MRE9                          MRE8
                                                Z
CAGCACCCCATCCCGGGCCGTGAGCAGAGGGCTGGGTTGCACACGCACTCCGGGAAAGGTGAAGCTGCGGTCAA   -300
          AP2 SmaI    BLE3                   BLE2         AP3
CGGTGCGAGGGGGCACGGGGAAGTGCCGGGAATTGGAGAGGGGGCGGGAAGGCTACAGGGAAGTCCAATACAGCC   -225
                                            SP1                         CCAAT
                                                       Z        DS
GGGCGCACAGCCTCGCTCGAGCGATCCGGAGCCAAAGAGCCTCCCCCGCCGTGCGCACCGAGCGGCGCTCCGGGA   -150
   MRE7                                     I  SP1 MRE6                 II

GTCGTGCGCCCGGCCCGCGCGGTGAGCTCAGCACGGCACGGACTGCGGGCCGTCGCCAGAGCTGGGGCGGGTGCA   -75
      MRE5              AP1    MRE4                      MRE3       SP1  BLE1

AGGGCCGGGGCGGCGCCTTTGCGCCCGGGCTCCCTCCCGCGGCAGTTAAAAGGAGCCGGCGCCCGCCTGGGCTCCA   1
  MRE2  SP1 GC  MRE1     SmaI            TATA                         CAP ↑
GCGCGCCTTCCACGAGCCTTGCGGAGTCTTCGCCTCTCACTTTGAACGTCGGACCTCACCTCGCTCGAAATGGAC   76
                                                                      MetAsp
CCCAACTGCTCCTGCGCCACAGGTAAGGGACGCTGGCTGTGGCTTGGGATGCGATAGCCCTGGTGGGGGAAG     151
ProAsnCysSerCysAlaThrG
AGGCCACGTTCAGGTCGCTGAGCTGAGAGCGGTTCCAGCTTCCCGGGGCTTCTGGTTTCTTCGCCAGAGCCGAGG   226
                                         SmaI
TCCCGGGGCTCCGGGCTGGCCCGCCTGACGTCATCCAGCAGGTCGGGGCGAGGGGTGGGGCTGGTGGCTGTGA    301
 SmaI
CCCCGCGCTTCGTGCAGGCCTCCGAGGACCCCGGCTGGCTGGAAGCAGAGAACCTTGCCTTTTCTGAGTTCAGG   376
               StuI
ACATAAGGTCCTTGCCCCTGGTCCCAGTATTCCTGGACTGGGCTGGATAAGACCAGGAGGACCCGTGCTCTTCC   451
CGGAGTGAGGGGACAGAGTGAGGTTTCTCTTTCTTTTCCCTGAGCGGGACAGGAACTCTCGGGTCTGATCTGCAT  526
CACACCCATCGCTCATTGCTGGCCTTTCTCCTCCTCCCAGGCAACTCCTGCACCTGTGCCAGCTCCTGCAAATGC  601
              *
                                        lyAsnSerCysThrCysAlaSerSerCysLysCys
AAAGAATGCAAATGCACCTCCTGCAAGAAGAGTGAGTGTGGCTTTCCCTGGAGACCGGGCTGGGGGCGGGCGG    676
LysGluCysLysCysThrSerCysLysLysS
ACCCAAGGCCAGTGAAGGTGAAGGAGTCGGGGGCTGAGCAGGCAGGAACTGGCCGGTGCCCAGCTTCCCCTCAGC  751
CCCTGATTTGGGGTCAGAGCTCCAAAGGCCATGGAGAAGAGTGTGTCCCAACTTCCCAAGATAGACGGCCCAGAG  826
GGGACAGCCACTGTTGTCAGGGGCAGAGCCAGAACTTGAACCTTTGTCTCCTGTGTCCTGATGGAGCCTTCCAAC  901
CCACCCAGGTCTGAAGCACTGGATGCACTTGACCTTGGGGACCACGTGTGGCTTCTGAGCTGCAGGGTCGACCTT  976
                                                            PstI
TTCCTATGTAGGCTCCCAGGGCTGCTTTGGAGGGGGTAGACCCCTGTCTCATCCTGAGGAACTCCTCACATCCCT  1051
GGACAGTTTGGTGGCTCCATTGCTCCCAGAGGTCCTGTTGCTGCCCCTCTGTGTGACTCCCAACCCAAGCATCCA  1126
CCAGCTGTGTGCCTGACCGGGCATGCCAGCTAGAACTAGGGCTCCTGCCGGGCTGGGGGCCGAGCCTGTGCCAGG  1201
                      SphI
GAGGGGGCCAGTCAAGTCAGCTCTGACCTCTTCCCTCTCTCTCTCCTCTTCCCAGGCTGCTGCTCCTGCTGCCCC  1276
                         *
                                                      erCysCysSerCysCysPro
GCCGGCTGCACCAAGTGTGCCCAGGGCTGCATCTGCAAAGGGGCATCGGACAAGTGCAGCTGCTGCGCCTGACGC  1351
AlaGlyCysThrLysCysAlaGlnGlyCysIleCysLysGlyAlaSerAspLysCysSerCysCysAla
CCGCAGAGCCCGGCTCCGGATGTGAAATAGTGCACTGTGTACAAACCTGGATTTTTTTTTTTTTCATACGAGCC   1426
                       III                             Poly T
TCATTTGCTACATTCCTTTTTCTATGAGAGATGTGGATGACAATAAACCCATTCAGACTTTCTGTGGCTCCAGGT  1501
                                         Poly A
GCCTTTGGTGTGTGGGAACATGGACTTGGTTGTAGTCGGGATGTAGAAGAAGTCAGTTGTGGGGAGGTGACACTG  1576
GGAGTGGGGGTCCGGGTACAAGGTGTTCTAGGTGTGCTGCAGCCTGGTCTAGCCTGGGCTGTCCTTTGTGGGCTG  1651
                   GRE                 PstI
TGGTGGGGAATTC
       EcoRI
```

FIGURE 2. DNA sequence of the MT-I gene locus. Numbering is from the proposed mRNA CAP site. Several proposed regulatory elements with the 5' promoter region and the conserved elements in the 3' untranslated region which are discussed in the text are underlined and labelled.

A

```
                GGATCCAGACTGCCTGGTTCAGTTTCTGGCTCTGCTACTCATTAGCTGTGTGACTGGGAAAGGC  -901
                BamHI
AATCTTTCTGTACTTCAGTTGTTTCCTTATATACAAATAGAGATAAGGACTTTATCCGAGAGTTATGAGAATTAA  -826
GTGCATTAATATTTTTATTTATTTATTTATTTATTTATTTGACAGGCAGAGTGGACAGTGAGAGAGAGAGA  -751
GACAGAAAGAAAGGTCTTCCTTTGCCGTTGGTTCACCCTCCAATGGCCGCCGCGGCCGGCGCACCGCGCTGATCC  -676
GGTGGCGGATGCCAGGAGCCTGGTGCTTTTCCTGGCCTCCCATGGGGTGCAGnnnnCAAGCACCTGAGCCATCCT  -601
CCACTGCACTCCCTGGCCACAGCAGAGAGCTGGCCTGGAAGAGGGGCAACCGGGACAGAATCCGGCGCCCCGACC  -526
GGGACTAGAACCCGGAGAGCTGGCACCGCTAGGCGAGGATTAGCCTAGTGAGCCGCGGTGCCGGCCTAAGAGCAT  -451
TAATATTTAAAGTGTTGAAAAAAAAAAAAAAAAAAAACTAAAATGGACCGGCGCCGTGGCACAGTAGGCCAAGC  -376
                                                        AvaII
CTCTGCCTGTGGGgCTGGCATCTCATATGGCGCCAGTTCATGTGGCTGCTCCTCTTCTCATCCAGCTCTCTGCT  -301
ATGGCCTGGGAAGGCAGTAGAAGATGCCCAAGTGCTTGGGCCCCTGCACCTGCATGTGAGACCTGGAAGAATTTC  -226
CTTGCTCCTGGCTTCGTATTGGCCCAGCTCCAGCTGTTGCGGCCATTTGGGGAGTGAACCAGTGGATGGAAGACC  -151
                                      PvuII
TTTCTCTCTGTCTCTCTCCTCTGTAACTTTACCTCTCAAATAAATGAATAAAATCTTTTAAAAATATTTGAAGTG  -76
TTTAACACACCTTCCAACTGCCGGACTGACTCTTCACCTTTCGCTTTGTATTTCCAGCTTcacCTGGGCTCAAAA  -1
                                                                           M
TGGAGCCCAGCTGCTCCTGCATCACCAGTGACTCCTGCACCTGTGCCAGCTCTTGCACATGCATGTCCTGCAGGA  75
etGluProSerCysSerCysIleThrSerAspSerCysThrCysAlaSerSerCysThrCysMetSerCysArgL
          PvuII                                                     PstI
AGAGCTGCTGCCCAGTGGGCTGTGTCCAGTGTGCCCAGGGCTGCATCTGCAAAGGGGCATCAAACAAATGCAGCT  150
ysSerCysCysProValGlyCysValGlnCysAlaGlnGlyCysIleCysLysGlyAlaSerAsnLysCysSerC
                                                                     PvuII
GCTGCGCCCCTGAGAGCCTGGCTCCCACATGCAAATAGAGCAACGTGTGCAAACCTGGATTTATACCCCCATACA  225
ysCysAla  Ter
ACCTGACTTCTATGCTGCATTCCTTCTTGTGTGAAATATGTGAATGAGAATACAAGTTGTTGATTTTAAAAGAAT  300
                                                                AhaIII
AAAGTGTTTAGAACATGGTCtGGGGCCAATGTCGTGGCATAGCTGGTTTAGCCACTACcTGCGATGCTGGCATCC  375
CACATGGGTACTGGTTTGAGTCCTGGCTGCTCCACTTCCAAGCCAGCTCCTGCTAATGCACCCAGGCCTGGCTGT  450
TGCAGCCATTTGGGGAGTGAACCATAGAATGCAAGAAGATGGCTCAGTGCTTGGGCACTTGCACCCAAGTGGAGA  525
CCCAGATGCAGCTCCTGGTTCCTGGcTTTGGCTTGGCCCAGCCCTGGcCATTGCAGCCAGTTGGGGAATGAACCA  600
GCAGCTTGGAAGATCTCTCTGTCTCTCTCTCTCTCTGTAACTCTTTCAAATAAGCAAATCTGGCCGGnGCCGC  675
            BglII
GGCTCAACAGGCTAATCCTCCCTAGCGGTGnCGGcACACCGGGTTCTAGTCCCGGTTGCCCCTCTTCCAGGCCAG  750
CCCTCTGCTGTGGCCCGGGAGTGCAGTGGAGGATGGCCCAGGTGCTTGGGCCCTGCACCCCATGGGAGACCAGGA  825
AAAGCACCTGGATCC
           BamHI
```

B

```
... ATTT...GA...AA...CT...AAAT...CCTGGGCTC...ATG...GCCCCTGA...CCTGGCTC...AAATAG...CT...
                     →           ———————→                      ————————→           →
     A     B    C    DR2    D       DR1       1      53  Ter1      DR1      62  Ter2 DR2
```

FIGURE 3. A: DNA sequence of the MT-II pseudogene (MT-IIψ) locus. Numbering is from the initiation codon. The direct repeats flanking the gene and the repetitive squences which are discussed in the text are underlined. B: Schematic representation of the MT-II pseudogene locus. The coding region is boxed in. Numbering refers to codon numbers. Ter 1 and Ter2 are termination codons. Direct repeats are indicated by arrows. A, B, C and D designate repetitive sequences.

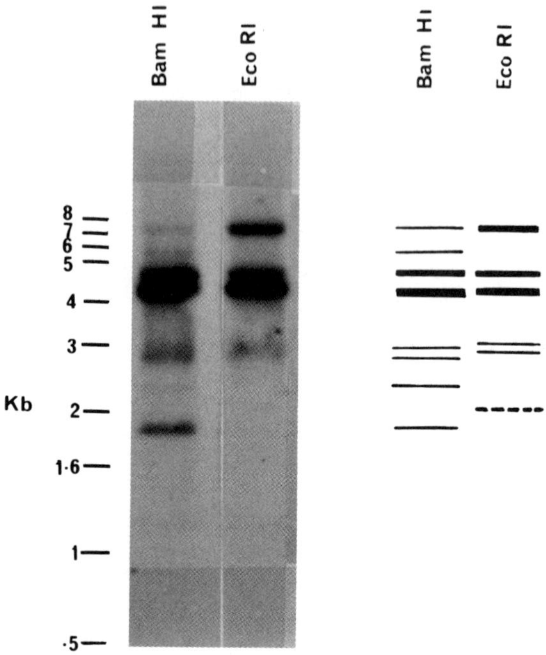

FIGURE 4. Southern genomic blots of rabbit DNA using as a probe a 1.4 Kb fragment that contains the second and third exon of the rabbit MT-I. A: BamHI digest; B: EcoRI digest.

DISCUSSION

The MT-I Rabbit Gene.

<u>Exons and introns</u>. We deduced the intron-exon junctions of the gene by comparing the DNA sequence and the amino acid sequence of the rabbit metallothionein protein(12). The rabbit MT-I, the mouse(13,14), human(15,16) sheep(17) and rat(18) MT genes share similar features: two introns interrupt three exons at the same amino acid positions. Exon 1 of the gene encodes the 5' untranslated region and amino acids 1 to 9 1/3; exon 2 encodes amino acids 9 2/3 to 30 1/3; exon 3 encodes amino acids 31 2/3 to 61 and the 3' untranslated region. Introns 1 and 2 are 468 and 624 base pair long respectively. The

two introns obey the GT-AG rule. The sequences of the exon-intron boundaries (or donor sequences), the intron-exon boundaries(20) (or receptor sequences) and the putative lariat branch points(21) (indicated with asterisks in fig. 2) all conform well with the consensus sequences for these structures.

The 3' untranslated region. Two interesting sequences present here have their counterparts in other mammalian MT genes. First, TGTGAAATA 23 bp after the termination codon (labelled element III in fig.2) is very similar to the sequence TGTAAATA present in the untranslated region of the mouse MT-II gene(14) and in the Chinese hamster MT-I and MT-II cDNAs(22). Second, an oligothymidylate sequence (Poly T in fig.2) is present after the termination codon. This sequence is also present in the 3' untranslated region of the mouse MT-II gene(14) and the monkey MT cDNAs(23). The functions of both sequences are unknown. The expected polyadenylation signal (Poly A in fig.2) is 119 bp from the stop codon.

The 5' promoter region. The 5' flanking region of the gene shows the typical features of eukaryotic genes. These features include the CAP site, TATA box and CCAAT box. The transcription CAP site (indicated by an arrow in fig. 2) was identified by comparison with other MT genes(13,14,15, 16,17,18). The CTGGGCTCCAGCGCGCCTTC sequence of the CAP site agrees well with the consensus sequence of CTRNRCYCCAYCACGCCTYC (R=purine)(17). Because in most eukaryotic genes the TATA box (consensus sequence 5'-TATA^{T}A^{T}-3'(24)) is centered 25 to 30 bp upstream from the CAP site, we suggest that the sequence TTAAAA (labelled TATA) centered at 23 to 25 bp, shares similar functions here. The sequence GTCCAATAC labelled (CCAAT) has 6 out of 9 bases in common with the CCAAT box consensus sequence GGTCAATCT(25).

Comparison of this promoter sequence with the promoters of other MT genes identified a number of oligonucleotide sequences that are recognized by *trans*-acting proteins involved in either basal level transcription or in the induction by metals, glucocorticoids and alfa interferon. Among them are sequences that are similar to the consensus sequence CTNTGCRCNCGGCCC (R=purines, N=any base) derived by Stuart *et al*.(26) for the GC rich metal regulatory elements (MRE). Here, 11 sequences (underlined and labelled MRE1 to 11, fig.3) have 9 or more bases identical to these 15 bases. Some of them are similar to the consensus sequence in both orientations. A consensus sequence (GGGCGYGTG) for basal level (BLE) have been derived(17). Three such similar BLEs (BLE1 to 3, fig.2) are detected here with BLE2 in the reverse orientation. Their sequences differ from the consensus in only 1 out of 9 bases. MRE1, MRE2 and BLE1 are highly conserved elements of functional MT genes. They are present in the rat(18), mouse(13,14), human(15,16), sheep(17) and rabbit MT promoters. The number of copies and locations of MREs and BLEs distal to the conserved region of MRE1, MRE2 and BLE1 are more variable among MT genes. This rabbit MT promoter contains potentially more copies of MREs than any previously isolated MT promoter.

Elements similar to the glucocorticoid and the interferon regulatory sequence consensus (GGTACANNNTGT TCT)(27) and (TTCN^{G}NACCTCANGCGTTTCTC^{C}TCTCT) (28) are labelled GRE and alfa-interferon (fig.3). The GRE located after the polyA signal is identical to the glucocorticoid regulatory sequence consensus. Its location suggests that it either regulates MT1 or it belongs to another gene promoter downstream of MT-I under the control of glucocorticoid. Element Alpha Interferon and its consensus sequence have 19 bases in common. Inside has been inserted the sequence AAAGAGAA. When taken in the reverse orientations, this insert and its neighboring bases have a 70% similarity to the interferon regulatory element consensus sequence in the reverse orientation:

TTCN$^{G}_{C}$N ACC	TCANGCGTTTCT$^{C}_{T}$CTCTCT	CONSENSUS
TCCTGAAAAGAGAAACCATCA	CTTTCTGCTTTTTAAGGTAACACA	ELEMENT ALPHA INTERFERON
AGA$^{G}_{A}$AAGAGAAAC	GCN TGAGGTN$^{C}_{G}$NGAA	CONSENSUS IN REVERSE ORIENTATION

Four eukaryotic nuclear transcription regulatory factors: Sp1(29), AP-1(30), AP-2(31) and AP-3(32) bind to specific enhancer sequences of simian virus 40 and MT genes(33). Factor Sp1 which binds to GGGCGG or CCGCCC in the reverse orientation(29) is repeated four times (underlined and labelled Sp1 in fig.2). AP-1 which binds to the consensus sequence ^{C}TGACT^{C}A(30) is repeated twice (labelled AP-1 in fig.2). Factor AP-2 binds to the consensus sequence CC^{C}C^{A}GCC(31). Two sites of similar sequences can be identified in the rabbit MT-I gene (underlined and labelled AP-2 in fig.2). A consensus sequence has not yet been determined for the factor AP-3. However, AP-3 has been shown to protect the core sequence TGTGGAAAGTCCCA of the simian virus 40 enhancer from nuclease digestion(32). A sequence (underlined and labelled AP-3 in fig.2) in the rabbit MT-I gene contains a similar core sequence.

Besides these elements known for their specific functions, two conserved sequences, a region of diad symmetry and 4 regions of alternating purine and pyrimidine are present in this promoter. Element I (fig.2) has 12 bases identical to the sequence CCAAAGGGNNGGTCCCGC which is conserved in all the MT-I genes but not the human MT-IIa gene(19). Element II is similar to the conserved promoter sequence RGCNGNNYTCCNGGAAYY identified by Peterson and Mercer(17). As in the mouse MT-I(13), sheep MT-Ia(17), human MT-Ia(15) and human MT-II genes(16), MRE1 is in a region of imperfect diad symmetry (overlined and labelled DS). However, this diad symmetric structure is not observed in the mouse MT-II promoter(14). The 4 regions of alternating purine and pyrimidine are overlined and labelled Z. Although no function has yet been identified for these 7 sequences, the fact that they are also detected in most other mammalian metallothionein promoters suggests that they are recognized by not yet identified <u>trans</u>-acting factors.

<u>The rabbit MT-II pseudogene (MT-IIψ)</u>. As shown in fig.3, there are 62 codons between the Met codon and the Ter2 codons. The first 53 codons have a high degree of homology with the other 53 codons of other metallothioneins while the last 9 codons (from 54 to 62) have no homology. This is reminiscent of the situation which prevails between beta hCG, FSH and LH (37) where the ancestral termination codon of beta hCG was put out of phase by a mutation resulting in the addition of about 30 amino acids to hCG. By analogy we shall show that most probably 1) Ter1 was the original termination codon and not Ter2 2) the coding region was the object of at least two recombination events and other additions or deletions 3) this gene is a pseudogene which has been integrated in the rabbit genome at the DR1 site.

Comparison of the similarities between the MT-I and MT-IIψ gene reveals 3 regions of very high homology separated by two deletions resulting from two recombination events. The two deletion removed 5 codons (21 to 26) and 3 codons (34 to 37) respectively from the MT-1 gene by the slipped mispairing mechanism (34) (fig.5). The sequence TGCAAATGC is present twice in this gene from codons 19 to 21 and from 24 to 26. This therefore promotes a slip during recombination and the loss of 5 codons for MT-IIψ. Similarly the sequence CTGCTGC is the sequence for codons 32 to 34 and for codons 35 to 37. A slip and a recombination delete 6 bases. Thus the last codon (no.61) of MT-1 correspondonds to codon 53 of MT-IIψ placing the ancestral termination codon at Ter1. After codon 53, two C are present before Ter1 ATG. This

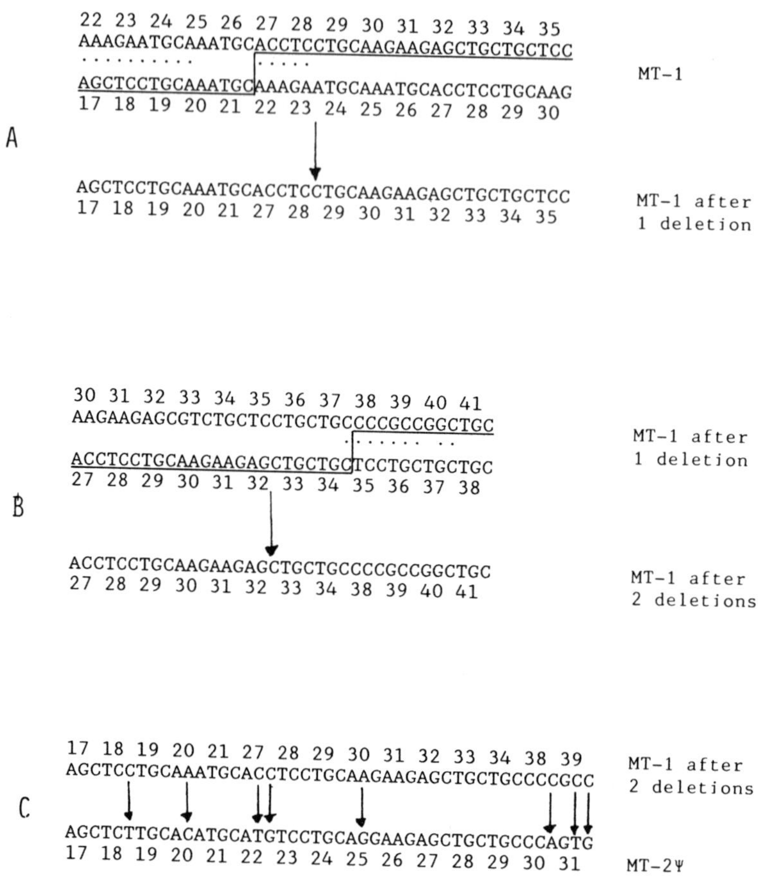

FIGURE 5. Possible recombination events of the rabbit MT-I cDNA for the deletion of direct repeat sequences to give the pseudogene MT-2ψ. The two upper sequences in part A and B of this figure depict a region of the MT-I cDNA (numbers refer to codons from the initiation codon ATG); the lower of the two sequences are the upper sequence shifted 5 or 3 codons to the right. Dotes indicate the alignment of identical nucleotides: a row of these indicate a region of direct repeat in the cDNA; the 2 predicated sequences which would occur after deletions caused by slipped mispairing mechanism(17) are also given. Part C of the figure shows the substitutions that could have taken place for the evolution of the pseudogene MT-2ψ from the recombination deleted cDNA from the region shown.

insertion of two bases before Ter1 or some other event put Ter1 out of phase and promote Ter2 as a second possible termination codon. Finally, either before or after these three genetic events, other point mutations occurred.

Similar to most-processed pseudogene(35), the rabbit MT-IIψ shows no intervening sequences, no promoter sequences and is flanked by direct repeats (8 and 9 base pair: CCTGG(G)CTC (labelled DR1, fig.2B)). In contrast to most other processed pseudogenes, the sequence homology between MT-IIψ and its functional counterpart does not cease at the points corresponding to the beginning and end of the transcript from its functional gene. In fact, sequence homology stops almost immediately 5' and 3' to the coding region. Moreover, the 3' poly A sequence present in most processed pseudogene is absent and the flanking direct repeats are not rich in adenosyl residues. This absence of adenosyl residues in DR1 direct repeats flanking immediately the coding sequence suggests that the mechanism of insertion of mRNA or cDNA in the formation of MT-IIψ may differ from that of other processed pseudogenes. The insertion of the mRNA in cDNA in the genome might have occurred at the sites in the middle of the mRNA or cDNA close to the beginning and the end of the coding sequence instead of the ends of the mRNA or cDNA transcript. Alternatively, further mRNA splicing might have taken place removing the 5' and 3' noncoding regions together with the poly A tract before the insertion of the mRNA or its cDNA transcript into the genome. The possibility of the 5' and 3' noncoding regions being removed by deletions after the insertion of the mRNA or cDNA is possible but unlikely because this process will predict the absence of the direct repeats as well.

Several repetitive sequences are found 5' and 3' to the coding region (labelled A,B,C and D, fig.3B). The second direct repeat (DR2, fig.3B) probably mark the sites of a second integration event which occurred in this region of the genome before or after the DR1 integration since repetitive CT sequences have been implicated as sites of DNA recombination and integration(36).

Nucleotide and Amino Acid Homologies. From the isolation of rabbit MT genes, the evolution relationship of MTs can be examined (Tables 1 and 2). Table 1 shows the nucleotide and amino acid homologies between the coding sequence of rabbit MT-IIψ and other cloned mammalian MT genes or cDNA. Table 2 shows the nucleotide and amino acid homologies between the coding sequence of rabbit MT-I and other cloned mammalian MT genes or cDNA. Similar to the method of analysis of other pseudogenes, deletions and insertions are not included in the calculations. As can be seen from Table 1, MT-IIψ has a higher amino acid homology

TABLE 1

	Nucleotide homologies	(%)	Amino acid homologies	(%)
Rabbit MT-I(16)	136/159	(86)	39/53	(74)
Chinese hamster MT-I(24)	128/159	(81)	38/53	(72)
Rat MT-I(23)	128/159	(81)	40/53	(75)
Mouse MT-I(21)	125/159	(79)	38/53	(72)
Sheep MT-Ia(20)	133/159	(84)	40/53	(75)
Monkey MT-I(25)	130/159	(82)	39/53	(74)
Human MT-Ia(18)	134/159	(84)	38/53	(72)
Rabbit MT-II(27)			41/53	(77)
Chinese hamster MT-II(24)	123/159	(77)	35/53	(66)
Mouse MT-II(22)	127/159	(80)	37/53	(70)
Monkey MT-II(25)	137/159	(86)	40/53	(75)
Human MT-IIa(19)	140/159	(88)	41/53	(77)

Nucleotide and amino acid homologies between the 53 codons of the rabbit MT-II with other cloned mammalian MT genes or cDNA. Deletions and insertions are not included in the calculations.

with rabbit MT-II than with rabbit MT-I, confirming the identification of MT-IIψ as a derivative of rabbit MT-II mRNA. As expected, pseudogene MT-IIψ shows less constrain on replacement site substitutions than the functional rabbit MT-I gene. For MT-IIψ, 55% of the nucleotide substitutions when compared with other MT sequences results in amino acid replacement while for rabbit MT-I, only 40% of the nucleotide substitutions results in amino acid replacement. For the 53 amino acids compared, there are 12 substitutions between MT-IIψ and its parent MT-I. Evidence shows that there is a rapid and constant rate of divergence of pseudogene sequences of 400 PAMs (1 PAM = 1 accepted point mutation per 100 residues per 100 million years)(18). This gives the approximate time of divergence of this pseudogene from its parent MT-IIψ gene of 5.7 million years. This estimate is probably conservative since we have ignored independent mutational events at a single site or superimposition. The rabbit MT-II protein isoform sequenced so far differs from other mammalian MTs in having an alanine residue inserted between residues 8 and 10(12). Because MT-IIψ DNA does not have this insertion, the insertion must have occurred very recently

TABLE 2

	Nucleotide homologies	(%)	Amino acid homologies	(%)
Chinese hamster MT-I(24)	162/183	(89)	52/61	(85)
Rat MT-I(23)	160/183	(87)	52/61	(85)
Mouse MT-I(21)	156/183	(85)	50/61	(82)
Sheep MT-Ia(20)	160/183	(87)	50/61	(82)
Monkey MT-I(25)	163/183	(89)	54/61	(89)
Human MT-Ia(18)	163/183	(89)	53/61	(87)
Rabbit MT-II(27)			54/61	(89)
Chinese hamster MT-II(24)	156/183	(85)	51/61	(84)
Mouse MT-II(22)	157/183	(86)	50/61	(82)
Monkey MT-II(25)	164/183	(90)	54/61	(89)
Human MT-IIa(19)	168/183	(92)	56/61	(92)

Nucleotide and amino acid homologies between the coding sequence of rabbit MT-I with other cloned mammalian MT genes or cDNA. Deletions and insertions are not included in the calculations.

(less than 5.7 million years ago). All mammals examined contain two major forms of metallothionein. Therefore, it has been suggested that these isoforms have arisen from a gene duplication prior to the mammalian radiation(2). This suggestion implies that interspecies homology of one isoform should be greater than the intraspecies homology of the two isoforms. Table 1 shows that the amino acid sequence of rabbit MT-I is more similar to that of rabbit MT-II than to that of other rodent MT-Is. Table 2 shows that rabbit MT-IIψ is more similar in coding sequence and amino acid sequence to rabbit MT-I than to other rodent MT-IIs. Thus, similar to the primate MTs the rabbit MT-I and MT-IIψ sequences do not substantiate the hypothesis that the two isoforms have arisen from a gene duplication prior to mammalian radiation. If the two isoforms did arise from gene duplication before mammalian speciation, a gene conversion between MT-I and MT-II or an evolution of MT-I and MT-II from one isoform with the inactivation of the other isoform might have taken place in the rabbit to make the two isoforms more similar. Interestingly enough, rabbit MT-I shows higher similarity in coding sequence and amino acid sequence to primate MT-Is and MT-IIs than to rodent MT-Is. Rabbit MT-IIψ also shows higher similarity in coding sequence and amino acid sequence to primate MT-Is and MT-IIs than to rodent MT-IIs. The explanation to this is not apparent.

SUMMARY

In summary, two members of the metallothionein New Zealand rabbit family have been isolated and sequenced. A MT-I coding and a MT-IIψ pseudogene have been cloned and characterized from a MT gene family. The MT-I gene possesses all the structures required for basal transcription, induced transcription as well as mRNA processing. The promoter region, with its numerous potential regulatory sequences, represents one of the more complete and extensive promoters isolated so far. About 30 possible regulatory sites have been identified. This promoter should be an excellent model to study gene regulation and to test the functions of the metal regulatory elements. The MT-IIψ processed pseudogene has been isolated and shown to have different features from most other processed pseudogene. Our analysis indicates, for the first time, a mechanism of formation of processed pseudogene in which a part of the transcript (only the coding sequence) is integrated into the genome.

ACKNOWLEDGMENTS

We thank F. Denis and S. Lewandowski for their help. Supported by MRC grant MT4860.

REFERENCES

1. Hamer DH (1986). Metallothionein. Ann Rev Biochem 55:913.
2. Beach LR, Palmiter, RD (1981). Amplification of the metallothionein-I gene in cadmium-resistant mouse cells. Proc Natl Acad Sci 78:2110.
3. McKnight GS (1978). The induction of ovalbumin and conalbumin mRNA by estrogen and progesterone in chick oviduct explant cultures. Cell 14:403.
4. Huynh TV, Young RA, Davis RW (1984). In Glover D (ed): "DNA cloning techniques: practical approach", IRL Press, Oxford.
5. Scherer G, Telford J, Baldari C, Pirrotta V (1981). Isolation of cloned genes differentially expressed at early and late stages of Drosophila embryonic development. Devel Biol 86:438.

6. Benton WD, Davis RW (1977). Screening gt recombinant clones by hybridization to single plaques in situ. Science 196:180.
7. Mbikay M, Maiti IB, Thirion JP (1981). Cloning and sequencing of cDNA for mouse liver metallothionein-I. BBRC 103:825.
8. Vieira J, Messing J (1982). The pUC plasmids, and M13mp7-derived system for insertion mutagenesis and sequencing with synthetic universal primers. Gene 19:259.
9. Yanisch-Perron C, Vieira J, Messing J (1985). Improved M13 phage cloning vectors and host strains: nucleotide sequences of the M13mp18 and pUC19 vectors. Gene 33:103.
10. Sanger F, Coulson AR, Barrell GB, Smith AJH, Roe BA (1980). Cloning in single-stranded bacteriophage as an aid to rapid DNA sequencing. J Mol Biol 143:161.
11. Southern EM (1975). Detection of specific sequences among DNA fragments separated by gel electrophoresis. J Mol Biol 98:503.
12. Wagner G, Neuhaus D, Worgotter E, Vasak M, Kagi JHR, Wuthrich K (1986) Sequence-specific ^{1}H-NMR assignments in rabbit-liver metallothionein-2. Eur J Biochem 157:275.
13. Glanville N, Durnam DM, Palmiter RD (1981). Structure of mouse metallothionein-I gene and its mRNA. Nature 292:267.
14. Searle PF, Davison BL, Stuart GW, Wilkie TM, Norsted G, Palmiter RD (1984). Regulation, linkage, and sequence of mouse metallothionein I and II genes. Mol Cell Biol 4:1221.
15. Richards RI, Heguy A, Karin M (1984). Structural and functional analysis of the human metallothionein-IA gene: differential induction by metal ions and glucocorticoids. Cell 37:263.
16. Karin M, Richards RI (1982). Human metallothionein genes-primary structure of the metallothionein-II gene and a related processed gene. Nature 299:797.
17. Peterson MG, Mercer JFB (1986). Structure and regulation of the sheep metallothionein-Ia gene. Eur J Biochem 160:579.
18. Andersen RD, Birren BW, Taplitz SJ, Herschman HR (1986). Rat metallothionein-I stuctural gene and three pseudogenes, one of which contains 5'-regulatory sequences. Mol Cell Biol 6:302.
19. Schmidt CJ, Jubier MF, Hamer DH (1985). Structure and expression of two human metallothionein-I isoform genes and a related pseudogene. J Biol Chem 260:7731.

20. Breathnach R, Chambon P (1981). Organization and expression of eucaryotic split genes coding for proteins. Ann. Rev Biochem 50:349.
21. Ruskin B, Krainer AR, Maniatis T, Green MR (1984). Excision of an intact intron as a novel lariat structure during pre-mRNA splicing *in vitro*. Cell 38:317.
22. Griffith BB, Walters RA, Enger MD, Hildebrand CE, Griffith JK (1983). cDNA cloning and nucleotide sequence comparison of Chinese hamster metallothionein I and II mRNAs. Nucleic Acids Res 11:901.
23. Schmidt CJ, Hamer DH (1983). Cloning and sequence analysis of two monkey metallothionein cDNAs. Gene 24:137.
24. Breathnach R, Benoist C, O'Hare K, Gannon F, Chambon P (1978) Ovalbumin gene: evidence for a leader sequence in mRNA and DNA sequences at the exon-intron boundaries. Proc Natl Acad Sci USA 75:4853.
25. Nishide T, Nakamura Y, Emi M, Yamamoto T, Ogawa M, Mori T, Matsubara K (1986). Primary structure of human salivary-amylase gene. Gene 41:299.
26. Stuart GW, Searle PF, Palmiter RD (1985). Identification of multiple metal regulatory elements in mouse metallothionein-I promoter by assaying synthetic sequences. Nature 317:828.
27. Klock G, Strahle U, Schutz G (1987). Oestrogen and glucocorticoid responsive elements are closely related but distinct. Nature 329:734.
28. Friedman RL, Stark GR (1985). -Interferon-induced transcription of HLA and metallothionein genes containing homologous upstream sequences. Nature 314:637.
29. Lee W, Haslinger A, Karin M, Tjian R (1987a). Activation of transcription by two factors that bind promoter and enhancer sequences of the human metallothionein gene and SV40. Nature 325:368.
30. Lee W, Mitchell P, Tjian R (1987b). Purified transcription factor AP-1 interacts with TPA-inducible enhancer elements. Cell 49:741.
31. Mitchell PJ, Wang C, Tjian R (1987). Positive and negative regulation of transcription *in vitro*: enhancer-binding protein AP-2 is inhibited by SV40 T antigen. Cell 50:847.
32. Chiu R, Imagawa M, Imbra RJ, Bockover JR, Karin M (1987). Multiple cis- and trans-acting elements mediate the transcriptional response to phorbol esters. Nature 329:648.

33. Fromm M, Berg P (1983). Simian virus-40 early-region and late-region promoter functions are enhanced by the 72-base pair repeat inserted at distant locations and inverted orientations. Mol Cell Biol 3:991.
34. Albertini AM, Hofer M, Calos MP, Miller JH (1982). On the formation of spontaneous deletions: the importance of short sequence homologies in the generation of large deletions. Cell 29:319.
35. Vanin EF (1985). Processed pseudogenes: characteristics and evolution. Ann Rev Genet 19:253.
36. Huberdeau D, Sylla BS, Herring-Gillam E, Bourgaux-Ramoisy D, Bourgaux P (1985). An excision event that may depend on patchy homology for site specificity. Mol Cell Biol 5:2608.
37. Fiddes JC, Goodman HM (1980). The cDNA for the -subunit of human chorionic gonadotropin suggests evolution of a gene by readthrough into the 3'-untranslated region. Nature 286:684.

Metal Ion Homeostasis: Molecular Biology and Chemistry, pages 49–57

DNA METHYLATION IS CORRELATED TO REGULATION OF HUMAN MT-IF, MT-IG AND MT-IIA GENES[1]

Nadia Jahroudi and Lashitew Gedamu

Department of Biological Sciences,
University of Calgary, Calgary, Alberta, Canada T2N 1N4

ABSTRACT We have studied the expression of three human metallothionein genes, MT-IF, MT-IG and MT-IIA in human hepatoblastoma (HepG2) and lymphoblastoid (Wi-L2) cell lines. All three genes are expressed in the HepG2 cell line in response to heavy metals. In addition MT-IIA is expressed at basal level and is induced by glucocorticoid hormone. In the Wi-L2 cell line, MT-IF and MT-IG genes are not expressed; however MT-IIA is expressed in response to heavy metals but not to glucocorticoid hormones. These observations suggest that the MT-I genes are expressed in a cell type specific manner and the induction of the MT-IIA gene by glucocorticoids is cell type specific. Transfection of Wi-L2 cells with appropriate constructs containing MT promoters fused to CAT gene (MT-CAT) resulted in CAT expression in response to inducers indicating the presence of active transacting factors in this cell line. DNA methylation appears to be involved in the regulation of these genes since azacytidine treatment of Wi-L2 cells resulted in the expression of MT-IF and MT-IG genes in response to heavy metals and the MT-IIA gene in response to dexamethasone induction. In addition, azacytidine treatment of HepG2 cells resulted in basal level expression of the MT-IF and MT-IG genes in this cell line. Analysis and comparison of the methylation pattern of MT-IF gene in HepG2 and Wi-L2 cells indicate that the promoter of this gene is methylated in a CG element located in at least one of the HpaII sites in the Wi-L2 cell line.

INTRODUCTION

Metallothioneins are a family of low molecular weight, cysteine-rich proteins (1). They are expressed in many different tissues and cell types when exposed to heavy metals, glucocorticoid hormones and a variety of other agents (1,2). In humans, a single MT-II polypeptide and at least five MT-I variants appear to be expressed at the protein level (3). The induction of these proteins is regulated mainly at the level of transcription. The genes coding for the human MT proteins are part of a multigene family which includes one MT-II gene, MT-IIA (4) and five functional MT-I genes [MT-IA (5), MT-IB (6), MT-IE (7), MT-IF (7,8) and MT-IG (9)]. Several cis-acting regulatory elements have been identified in the promoter region of MT genes (8,12). Repeated copies of a short sequence called the metal-responsive element (MRE) are responsible for heavy metal induction of these genes (10). In addition to the MREs, a glucocorticoid responsive element (GRE) and two copies of basal level enhancer elements (BLE), which are responsible for glucocorticoid induction and basal level expression of MT-IIA gene respectively (12,13), have also been found in the MT-IIA promoter.

The expression of some of the human MT genes has been studied. For example, MT-IA and MT-IIA genes are shown to be expressed in all cell lines (5,14). Differential regulation in response to different inducers has been reported for some of the members of this gene family (15). Furthermore, cell-type specific expression of the MT-IB, MT-IE and MT-IF genes has been reported (6,8,14) and DNA methylation is correlated to the cell type specific expression of MT-IB and MT-IE genes (6,14). In this report we have examined the expression of three human metallothionein genes, namely MT-IF, MT-IG and MT-IIA, in human hepatoblastoma (HepG2) and lymphoblastoid (Wi-L2) cell lines. We demonstrate that DNA methylation is involved in the regulation of expression of these genes.

MATERIALS AND METHODS

Cell Culture and Northern Blot Analysis of MT mRNA

Human hepatoblastoma (HepG2) and lymphoblastoid (Wi-L2) cells were maintained in modified Eagle's medium (GIBCO Laboratories) supplemented with 10% and 7.5% fetal calf serum, respectively (8). Cells were induced with 2 μM $CdCl_2$, 100 μM $ZnCl_2$, 150 μM $CuCl_2$ and 10 μM dexamethasone (Dex) under optimal conditions. In some cases, cells were treated with 5-azacytidine (8 μM) for 72 h prior to metal exposure. Total RNA was isolated by the guanidinium isothiocyanate extraction method (16) and Northern blot analysis was performed using MT specific RNA probes (15).

Transfection and CAT Assay

The expression vector, MT pGEM2cat, was constructed by blunt end ligation as follows: (i) The parent plasmid (pGEM2cat) was made by inserting the HindIII-BamHI fragment containing the CAT structural gene and the SV40 early region polyadenylation site from pSVOcat (17) into the PvuII site of pGEM-2 (Promega Biotech). (ii) The promoter region of MT-IIA (BamHI to HindIII, +75 to -770), MTIG (AvaI to ApaI, +64 to -635) or MT-IF (the first nucleotide before the translation start site to XbaI, +71 to -440) were inserted into the SmaI site, in the polylinker region of pGEM2cat. The orientations of insertions were determined by the dideoxy nucleotide sequencing method (18). Transfection was performed using DEAE-dextran as described by Grosschedl and Baltimore (19). After 30 h of transfection, cells were induced with heavy metals or Dex for 12 h, harvested and CAT activity was assayed (17).

RESULTS AND DISCUSSION

MT-IF and MT-IG Genes Are Expressed in HepG2 Cells But Not in Wi-L2 Cells

Based on the results of the time course studies, optimal conditions for accumulation of MT mRNA in response to heavy metals (cadmium, zinc and copper) and Dex were chosen to investigate the expression of MT genes in human hepatoblastoma (HepG2) and lymphoblastoid (Wi-L2) cells. Specific probes generated from the corresponding MT genomic clones were used to study the level of MT-IF, MT-IG and MT-IIA mRNAs in these cell lines. These probes contained the 3'untranslated region and part of the 3' flanking region of each gene and were shown to be specific when used either as RNA or DNA probes. The results of Northern blot analysis indicated that in HepG2 cells heavy metals led to induction of MT-IF (Fig. 1A) and MT-IG (Fig. 1B) genes. There was no detectable basal level accumulation of these mRNAs and were not expressed in response to Dex. Similar experiments in Wi-L2 cells have indicated that neither heavy metals nor Dex cause the induction of these genes (Fig. 1D and 1E). The results of these experiments suggest that the expression of MT-IF and MT-IG genes is regulated in a cell-type specific manner.

MT-IIA Gene Induction by Glucocorticoid Hormone is Cell Type Specific

Studies on the expression of MT-IIA gene in HepG2 and Wi-L2 cells have shown that this gene is expressed at the basal level and induced by heavy metals in both cell lines. However, Dex induces MT-IIA gene only in HepG2 cells and not in Wi-L2 cells (Fig. 1C and 1F). These results suggest that, although the expression of MT-IIA gene is not cell type specific, its induction in response to Dex is regulated in a cell type specific manner.

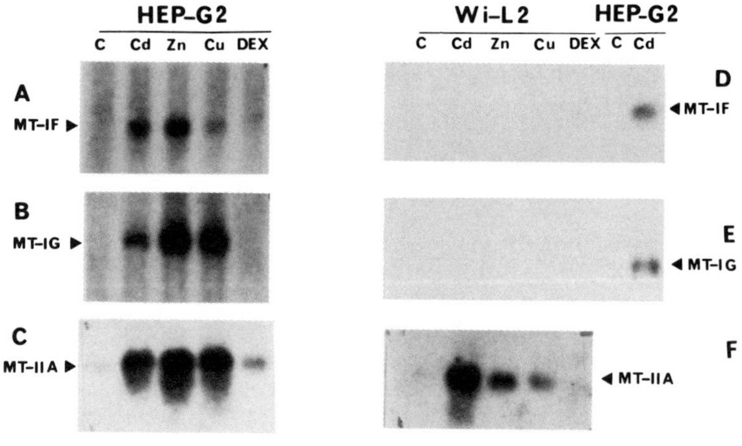

Figure 1. Expression of MT-IF, MT-IG and MT-IIA in HepG2 and Wi-L2 cells. Total nucleic acids were hybridized with the MT-IF (A,D), MT-IG (B,E) and MT-IIA (C,F) probes. Panels D and E, uninduced (C) and Cd induced (Cd) HepG2 cells.

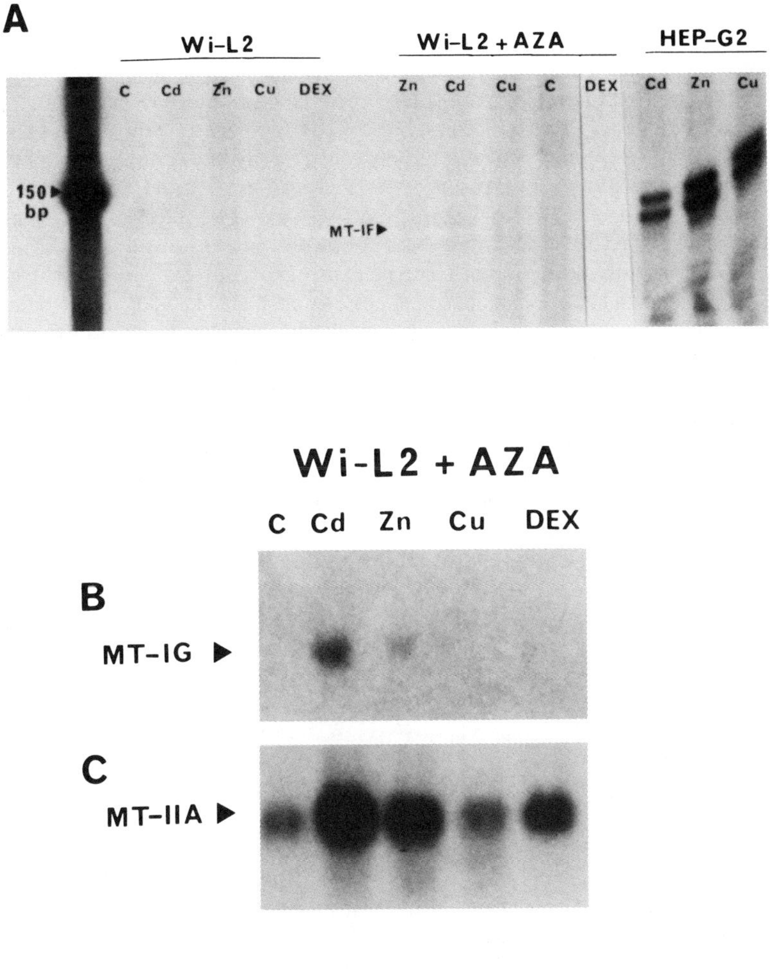

Figure 2. Expression of MT-IF, MT-IG and MT-IIA genes in aza treated Wi-L2 cells. (A) RNase protection analysis. Total RNA (20 µg) was hybridized with the MT-IF specific RNA probe (5 X 10^4 cpm), and processed as in Methods. The 150 bp DNA fragment marker is generated from a HpaII digested pBR322. (B,C) Northern blot analysis of total nucleic acids hybridized with the MT-IG (B) or MT-IIA (C) probe.

Induction of MT Genes After Treatment of the Cells With 5-azacytidine

In order to understand the nature of cell type specific expression of MT-IF and MT-IG genes and cell type specific induction of MT-IIA gene, the possible role of DNA methylation in MT gene regulation was investigated. Thus, Wi-L2 cells were exposed to heavy metals after azacytidine (aza) treatment and RNA was analysed as in Methods. Under these conditions, MT-IF (Fig. 2A) and MT-IG genes (Fig. 2B) were expressed in response to Cd and the MT-IIA gene was also significantly induced by Dex over basal level (Fig. 2C). The expression of MT-IF and MT-IG genes in aza treated Wi-L2 cells indicates that DNA methylation is correlated with cell-type specific expression of these genes. Futhermore, such DNA modification is also shown to be involved in cell type specific induction of MT-IIA gene in response to Dex. The possible role of DNA methylation in the regulation of basal level expression of MT-IF and MT-IG genes in HepG2 was also investigated. The exposure of HepG2 cells to aza resulted in basal level expression of both genes (data not shown), further supporting the important role of DNA methylation in the regulation of MT gene expression. Treatment of cells with aza may result in hypomethylation of the MT genes or the genes coding for trans-acting factors which are necessary for the activation of MT genes. These results cannot exclude either possibility.

Transfection of MT-CAT construct into Wi-L2 cells

Introduction of the human MT promoter gene fused to a reporter gene into Wi-L2 cells provides a useful technique to investigate whether or not trans-acting factors involved in MT gene induction are present in this cell line. MT-CAT constructs were, therefore, transfected into Wi-L2 cells and CAT activity was assayed. MT-IFcat plasmid was expressed in response to Cd and Zn induction. However, there was no increase in CAT activity above the basal level in Dex or Cu induced cells (Fig. 3A). A similar pattern of CAT activity was observed in MT-IG cat transfected Wi-L2 cells (Fig. 3B). The MT-IIAcat construct showed a high level of CAT activity in response to Dex as well as heavy metals (Fig. 3C). The pattern of expression of a MT-CAT gene in transformed cells was similar to that of each corresponding MT gene in aza-treated Wi-L2 cells (Figs. 2A-C). These results

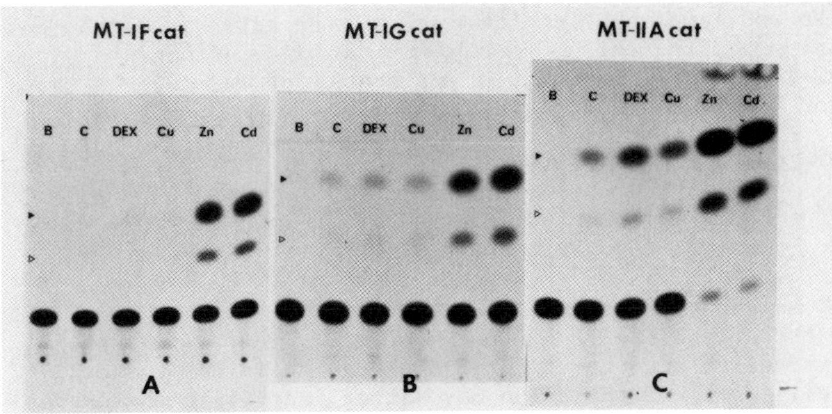

Figure 3. Expression of MTcat constructs in transfected Wi-L2 cells. Expression of MT-IFcat, MT-IGcat and MT-IIA-cat recombinant genes in response to no inducer (C), Dex, Cu, Zn and Cd. CAT activity was assayed by separating [^{14}C]-chloramphenicol and its acetylated forms [1- and 3-acetyl chloramphenicol (17)] using thin layer chromatography. B represents Wi-L2 cells transfected with pGEM2-cat.

demonstrate that trans-acting factors necessary for induction of MT-IF and MT-IG genes by heavy metals are present and functional in the Wi-L2 cell line. The observation that Dex induction resulted in a significant increase in CAT activity, as directed by the GRE-containing MT-IIA promoter, indicates that glucocorticoid receptors are functional in these cells. Therefore, these observations support the hypothesis that 5-azacytidine treatment of Wi-L2 cells results in activation of MT genes as a consequence of their hypomethylation. In particular, similarity in the expression pattern of these MT genes in aza-treated Wi-L2 cells and cells transfected with MT-cat gene constructs suggest that methylation of the promoter region of these genes controls their regulation.

We investigated the methylation pattern of the promoter region of the MT-IF gene in HepG2 (expressing) and Wi-L2 (non-expressing) cell lines using HpaII and MspI restriction enzymes. Our studies indicate that a 167 bp fragment is generated by digestion of HepG2 genomic DNA with both HpaII and MspI. However, this fragment is only present in MspI digested Wi-L2 DNA (data not shown). The lack of appearance of this band from HpaII digested Wi-L2

DNA indicates that at least one of the sites in either end of this fragment is methylated in the promoter of the MT-IF gene. The methylation status of other sites have not yet been identified. Similar experiments will be carried out to investigate the methylation patterns of MT-IG and MT-IIA genes.

CONCLUSION

The results of these studies indicates that DNA methylation of MT genes regulates cell type specific expression and induction of these genes in HepG2 and Wi-L2 cell lines. Methylation may either inactivate the expression of the gene by affecting chromatin structure or interfere with binding of trans-acting factors to cis-acting DNA regulatory elements.

ACKNOWLEDGMENTS

This work is generously supported by the Medical Research council of Canada. The MT-IIA gene is kindly provided by Dr. Dean Hamer. We would like to give special thanks to Janet Price-Haughey for her critical reading and helpful suggestions in preparing the manuscript and Norma Herrington for her excellent typing.

REFERENCES

1. Kagi, J.H.R., and Kojima, Y. Metallothionein II, Birkhauser, Basel (1987).
2. Hamer, D.H. Annual Rev. Biochem. 55 (1986).
3. Kagi, J.H.R., Vasak, M., Lerch, K., Gilg, D.E.O., Hunziker, P., Bernhard, W.R., and Good, M. Environ. Health Perspect. 54, 93-103.
4. Karin, M., and Richards, R.I. Nature 299, 797-802 (1982).
5. Richards, R.I., Heguy, A., and Karin, M. Cell 37, 263-272 (1984).
6. Heguy, A., West, A., Richards, R.I. and Karin, M. Mol. Cell. Biol. 6, 2149-2157 (1986).
7. Schmidt, CLJL, Jubier, M.F., and Hamer, D.H. J. Biol. Chem. 260, 7731-7737 (1985).
8. Varshney, U., Jahroudi, N., Foster, R., and Gedamu, L. Mol. Cell. Biol. 6, 26-37 (1986).

9. Foster, R., Jahroudi, N., Varshney, U., and Gedamu, L. J. Biol. Chem. In press (1988).
10. Stuart, G.W., Searle, P.F., Chen, Y.N., Brinster, R.L. and Palmiter, R.D. Proc. Natl. Acad. Sci. USA 81, 7318-7322 (1984).
11 Stuart, G.W., Searle, P.F., and Palmiter, R.D. Nature 317, 828-831.
12. Karin, M., Haslinger, A., Heguy, A., Dietlin, T., and Cooke, T. Mol. Cell. Biol. 7, 606-613 (1987).
13. Karin, M., Haslinger, A., Holtgreve, H., Richards, R.I., Krauter, P., Westphal, H.J., and Beato, M. Nature 308, 513-519 (1984).
14. Schmidt, C.J., and Hamer, D.H. Proc. Natl. Acad. Sci. USA. 83, 3346-3350 (1986).
15. Sadhu, C., and Gedamu, L. J. Biol. Chem. 263, 2679-2684 (1988).
16. Kaplan, B.B., Bernstein, S.L., and Gioio, A.E. Biochem. J. 183, 181-184 (1979).
17. Gorman, C.M., Moffat, L.F., and Howard, B.H. Mol. Cell Biol. 2, 1044-1051 (1982).
18. Begin, M.D., Gibson, T.J., and Hong, G.F. Proc. Natl. Acad. Sci. USA. 80, 3963-3965 (1983).
19. Grosschedl, R., and Baltimore, D. Cell 41, 885-897 (1985).

Metal Ion Homeostasis: Molecular Biology and Chemistry, pages 59–69

EFFECT OF HEAVY METALS ON PROTEIN BIOSYNTHESIS OF TWO DIFFERENTIATED HUMAN HEPATOMA CELL LINES[1]

P. Remondelli, M.C. Pascale and A. Leone

Istituto di Biochimica Cellulare e Molecolare, University of Naples, Naples, Italy 80131

ABSTRACT Exposure for ten hours to high concentrations of copper, zinc and cadmium induced the synthesis of several proteins in HepG2 and Hep3B, two differentiated human hepatoma cell lines. Metallothioneins and the heath shock Hsp70 appeared in both cell lines after induction. Other proteins were synthesized after exposure to heavy metals in a cell line specific manner. One of them was also induced following exposure to acute phase reactants.

INTRODUCTION

In mammals the liver plays a major role in the regulation of the homeostasis of heavy metals, balancing the excretion of toxic amounts of these ions in form of bile coniugates with the distribution of these elements to the other tissues via the secretion of specific serum protein carriers (i.e. ceruloplasmin).

Differentiated liver cell lines provide an excellent system to study the biochemical and molecular basis of the metabolism of metal ions *in vitro*. Our experiments have

---

This work was supported by a grant of Associazione Italiana per la Ricerca sul Cancro

been focused on the effect of copper, zinc and cadmium on protein biosynthesis of two well characterized human hepatoma cell lines, HepG2 and Hep3B (1). These cell lines secrete twenty serum proteins and have been successfully used to study liver tissue specific gene regulation (2).

The results presented show that apart metallothioneins and Hsp70, other proteins previuosly not characterized are induced by metals in these cells. Moreover, one of these proteins was also made in response to inflammatory agents.

METHODS

HepG2 and Hep3B cell lines were respectively cultured at 37°C in Eagle Minimal Essential Medium and Dulbecco Modified Medium supplemented with 10% Fetal Bovine Serum, Penicillin 60 u/ml and Streptomycin 100 u/ml in 5% CO_2 atmosphere. In a typical experiment, cells were grown in Costar multiwells at approximately full confluency, induced for ten hours with different concentrations of zinc sulfate, copper sulfate and cadmium chloride and labeled the last ninety minutes with 30 µCi/ml ^{35}S-methionine or ^{35}S-cysteine.

Before the pulse, cells were starved for one hour in full medium lacking methionine or cysteine. Heath shock was obtained shifting the temperature to 42°C for two hours. Acute phase response was induced adding to the cells for thirtysix hours 10% v/v of LPS monocyte activated supernatant (3). Same amounts of labeled cell lysates or radioactive media concentrated by trichloroacetic acid precipitation were reduced, alkylated (4) and analyzed by 20% SDS polyacrilamide gel electrophoresis (5). After the run, gels were processed for fluorography (6). Radioactive molecular wheight markers were ^{35}S-methionine labeled Sindbis virus proteins (7).

RESULTS

Increasing concentrations of copper, zinc and cadmium were added for ten hours to HepG2 and Hep3B cell lines and ^{35}S-methionine and ^{35}S-cysteine labeled proteins were analyzed by SDS polyacrilamide gel electrophoresis.

Two major proteins were synthesized in both HepG2 and Hep3B cell lines in response to metals: metallothioneins (MT) and the heath shock protein Hsp70. The cysteine rich MT appeared to be strongly induced by zinc and cadmium in both cell lines and poorerly by copper in HepG2 (Fig.1, panel A and B). In our experimental conditions we were not able to detect MT synthesis in Hep3B cell line after copper induction; experiments are in progress to clarify this finding.

We also noted that high concentrations of copper and cadmium were inhibiting the incorporation of labeled aminoacids in neosynthesized proteins, while zinc had a less pronounced effect. Maximal induction of MTs corresponded to approximately fifty per cent reduction of total protein synthesis (data not shown). These observations paralleled the appearance in both cell lines of the heath shock Hsp70 protein in response to all three metals (Fig.1, Fig.2, Fig.4 and Fig.5, panel A and B); again, zinc showed a reduced effect on the induction of Hsp70 biosynthesis.

Other proteins were synthesized in a cell line specific manner. In Hep3B cells a 15 Kd MW cysteine rich protein (Fig.1, panel B) and a 28 Kd MW methionine rich protein (Fig.2, panel B) appeared after copper and cadmium induction. A 28 Kd MW methionine rich protein was also present in HepG2 cell line after exposure to the same metals and, weakly, to zinc (Fig.2, panel A). This protein comigrate with the 28 Kd MW specie synthesized in the Hep3B cells (data not shown).

We also studied the effect of metals on protein secretion of our cell lines. At 250 µM zinc chloride the secretion of a 47 Kd MW protein was strongly enhanced in HepG2 cells (Fig.3, panel A). Lower amounts of the same protein were also present in the media of cells induced with 200 µM

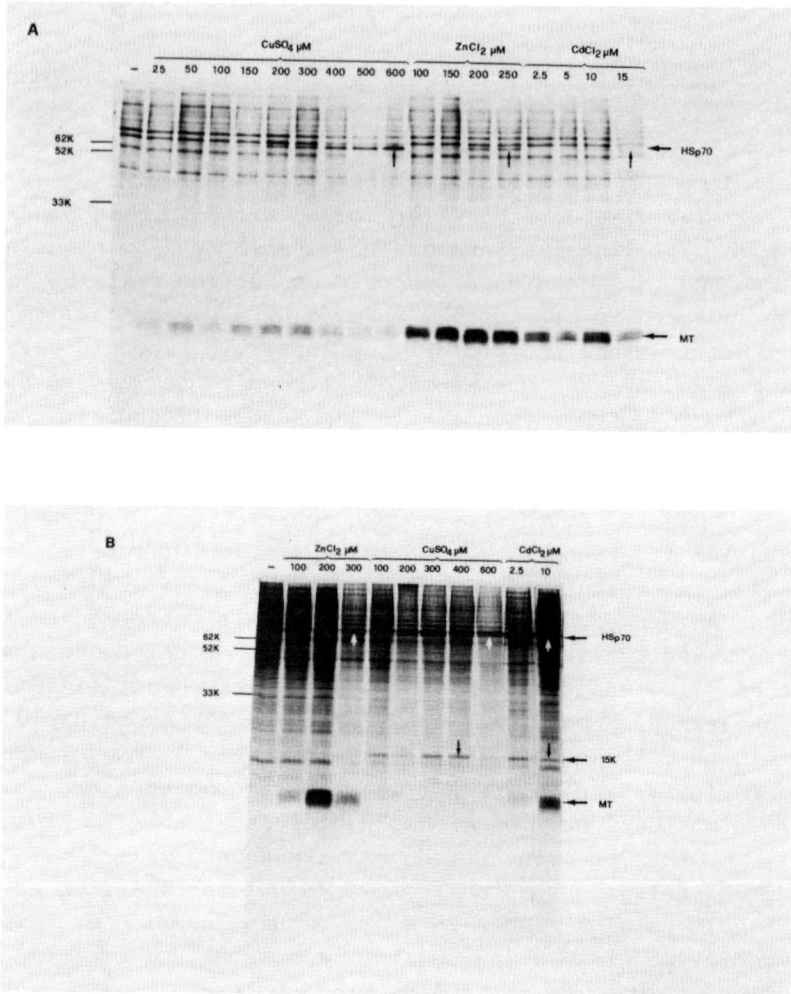

FIGURE 1. Effect of different concentrations of metals on protein profiles of ^{35}S-cysteine labeled extracts of HepG2 (panel A) and Hep3B (panel B) cell lines, analyzed by SDS polyacrilamide gel electrophoresis. MT: metallothioneins. Upward arrows indicate Hsp70 protein, downward arrows the 15 Kd protein.

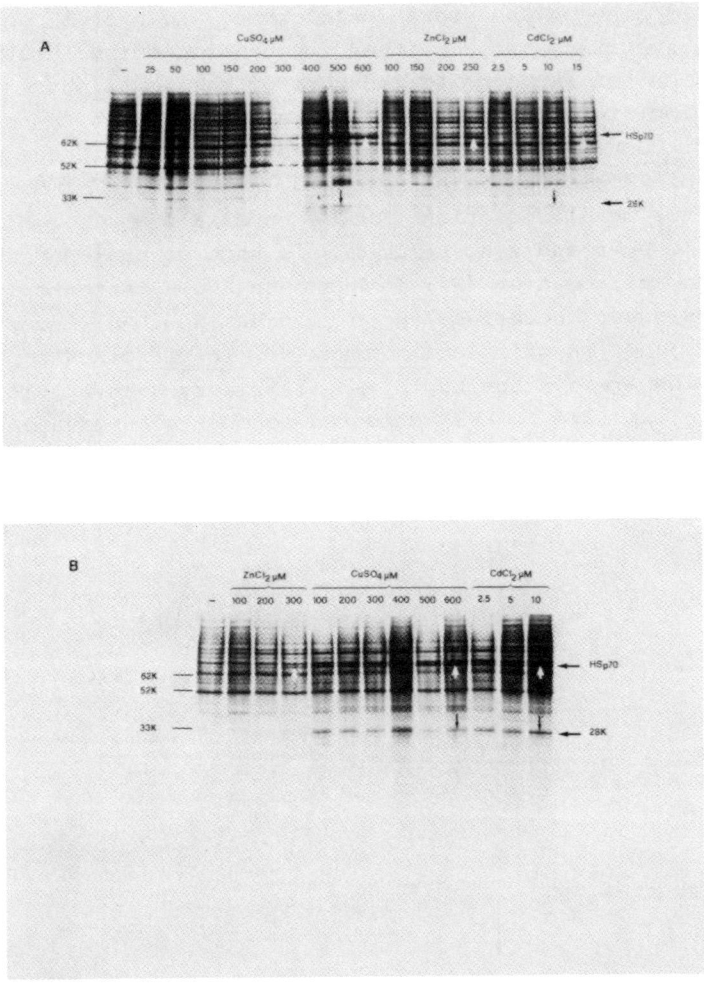

FIGURE 2. Effect of different concentrations of metals on protein profiles of ^{35}S-methionine labeled extracts of HepG2 (panel A) and Hep3B (panel B) cell lines, analyzed by SDS polyacrilamide gel electrophoresis. Upward arrows indicate Hsp70 protein, downward arrows indicate the 28 Kd proteins.

$ZnCl_2$, but not with 100 uM $ZnCl_2$ (data not shown). The effect of zinc on the secretion of the 47 Kd protein seemed quite specific, since neither copper or cadmium, heath shock treatment or exposure to inducers of the acute phase response stimulated the appearance of this protein in the same cells. Conversely, we did not observe changes in the pattern of labeled proteins released in the media of Hep3B cells after metal induction (Fig.3, panel B).

Copper and zinc serum levels undergo profound changes during inflammation (8); it has been shown that the inflammatory agent bacterial lipopolysaccharide (LPS) induces mouse MT gene trascription *in vivo* (9). *In vitro* acute phase proteins are synthesized by mouse primary hepatocytes following exposure to LPS activated monocytes supernatant (10).

Therefore, we used this experimental system to compa-

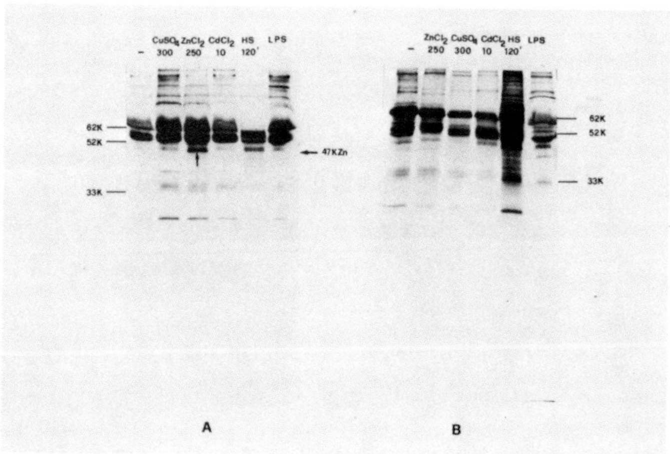

FIGURE 3. Effect of different concentrations of metals on profiles of ^{35}S-methionine labeled secreted proteins of HepG2 (panel A) and Hep3B (panel B) cell lines, analyzed by SDS polyacrilamide gel electrophoresis. Upward arrow indicates the 47 Kd protein.

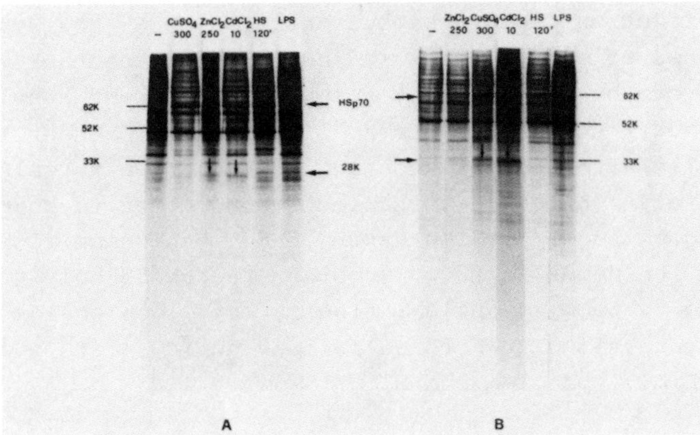

FIGURE 4. Effect of metals, heath shock and inflammatory stress on protein profiles of ^{35}S-cysteine labeled extracts of HepG2 (panel A) and Hep3B (panel B) cell lines, analyzed by SDS polyacrilamide gel electrophoresis. Downward arrows indicate the 15 Kd protein.

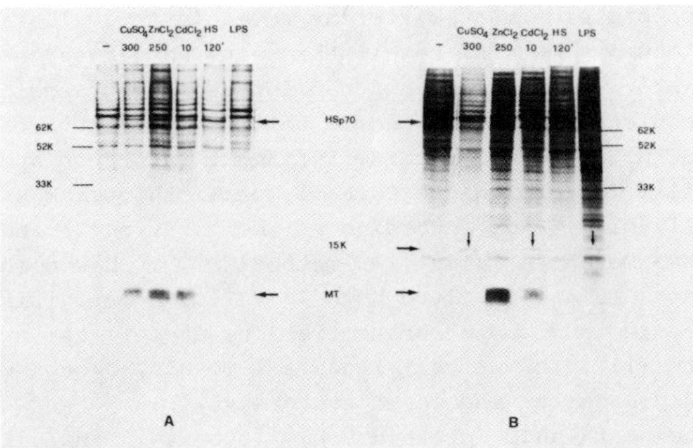

FIGURE 5. Effect of metals, heath shock anf inflammatory stress on protein profiles of ^{35}S-methionine labeled extracts of HepG2 (panel A) and Hep3B (panel B) cell lines, analyzed by SDS polyacrilamide gel electrophoresis. Downward arrows indicate the 28 Kd proteins.

re the effect on protein biosynthesis of HepG2 and Hep3B cell lines of heavy metals and heath shock together with inducers of the inflammatory response. Different results were obtained: the 15 Kd cysteine rich protein present in Hep3B cells after copper and cadmium induction was also synthesized after exposure to LPS activated monocyte supernatant (Fig.5, panel B); the Hep3B 28 Kd methionine rich protein and the HepG2 28 Kd polypeptide responded only to metals (Fig.4, panel A and B); finally metallothioneins were not synthesized in both cell lines during acute phase response (Fig.5, panel A and B).

DISCUSSION

Using two differentiated human hepatoma cell lines, we showed that copper, zinc and cadmium induced the synthesis of different proteins. Metallothioneins and Hsp70 appeared in both cell lines after ten hours induction. These two groups of proteins play different roles in the cellular response to heavy metals. Metallothioneins have been shown to bind tightly copper, zinc and cadmium and are trascriptionally regulated by these ions as well by glucocorticoids, interleukin 1, phorbol esters, inflammatory stress and others (11). The biological role of metallothioneins has not been fully clarified; according to the different roles played by each metal in mammallian metabolism, it has been proposed that MTs are involved both in cellular detoxification (i.e. cadmium and copper sequestration) as well as in zinc transport (11). In our cell lines all metals, except copper in Hep3B cells, induce MT synthesis.

The heath shock protein Hsp70 is part of the cellular response to several stress: heath, metal poisoning, exposure to amino acid analogue, etc. (12). Therefore its appearance seemed connected to the toxic role played by copper, cadmium and zinc on the cells.

We also found that metals did not induce the synthesis of the same protein(s). Thus, the 15 Kd cysteine rich and the 28 Kd methionine rich proteins present in Hep3B cells

TABLE 1
SUMMARY OF INDUCED PROTEINS

Protein		Cu	Zn	Cd	HS	LPS
HepG2	28 Kd	+	±	+	-	-
	47 Kd	-	+	-	-	-
Hep3B	28 Kd	+	-	+	-	-
	15 Kd	+	-	+	-	+

are synthesized in response to copper and cadmium, while zinc alone induced the secretion of the 47 Kd protein in HepG2 cells.

Table 1 summarizes the results obtained with different inducers in both cell lines. In our experimental conditions all proteins responded only to metals except the 15 Kd cysteine rich protein which shared a common response with acute phase reactants. Inflammation severely affects metal distribution (8): in particular copper levels raise in patients with cancer and other infectious diseases. Therefore, according to our results, the 15 Kd protein could represent a target for both stimuli in the Hep3B cell line.

It has been shown that mouse MTI gene is actively transcribed _in vivo_ after twelwe hours of inflammatory stress (9); no data are available on MT gene expression at longer times. Thus, we cannot exclude that the lack of MT synthesis observed in our cell lines after thirtysix hours exposure to LPS induced monocyte supernatant reflects a transient activation of MT transcription, followed by a decrease to basal levels; conversely, our _in vitro_ conditions mimic only in part the _in vivo_ inflammatory response.

ACKNOWLEDGMENTS

We thank dr. M. Terrana and dr. V. Colantuoni for providing us respectively the HepG2 and the Hep3B cell lines; dr. G. Scala for LPS induced monocyte supernatant; dr. S. Bonatti and dr. G. Migliaccio for helpful discussion.

REFERENCES

1. Knowels BB, Howe CC, Aden DP (1980). Human hepatocellular carcinoma cell lines secrete major plasma proteins and hepatitis B surface antigen. Science 209:497.
2. Ciliberto G, Dente L, Cortese R (1985). Cell specific expression of a transfected human α1-antitrypsin gene. Cell 41:531.
3. Scala G, Allavena P, Djeu JK, Kashara T, Ortaldo JR, Herbman RB, Oppenheim JJ (1984). Human large granular lymphocytes are potent inducers of interleukin 1. Nature 309:56.
4. Bonatti S (1983). Biosynthesis of Sindbis virus membrane glycoproteins _in vitro_. Meth Enzim 96:512.
5. Welch WJ, Sefton BM (1979). Two small virus specific polypeptides are produced during infection with Sindbis virus. J Virol 29:1186.
6. Bonner WM, Laskey RA (1974). A film detection method for tritium labeled proteins and nucleic acids in polyacrilamide gels. Eur J. Biochem 43:83.
7. Rice CM, Strauss JH (1981). Synthesis, cleavage and sequence analysis of DNA complementary to the 26S messanger RNA of Sindbis Virus. J. Mol Biol 150:315.
8. Durnam DM, Hoffman JS, Quaife CJ, Benditt EP, Chen HY, Brinster RL, Palmiter RD (1984). Induction of mouse MTI mRNA by bacterial endotoxin is indipendent of metals and glucocorticoid hormones. PNAS 81:1053.
9. Baumann H, Jahreis GP, Gaines KC (1983). Synthesis and regulation of acute phase plasma proteins in primary cultures of mouse hepatocytes. J Cell Biol 97:866.

10. Hamer DH (1986). Metallothioneins. Ann Rev Biochem 55:913.

11. Sorenson JR (1982). "Inflammatory diseases and copper: the metabolic and therapeutic roles of copper and other essential metalloelements in humans." Clifton: Humana Press.

12. Schlesinger MJ, Ashburner M, Tissieres A (1982) "Heath shock from bacteria to man". Cold Spring Harbor: Cold Spring Harbor Laboratory Press.

Metal Ion Homeostasis: Molecular Biology and Chemistry, pages 71–80

REGULATION OF METALLOTHIONEIN GENE EXPRESSION IN ZINC DEFICIENT RATS[1]

M.L.Scarino, G.Morelli, O.Capasso
and
M.C.Mellone*, A.Leone**

Istituto Nazionale della Nutrizione, Rome, Italy 00179

* Istituto Scienze Biochimiche and
**Istituto Biochimica Cellullare e Molecolare
University of Naples, Naples, Italy 80131

ABSTRACT We studied the effect of zinc administration on metallothionein (MT) mRNA induction in liver and kidney of normal and zinc deficient rats. In Zn-deficient rats the peak of accumulation after zinc injection is delayed compared to the control and the MT mRNA levels remain high even 24 hrs after the induction. On the contrary, transcription rates of MT genes are very similar in both groups and MT transcripts are present at high level at late times (24-48 hrs). In view of these results, a role for zinc in the control of MT mRNA stability is discussed.

INTRODUCTION

Zinc is an essential trace element involved in several processes, such as growth, development and differentiation. A wide number of enzymes need zinc for their functions. Typical signs of zinc

---

[1]This work was supported by CNR, Italy. Grant IPRA Subproject 3, paper 1832

deficiency in mammals are impairment of immune system, growth retardation and skin lesions (1).

The metallothioneins are a class of cysteine rich, small molecular weight proteins that are able to bind zinc and other heavy metals in all tissues (2). Although their biological function has not been fully elucidated, multiple roles have been suggested for these proteins in cell detoxification as well as in the control of the homeostasis of essential trace elements (3).We investigated the modulation of the expression of this gene family in zinc deprived rats after feeding a diet rich in a zinc chelating agent.

METHODS

Male Sprague-Dawley rats (200-220 g) were fed (30 days) a complete balanced diet (4) (normal group) or a diet supplemented with Na phytate (a zinc chelating compound) in a 80:1 molar ratio to zinc (zinc deficient group). Plasma zinc concentration, monitored during the last week of the experiment, was 1.1-1.5 µg/ml in normal and 0.1-0.4 µg/ml in zinc deficient group. Animals were injected intraperitoneally with a solution of zinc sulfate (1 mg Zn^{++}/100 g body weight)and sacrified at different times (0,1,3,6,10,24 and 48 hours). Total RNA was extracted with the guanidine-thiocyanate method (5) and 5µg of each sample analyzed by Northern and slot blot assays (6),using a monkey MT-I cDNA probe that recognizes both MT-I and MT-II mRNAs (7).Relative amounts were calculated as a ratio of ferritin mRNA identified on the same samples with a cDNA probe of rat ferritin light chain (8). Purification of nuclei and run-on transcription assay were performed according to Schibler et al. (9). Autoradiographic films were scanned with an LKB Ultroscan laser densitometer. Zinc content in the organs was measured with flame atomic absorption spectrometry using a Perkin-Elmer model 372 atomic absorption spectrophotometer.

RESULTS

In order to follow, in zinc deficient rats, the kinetics of accumulation of MT mRNA after zinc induction, we performed a Northern blot analysis of RNA purified from liver and kidney, the two organs principally involved in zinc metabolism. A typical result is shown in Fig.1:

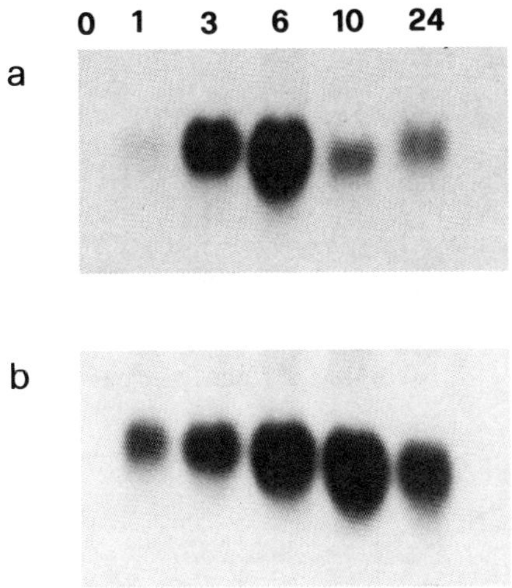

FIGURE 1. Induction of MT mRNA in liver of normal (Panel a) and zinc deficient (Panel b) rats following zinc administration. Lane numbers are hours of zinc induction.

in normal feeding conditions (Panel a), MT mRNA peaked at 6 hours and rapidly decreased to the basal level after 10 hours. In zinc depleted conditions (Panel b) the peak of MT mRNA was delayed and remained high even at late times after induction (10-24 hours). When animals from both groups were injected with saline(0.9% NaCl) no induction of MT mRNA was observed (data not shown). Similar results have been obtained from three independent sets of experiments using RNA slot blot analysis to quantitate the level of MT mRNA induction in liver (Fig.2a) and in kidney (Fig.2b).

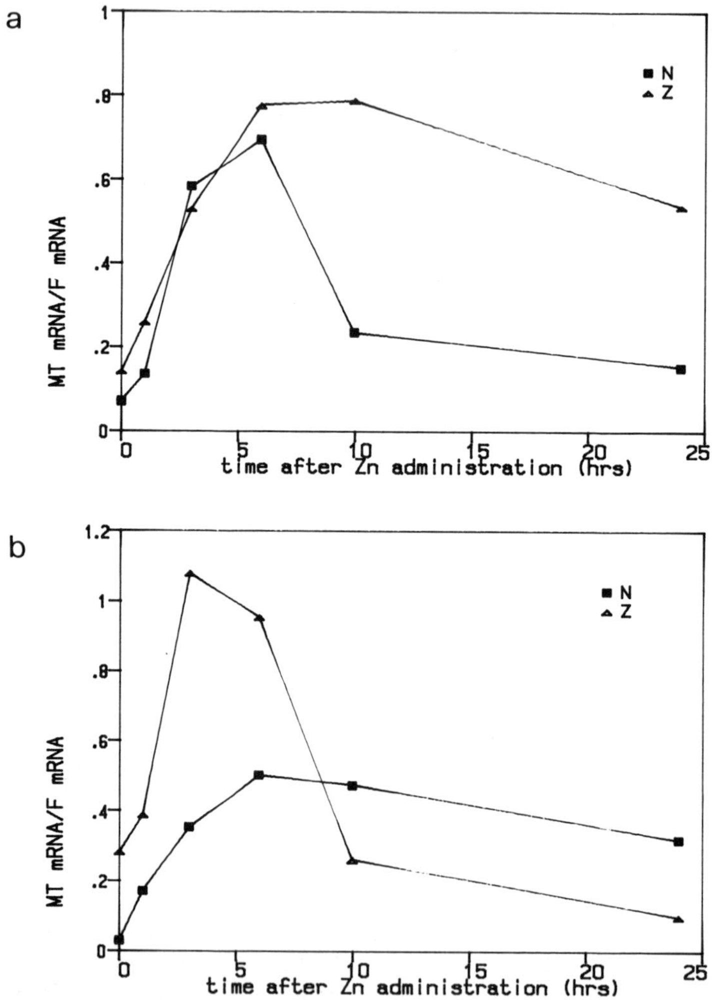

FIGURE 2. Kinetics of MT mRNA induction in liver (Panel a) and kidney (Panel b) of normal (N) and zinc deficient (Z) rats following zinc administration.

We found clear differences in the kinetics of induction in both organs of the two groups of animals. However, in normal animals the maximum level of accumulation at 6 hours was higher in kidney than in liver with the ratio: liver MT mRNA/kidney MT mRNA = 0.64. In zinc deficient animals this ratio was more than doubled, i.e. liver MT mRNA/kidney MT mRNA = 1.6.

To compare the accumulation of steady state mRNA levels with the transcription rates of MT genes we performed a nuclear run on assay on liver nuclei isolated from normal and zinc deficient rats (Fig.3).

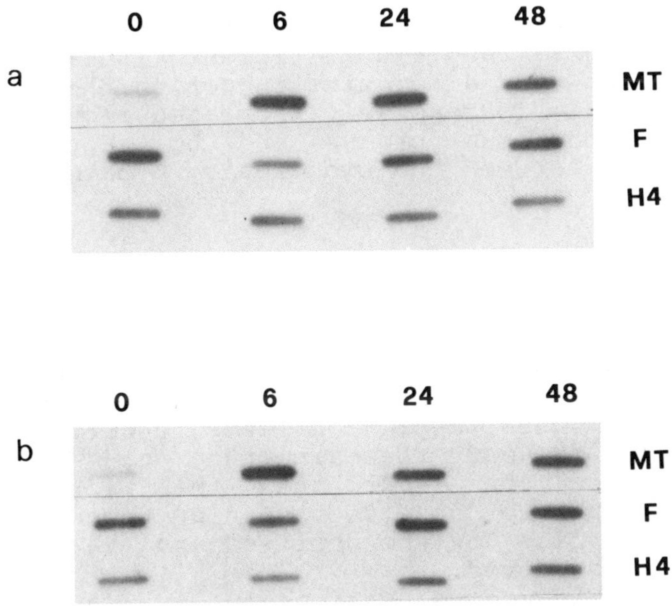

FIGURE 3. Run on assays in liver nuclei from normal (Panel a) and zinc deficient (Panel b) rats following zinc administration. Lanes 0, 6, 24, 48: hrs after zinc induction. MT: Metallothionein, F: Ferritin, H4: Histone H4

The basal level of transcription of MT genes was very low in non induced animals of both groups (lane 0, 3a and 3b). The transcription rate was greatly enhanced 6 hours after the Zn induction and did not decrease in both normal and zinc deficient rats even at late times after zinc injection (i.e. 24 and 48 hours). Interestingly this kinetics did not parallel the accumulation of MT mRNA in both groups as shown in the Northern blot experiment (Fig.1). Moreover MT mRNA levels at 48 hours returned to basal values in both sets of animals (data not shown).

To correlate MT mRNA induction and zinc accumulation, we analyzed zinc content in liver and in kidney (Fig.4) of treated animals. The zinc content before the zinc load (Fig.4, Panel a and b time 0) showed some differences between normal (hatched bars) and zinc deficient (solid bars) animals. However after zinc injection the content of this metal progressively increased from 1 to 24 hours in the two organs and no appreciable differences between the two dietary treatment were observed.

DISCUSSION

The results presented in this paper indicate that modulation of metallothionein gene expression is strongly influenced by zinc deficiency.In particular, this depletion seems to affect both the kinetics and the amplitude of MT mRNA induction in the liver and the kidney with although little or no effect on zinc accumulation in both organs of the deprived and the normal animals is observed.

Total zinc concentrations in liver and kidney indicate that zinc is very efficiently retained and accumulates in these tissues although clear signs of zinc deficiency are evident after one month of phytate diet (arrest in growth, skin lesions, drop in serum zinc concentration). However MT mRNA induction curves do not parallel the corresponding time dependent zinc accumulation in the liver of both groups of animals. It has been demonstrated, at least in hepatocytes (10),

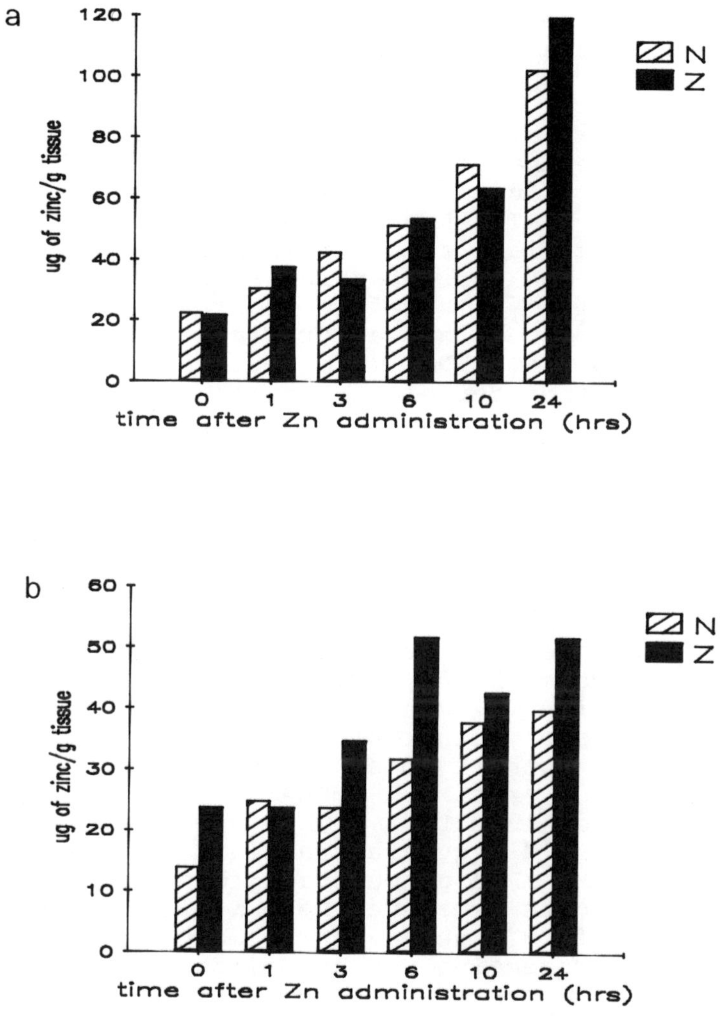

FIGURE 4. Zinc content in liver (Panel a) and kidney (Panel b) of normal (N, hatched bars) and zinc deficient (Z, solid bars) rats after zinc sulfate administration.

that zinc distributes in two separate intracellular pools and that the metallothionein protein content of hepatocytes is directly related to the flux of zinc into these pools. Our data raise the possibility that zinc deprivation affects the compartimentalization of this metal in liver.

The analysis of the MT gene transcription rates and the analysis of MT mRNA content in liver of normal induced animals evidence a clear discrepancy: MT RNA transcripts are still synthesized at late times of induction -i.e. 24,48 hours- with no corresponding accumulation of MT mRNA molecules. Therefore post-transcriptional events seem to play a crucial role on the stability of MT mRNA molecules. Conversely in liver of zinc deficient animals we found an increase of MT RNA transcription with a concomitant MT mRNA accumulation. These data suggest that zinc deprivation affects the stability of MT mRNA rather than its transcription rate; this process could be essential in the restoration of normal zinc distribution in both groups of animals.

Regulation of mRNA turnover can be modified by different mechanisms, as autoregulated changes of polyribosome bound mRNAs (11), modulation of mRNA stability (12) and others. Data on MT mRNA stability have been reported by Mercer and Wake (13). These authors observe a shortening in poly-(A) tails of rat MT-I mRNA after copper and dexamethasone induction. They suggest that this shortening could regulate translation of the newly synthesized MT mRNA as well as its stability.

The results presented here indicate that the induction of MT genes can be modulated at the level of processing or stability of MT mRNA. Experiments are now in progress to elucidate the molecular mechanism(s) that could control this modulation.

ACKNOWLEDGMENTS

We thank D.H. Hamer, N. Heintz and H.N. Munro for cDNA clones of Metallothionein I, Histone H4 and

Ferritin. We are grateful to P. Rami and R. Rami for technical assistance in animal care. We appreciate the help of F. Luti for RNA purification, A.M. Rauseo for word-processing the manuscript and S. Gaetani for helpful suggestions and discussion during this work.

REFERENCES

1. Hambidge KM,CAsey EC,Krebs FK (1986). Zinc.In Mertz W (ed):"Trace Elements in Human and Animal Nutrition - Fifth Edition" London: Academic Press,vol 2 p 1.
2. Hamer DH (1986).Metallothioneins. Ann Rev Biochem 55:913.
3. Karin M(1985). Metallothioneins:proteins in search of function. Cell 41:9.
4. Bises G,Mengheri E,Gaetani S (1987). Limphocyte subsets in zinc deficient and in protein malnourished rats.Nutr Rep Int 36:1371
5. Chirgwin JM,Przybyla AE,Mac Donald RJ, Rutter WJ(1979).Isolation of a biologically active ribonucleic acid from sources enriched in ribonuclease.Biochemistry 18:5294
6. Maniatis T,Fritsche F,Sambrook J (1982). "Molecular Cloning:A Laboratory Manual" Cold Spring Harbor:Cold Spring Harbor Laboratory
7. Schmidt CJ and Hamer DH (1983).Cloning and sequence analysis of two monkey metallotionein cDNAs.Gene 24:137
8. Brown JP, Leibold E and Munro HN (1982).Isolation of cDNA clones for the light subunit of rat liver ferritin: evidence that the light subunit is encoded by a multigene family.Proc Natl Acad Sci USA 80:1265
9. Schibler U,Hagenbuechle O,Wellauer PK and Pittet A-C (1983).Two promoters of different strenghts control the transcription of the mouse alpha-amylase Amy- 1a in the parotid gland and the liver. Cell 33: 501
10. Pattison SE and Cousins RJ (1986).Zinc uptake and metabolism by hepatocytes. Federation Proc 45:2805

11. Yen TJ, Gay DA,Patcher JS,Cleveland DW (1988).Autoregulated changes in stability of poliribosome-bound β-Tubulin mRNAs are specified by the first 13 translated nucleotides.Mol Cell Biol 8:1224
12. Jones TR and Cole MD (1988).Rapid cytoplasmic turnover of c-myc mRNA: requirement of the 3'untraslated sequences.Mol Cell Biol 7:4513
13. Mercer JFB and Wake SA (1985).An analysis of the rate of metallothionein mRNA poly(A)-shortening using RNA blot hybridization.Nucl Acids Res 13:7929

Metal Ion Homeostasis: Molecular Biology and Chemistry, pages 81–90

PURINERGIC AGONIST MEDIATED INDUCTION OF RAT HEPATIC ZINC METALLOTHIONEIN[1]

Frank O. Brady, Scott H. Garrett, and Koji Arizono[2]

Department of Biochemistry, School of Medicine, University of South Dakota, Vermillion, South Dakota 57069

ABSTRACT Purinergic agonists cause activation of either adenylate cyclase (P_1) or phospholipase C (P_2) by adenosine (AMP) and ATP (ADP), respectively. Rat liver possesses the A_2 subtype of the P_1 receptor and the P_2y subtype of the P_2 receptor. In vivo injections of ATP (10 μmoles/kg) or adenosine (100 μmoles/kg) were administered to male, Sprague-Dawley rats (250-300 g), and the levels of hepatic zinc metallothionein (MT) were assessed. ATP (single or three injections) did not elevate ZnMT levels above controls receiving saline injections, but adenosine (three injections at 0, 4, and 8 hr) did (3.4-fold induction). Various agonist analogues of adenosine are available, and these were studied in vivo. 2-Chloroadenosine (100 μmoles/kg) and 5-N-ethylcarboxamide adenosine (NECA) (10 μmoles/kg) were effective in vivo inducers at 11 hr after single injections with 7.3- and 5.5-fold inductions, respectively. 5'-Chloro 5'-deoxyadenosine (50 μmoles/kg) were modestly effective after a single injection (2.1-fold) and better after three injections (3.9-fold). A time course study with 2-chloroadenosine (100 μmoles/kg) revealed near maximal induction at

---

1 This work was supported by a grant from the Parson's Fund, USD-SM

2 Permanent address: Faculty of Pharmaceutical Sciences, Nagasaki University, 1-14 Bunkyo-machi, Nagasaki 852, Japan

11 hr with a continuing rise to 25 hr. A dose response curve (11 hr) revealed an $ED_{0.5}$ for 2-chloroadenosine of 20 μmoles/kg. Theophylline pretreatment can block the induction by 2-chloroadenosine. The in vivo studies have been confirmed in vitro, using isolated, cultured rat hepatocytes. The multihormonal modulation of the induction of rat hepatic ZnMT now includes purinergic agonists which activate adenylate cyclase.

INTRODUCTION

The hormonal induction of hepatic zinc metallothionein (ZnMT) is a complex and multifaceted process. We have demonstrated the ability of catecholamines, angiotensin II, and phorbol esters to induce ZnMT in vivo (1-4). Cousins and coworkers have shown the same ability for zinc, glucocorticoids, glucagon, cAMP analogues, and interleukin-1 (5). In this report we document the ability of adenosine and adenosine analogues to induce hepatic ZnMT, presumably via the cAMP pathway.

As shown in Figure 1, all of these inducers except zinc and glucocorticoids (interleukin-1 is unknown) cause their effects by communicating via their receptors to adenylate cyclase or to phospholipase C, resulting in the activation of protein kinases A and C and calmodulin-dependent protein kinase. Epinephrine, glucagon, and adenosine activate adenylate cyclase, raising cytosolic cAMP levels. cAMP or its synthetic analogues directly activate protein kinase A. Norepinephrine and angiotensin-II activate phospholipase C, raising the levels of cytosolic inositol trisphosphate (IP_3) and membrane bound diacylglycerol (DG). IP_3 liberates Ca^{2+} from endoplasmic reticulum stores, which in turn activates calmodulin-dependent protein kinase. DG and phorbol esters (TPA) activate protein kinase C. By an unknown mechanism(s) activation of these protein kinases results in increased transcription of ZnMT genes. Karin's group has demonstrated the involvement of transcription factors AP-1 and AP-2 in these processes (6). Zinc and glucocorticoids (dexamethasone) are primary inducers of ZnMT. Zinc and other metals act by an unknown mechanism, which culminates in an interaction with one or more metal regulatory elements in the promoter region of MT genes. Glucocorticoids act via the glucocorticoid receptor, which binds to a glucocorticoid regulatory element in the promoter region of MT genes.

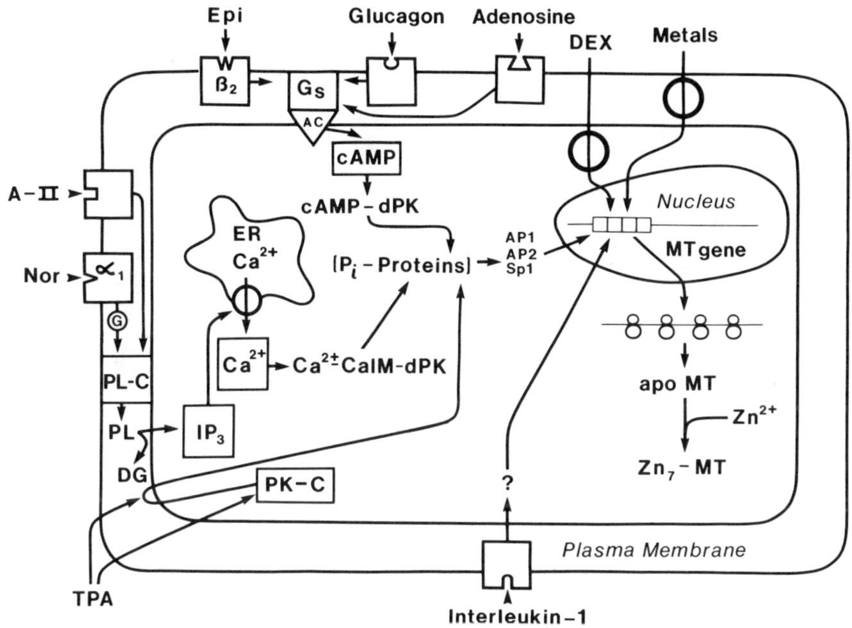

FIGURE 1. Induction of hepatocyte metallothionein

Purinergic receptor communication is illustrated in Figure 2 (7,8). Adenosine and AMP interact with P_1 receptors to inhibit (A_1 subtype) or to stimulate (A_2 subtype) adenylate cyclase. G proteins (G_i and G_s) mediate this communcation. Liver contains the A_2 subtype of adenosine receptors. A variety of synthetic agonists are available for adenosine receptors, and methylxanthines are effective antagonists of these receptors. ATP and ADP interact with P_2 receptors to activate phospholipase C. Liver contains the P_2y subclass. No synthetic agonists or antagonists are available for P_2 receptors, making any studies of the role/function of these receptors difficult.

METHODS

The use of rats, methods of injection, analysis of ZnMT, chemicals, and data analysis are similar to previous reports from our laboratory (9). Other details are presented in the legends of Figures and Tables.

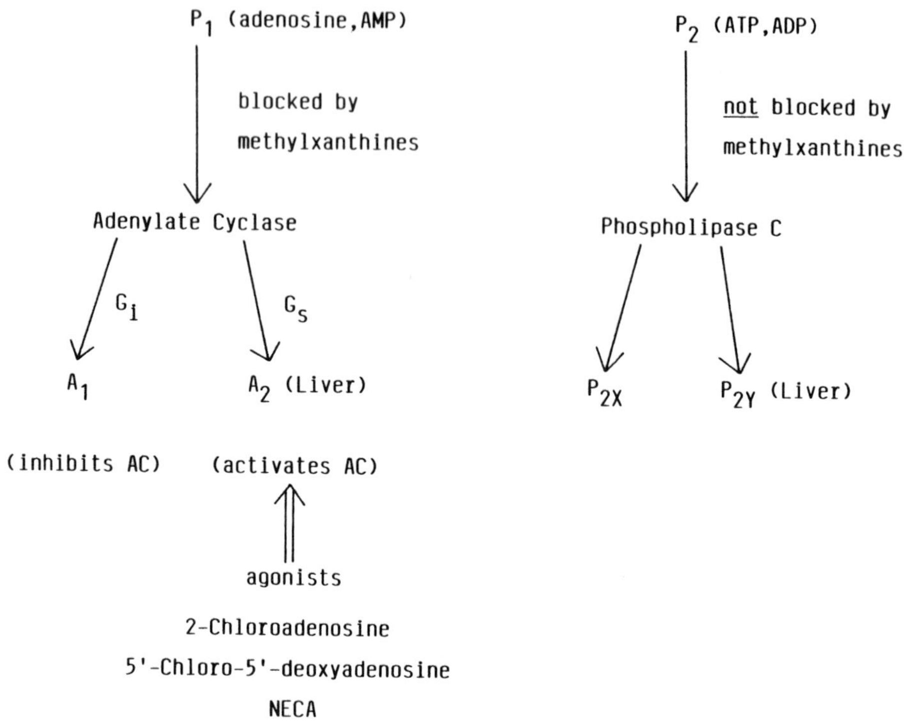

FIGURE 2. Purinergic Receptors Scheme

RESULTS

In our initial experiments we tested ATP, adenosine, and some adenosine analogues for their ability to induce rat hepatic ZnMT in vivo. As can be seen in Table 1, adenosine (three injections) and the three adenosine analogues (2-chloroadenosine; NECA, 5'-(N-ethyl) carboxamidoadenosine; and 5'-chloro-5'-deoxyadenosine) were effective inducers of ZnMT. The inability of a single injection of adenosine or single or multiple injections of ATP to induce ZnMT is probably due to their rapid uptake and metabolism by cells under in vivo conditions.

2-Chloroadenosine was selected as an inducer for further study. Figure 3 shows a dose response curve for the 11 hr induction of hepatic ZnMT. An effective dose for 50% of the maximal increase in MT ($ED_{0.5}$) was 22.0 μmoles 2-Cl-Ado/kg b.w. A dosage of 200 μmoles 2-Cl-Ado/kg b.w. was

TABLE 1

Effects of Purinergic Agonists on Rat Hepatic Zinc Thionein Levels

Compound	Zn in MT (µg/g liver)
Control	1.31 ± 0.05
ATP, 1x (0h)	1.76 ± 0.36
ATP, 3x (0, 4, 8h)	1.79 ± 0.32
Adenosine, 1x (0h)	1.33 ± 0.08
Adenosine, 3x (0, 4, 8h)	5.38 ± 0.36[a]
2-Chloroadenosine, 1x (0h)	9.53 ± 0.45[a]
NECA, 1x (0h)	7.14 ± 0.48[a]
NECA, 2x (0, 5.5h)	7.48 ± 1.29[b]
5'-Chloro-5'-deoxyadenosine, 1x (0h)	2.80 ± 0.63[b]
5'-Chloro-5'-deoxyadenosine, 3x (0, 4, 8h)	5.17 ± 0.57[b]

All compounds were administered ip in 1.0 ml saline at 0h. Rats were killed at 11h. Zn in MT was determined by AAS after Sephadex G-75 column chromatography of cytosol. At least 4 animals were in each group. Dosages were: ATP (10 µmoles/kg), adenosine (100 µmoles/kg), 2-chloroadenosine (100 µmoles/kg), NECA (10 µmoles/kg), and 5'-chloro-5'-deoxyadenosine (50 µmoles/kg). NECA is 5'-(N-ethyl) carboxamidoadenosine. Statistically significantly different from control levels at P < a, 0.001; b, 0.01, using t test.

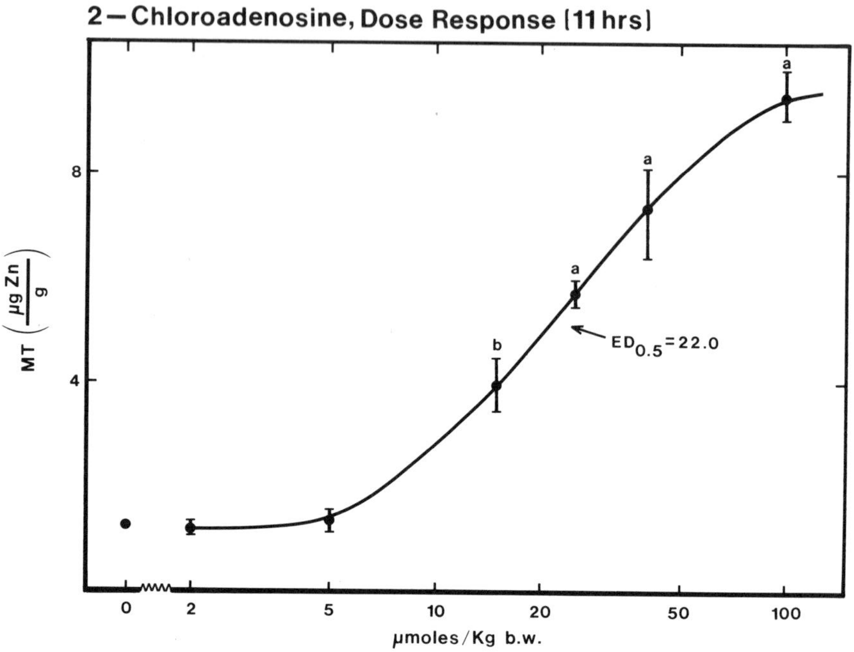

FIGURE 3

administered to one group of animals, but most of this group died before 11 hr. The time course for induction by 100 µmoles 2-Cl-Ado/kg b.w. is shown in Figure 4. This curve is somewhat unusual for an MT induction in two respects: no increase is seen at 4 hr, and a maximal induction is reached by 11 hr. Complete metal analyses (total and cytosolic zinc and serum zinc and copper) for these two experiments are presented in Table 2. Noteworthy in this data is the hypozincemic effect of 100 µmoles 2-Cl-Ado/kg b.w., especially at the 7 hr and 11 hr time points.

To establish that the induction of hepatic ZnMT by the adenosine agonist, 2-chloroadenosine, was being mediated by purinergic receptors, we tested the ability of two different methyl-xanthines, caffeine and theophylline, to block the induction by 2-Cl-Ado. As shown in Table 3, both caffeine and theophylline at 100 µmoles/kg b.w. were able to block completely the induction by 2-Cl-Ado at 25 µmoles/kg b.w., thus confirming the role of adenosine receptors in the induction of hepatic ZnMT.

DISCUSSION

Purinergic agonist mediated induction of rat hepatic zinc metallothionein, presumably via adenylate cyclase activation, has now been demonstrated under in vivo conditions. The extent of induction by 2-chloroadenosine is similar to that achieved by metals and by catecholamines (Figure 5). Preliminary in vitro results in our laboratory indicate that purinergic agonists are inducing hepatic ZnMT by direct action on hepatocyte receptors.

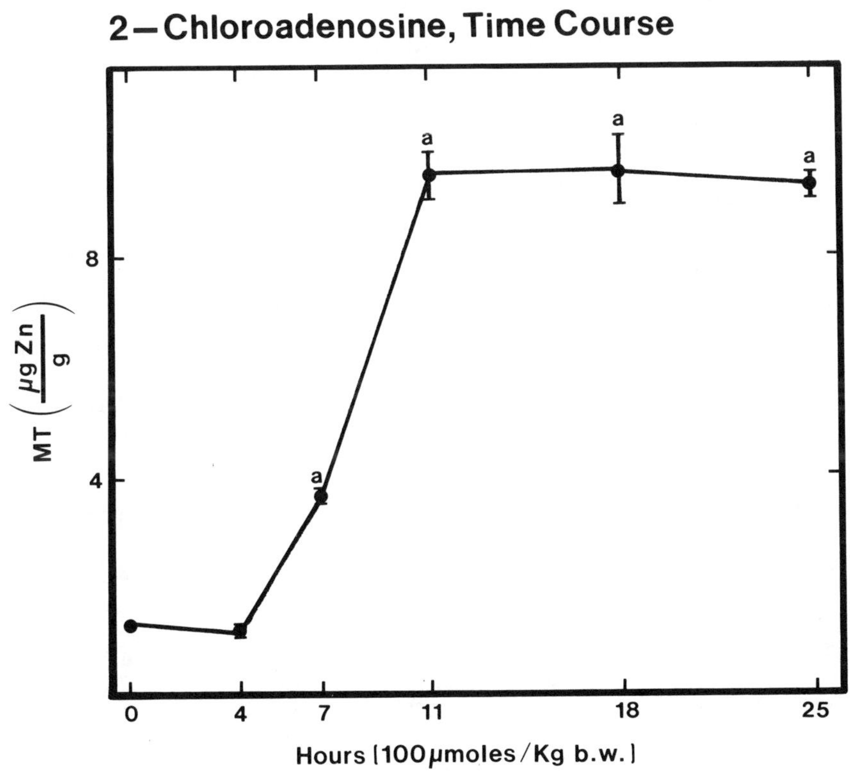

FIGURE 4

TABLE 2

2-Chloroadenosine, Metal Analyses
Time Course and Dose Response

Treatment	Total Zn (µg/g)	Cytosol Zn (µg/g)	Serum (µg/ml) Zn	Cu
100 µmoles/kg, hr				
0	29.69±0.66	15.00±0.73	2.26±0.21	1.06±0.04
4	36.44±1.71[b]	17.13±1.49	0.89±0.05[b]	0.61±0.03[a]
7	43.69±0.89[a]	20.27±0.53[a]	0.55±0.03[a]	1.05±0.14
11	42.38±1.30[a]	24.32±1.96[a]	0.68±0.07[a]	1.76±0.21[c]
18	44.75±0.81[a]	20.50±0.71[a]	0.82±0.10[a]	0.98±0.07
25	46.69±1.10[a]	20.79±0.71[a]	0.91±0.08[b]	1.12±0.06
11 hr, µmoles/kg				
0	29.69±0.66	15.00±0.73	2.26±0.21	1.06±0.04
2	38.50±1.63[a]	15.35±1.40	1.39±0.12[d]	0.89±0.03[c]
5	33.93±0.82[b]	15.25±0.54	1.06±0.04[b]	0.96±0.07
15	37.88±1.69[a]	17.47±0.23[d]	1.08±0.02[b]	0.96±0.02
25	40.22±0.94[a]	18.76±1.12[d]	0.92±0.08[a]	0.98±0.04
40	42.54±1.45[a]	20.17±1.01[b]	0.83±0.10[a]	1.00±0.03
100	42.38±1.30[a]	24.32±1.96	0.68±0.07[a]	1.76±0.21

TABLE 3. 2-Cl-Ado induction of MT. Antagonist blockade.

Treatment[1]	MT (µg Zn/g)	Relative Induction
Saline	1.31 ± 0.05	1.00
2-Chloradenosine	5.70 ± 0.23[a]	4.35
Caffeine	0.99 ± 0.28	0.76
Caffeine + 2-Chloroadenosine	1.00 ± 0.13	0.76
Theophylline	0.97 ± 0.29	0.74
Theophylline + 2-Chloroadenosine	0.64 ± 0.09[a]	0.49

[1] Dosages: 2-Chloroadenosine, 25 µmoles/kg; caffeine, 100 µmoles/kg; theophylline, 100 µmoles/kg, all given in 1.0 ml of sterile saline at time 0. Animals were killed at 11 hr. At least four rats (200-250g b.w.) were in each group

[a] P < 0.001, as compared to control

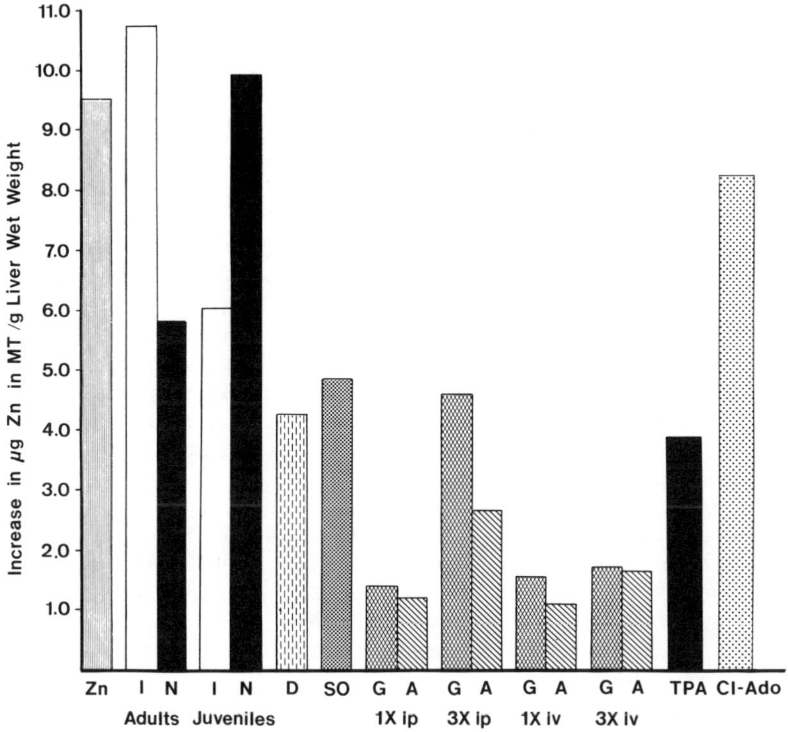

FIGURE 5. Increases in ZnMT 11 hr after various inductions.

Currently, we view ZnMT induction as a fundamental response by the liver to stresses on the animal. High levels of ZnMT may be necessary for the hepatocyte to make the large quantities of new macromolecules (DNA, RNA, and proteins) it needs, when changing its metabolism in response to stresses. Many of the enzymes involved in nucleic acid and protein biosynthesis are zinc requiring enzymes. The recent recognition of "zinc finger"-containing proteins as transcriptional regulators demonstrates another early need for zinc, when the hepatocyte changes its metabolism (10).

The induction of hepatic ZnMT by metals and a multitude of hormones remains an area of fruitful and interesting endeavor by a wide variety of biological scientists.

REFERENCES

1. Brady FO, Helvig BS (1984) The effect of epinephrine and norepinephrine on zinc thionein levels and induction in rat liver. Am J Physiol 247:E318.
2. Helvig BS, Brady FO (1984) Effects of glucagon, Arg-vasopressin, and angiotensin II on rat hepatic zinc thionein levels. Life Sci 35:2513.
3. Brady FO, Helvig BS, Funk AE, Garrett SH (1987) The involvement of catecholamines and polypeptide hormones in the multihormonal modulation of rat hepatic zinc thionein levels. In Kagi JHR, Kojima Y (eds): "Metallothionein II", Basel: Birkhauser Verlag, p 555.
4. Garrett SH, Arizono K, Brady FO (1988) Hormonal modulation of rat hepatic zinc metallothionein levels. In Hurley L (ed): "Trace Elements in Man and Animals-6", in press.
5. Cousins RJ (1985) Absorption, transport, and hepatic metabolism of copper and zinc: special reference to metallothionein and ceruloplasmin. Physiol Rev 65:238.
6. Angel P, Allegretto EA, Okino ST, Hattori K, Boyle WJ, Hunter T, Karin M (1988) Oncogene jun encodes a sequence-specific trans-activator similar to AP-1. Nature 332:166.
7. Satchell D (1984) Purine receptors: classification and properties. Trends Pharm Sci 5:340.
8. Stiles GL (1986) Adenosine receptors: structure, function and regulation. Trends Pharm Sci 7:486.
9. Brady FO, Webb M (1981) Metabolism of zinc and copper in the neonate. (Zinc, copper)-thionein in developing rat kidney and testis. J Biol Chem 256:3931.
10. Evans RM, Hollenberg SM (1988) Zinc fingers: gilt by association. Cell 52:1.

Metal Ion Homeostasis: Molecular Biology
and Chemistry, pages 91–99

THE METALLOTHIONEIN SYSTEM OF DROSOPHILA MELANOGASTER[1]

Nour-Eddine Erraïss, Philippe Silar,
Agnès Cadic-Jacquier, Raja Mokdad-Gargouri
and Maurice Wegnez

Laboratoire d'Embryologie Moléculaire et Expérimentale
Université Paris XI, Bâtiment 445, 91405 Orsay Cedex,
France

ABSTRACT The main cadmium-binding metallothionein (MT) of _D. melanogaster_ has an amino acid composition compatible with the structure of the gene **Mto**. Both _Drosophila_ MT genes, **Mto** and **Mtn**, can be induced by cadmium and copper, but not by zinc ions. The two genes most probably derive from a common ancestral MT gene, as inferred from the position of a single intron in their sequences. Structural characteristics of MT make difficult any statement concerning homology of similar segments found in MT from distantly related species.

INTRODUCTION

Two very distinct genes characterize the metallothionein (MT) system of _Drosophila_ _melanogaster_. The gene **Mtn** codes for a protein of 40 amino acids (1). We have cloned a cDNA from a cadmium-resistant cell line that corresponds to a 43-amino acid MT, **Mto** (2). The **Mtn** and **Mto** proteins, that have markedly different sequences, are only 26% isologous (2,3). In this paper, we further characterize **Mto** at both protein and gene levels.

---

[1]This work was supported by Research Grant 861027 from the "Institut National de la Santé et de la Recherche Médicale".

RESULTS AND DISCUSSION

The **Mto** Protein.

Induction of MT by cadmium, as shown in figure 1, is very effective in Drosophila. A calibration of the Sephadex G-25 column shown in figure 1 allows the quantification, in relative values, of the amount of MT in protein extracts (4). A 24-fold increase of the MT concentration relative to the basal level thus is observed after feeding Drosophila flies a medium containing 0.1 mM cadmium chloride (figure 1).

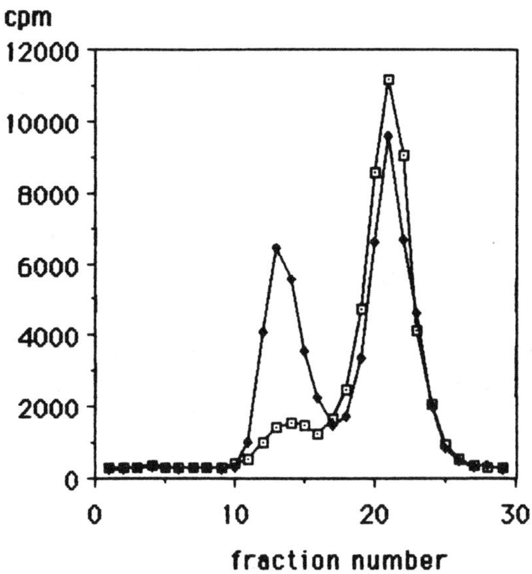

FIGURE 1. Quantification of MT in control (⊡) and 0.1 mM $CdCl_2$ induced flies (◆). Twenty µg of thermo-resistant proteins extracted from adult flies were dissolved in 200 µl of 100 mM ammonium formate, pH 8.8, 2.5% β-mercaptoethanol, 0.1 µCi $^{109}CdCl_2$ (5×10^{-3} µM) and fractionated on a Sephadex G-25 column (55 x 0.7 cm) equilibrated in 100 mM ammonium formate, pH 8.8, 0.1% β-mercaptoethanol. Elution fractions : 0.60 ml. Excluded proteins elute in fractions 11-16 and free $^{109}CdCl_2$ in fractions 19-24. The radioactive cadmium eluting in the excluded peak is bound to MT (4).

Purification and subsequent analysis of the protein binding ^{109}Cd in intoxicated Drosophila, either in larvae or in adults, shows a better agreement with the properties we assign to **Mto** than with those of **Mtn**. We have purified the cadmium-binding MT of Drosophila either by polyacrylamide gel electrophoresis (data not shown), or by DEAE-cellulose chromatography (figure 2).

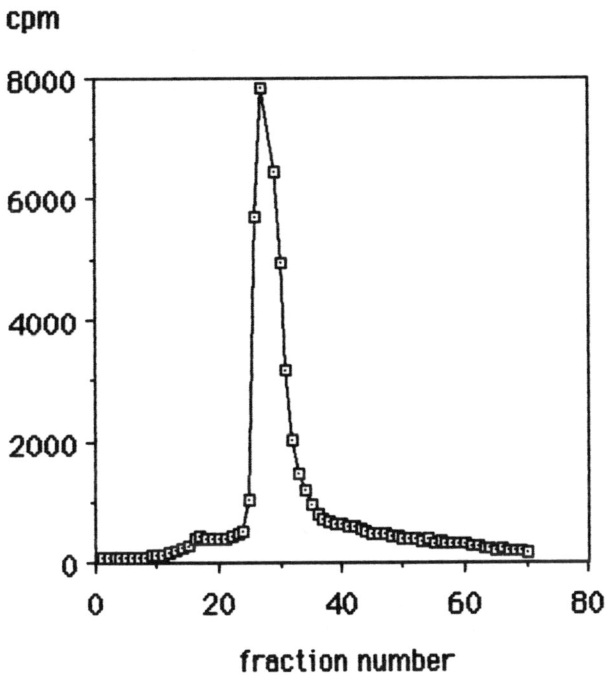

FIGURE 2. DEAE-cellulose purification of MT from 0.1 mM $CdCl_2$ induced larvae. Thermo-resistant proteins were first fractionated on a Sephadex G-75 column (4). The MT were acetone precipitated from the cadmium-binding fraction and loaded on a DE52 cellulose column equilibrated with 5 mM Tris-HCl, pH 8.8 (4). Elution was performed using a linear gradient of 5-500 mM Tris-HCl, pH 8.8.

The amino acid composition of the main MT eluting from the DEAE-cellulose column, as shown in table 1, agrees well with the amino acid composition of the deduced structure of **Mto** protein (2).

TABLE 1
AMINO ACID COMPOSITION OF LARVAL DROSOPHILA CADMIUM-BINDING METALLOTHIONEIN

Residue	Experimental[a]	Integral value	**Mto**[b]	**Mtn**[c]
Ala	2.4	2	2	3
Arg	0.2	0	0	0
Asx	5.9	6	8	3
Cys	7.2	7	12	10
Glx	3.9	4	4	2
Gly	4.6	5	4	7
His	0.1	0	0	0
Ile	0.3	0	0	0
Leu	0.5	0-1	0	0
Lys	4.9	5	6	5
Met	0.1	0	1	1
Phe	0.1	0	0	0
Pro	1.7	2	1	2
Ser	2.3	2	2	6
Thr	1.4	1	1	1
Tyr	0.1	0	0	0
Val	1.7	2	2	0

[a]Average of six analyses.
[b]Calculated from reference (2).
[c]Calculated from reference (1).

Valine is the best-suited amino acid to discriminate between both _Drosophila_ MT, since no valine should be found in **Mtn** (1), in contrast to **Mto** which should contain two valine residues (2). The main cadmium-binding MT from intoxicated _Drosophila_ larvae, as shown in table 1, contains two valine residues. The experimental values found for Asx, Glx and Ser also demonstrate that this protein most probably corresponds to **Mto** (table 1).

Mto Gene Transcription.

Cadmium, zinc and copper ions are all very efficient MT gene inducers in mammalian species (5). Surprisingly, zinc ion is able to induce neither **Mto** nor **Mtn** gene transcription in cell cultures, larval, or adult Drosophila stages (data not shown). On the contrary, cadmium and copper are good inducers of both **Mto** and **Mtn** genes. The main difference in their transcription properties is that **Mto** is best induced by cadmium while **Mtn** is induced more efficiently by copper (data not shown).

Mto Gene Structure.

Both Drosophila MT genes are localized on the third chromosome. The gene **Mtn** was shown to be located at position 85E 10-15 (1). We have located the gene **Mto** by in situ hybridization at position 92E (data not shown).

The dual MT system of Drosophila, composed of genes sharing a very low level of isology, poses an interesting evolutionary problem. Is it possible to state whether or not they are homologous, i.e. whether or not they derive from a common ancestral gene ? The determined sequences of the two genes favor the common origin hypothesis. Both **Mtn** (1) and **Mto** (unpublished results) are interrupted by a single intron. As shown in figure 3, this intron is much shorter in **Mto** (61 bp) than in **Mtn** (265 bp). Interestingly, both introns are located at exactly the same position when the genes are aligned according to the best protein alignment found with the program EPURE (2). This strongly suggests that the two genes are homologous. At the present time, the MT system of D. melanogaster is thus characterized by two genes, **Mto** and **Mtn**, resulting from a very ancient duplication of their ancestral gene. It is also characterized by several recent duplications of the **Mtn** gene that are more or less spread throughout the D. melanogaster populations (6).

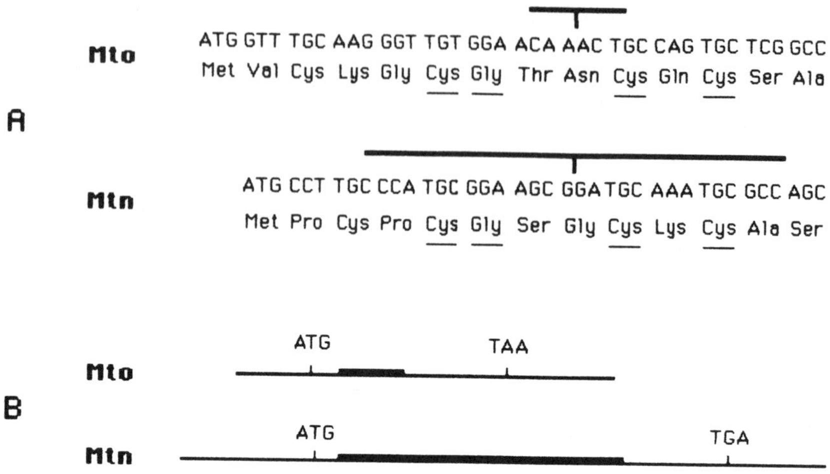

FIGURE 3. Comparison of **Mto** and **Mtn**. (A) The first 14 (**Mto**) and 13 (**Mtn**) amino acids of Drosophila MT, and their corresponding cDNA sequences, are aligned according to EPURE (2). Shared amino acids are underlined. The two upper lines indicate the position of the single intron characterized in the **Mto** gene (61 bp, unpublished results) and in the **Mtn** gene (265 bp, ref.1). (B) Alignment of **Mto** and **Mtn** non-processed messenger RNAs. The introns are drawn as a solid box.

Metallothionein Sequence Comparisons.

The unusual composition of MT considerably biases the search for homologous segments between sequences from distantly related species.The MT system of D. melanogaster illustrates the difficulty of comparing such proteins in an evolutionary perspective. Cysteine alone represents 30% of total residues in most MT. A mean value of 10% for each of the three amino acids glycine, lysine and serine is found when considering sequences from a large range of species (7).

Furthermore, no or very few histidine, isoleucine, leucine, phenylalanine, tyrosine and tryptophan residues are present in MT. These data lead us to question the significance of the sequence alignements commonly proposed for MT from species in far related groups. We also question the validity of assigning a common evolutionary origin to similar motifs found in MT from different species. We illustrate these two points in the following sections.

<u>Alignment of MT</u>. To assess the validity of alignments, we constructed random sequences having an amino acid composition similar to the mean value obtained when considering MT from different groups of species. When these random sequences are compared directly, a mean isology value of 12% is found. This value rises to 23% when searching for optimal alignments of segments 60 amino acid-long without allowing any gap. This is very similar to the 26% isology level found for the best **Mto-Mtn** alignment (2). These calculations demonstrate that no statistically significant alignments can be proposed when simply considering MT sequences from species in far-related groups.

<u>Evolutionary conserved motifs</u>. The motif **βcβc∂∂∂ckc** where **c** is cysteine, **k** lysine, **β** serine or threonine and **∂** any amino acid if not cysteine, has been "extracted" from MT sequence comparisons (7). The probability (P_n , n = number of mismatches) for a 10 amino acid segment from random MT-like proteins to match this sequence can be calculated as

$$P_0 = 2.78\ 10^{-6}$$
$$P_1 = 1.04\ 10^{-4}$$
$$P_2 = 1.57\ 10^{-3}$$
$$P_3 = 1.21\ 10^{-2}$$
$$P_{n\leq 3} = P_0 + P_1 + P_2 + P_3 = 1.38\ 10^{-2}$$

The probability of finding at least one such motif in any 60 amino acid-long MT, if allowing three mismatches, thus is 0.49. In the sea urchin the segment **ckc∂∂∂cβcβ** was also considered (7), which rises the probability to 0.76. Simulations with random sequences agree very well with these calculations.

It is clear that MT from all species share structural similarities, obviously related to their properties, i.e. binding of metal ions. These structural features most probably were conserved during evolution. Two ways for maintaining such structures are possible. The first would be the strict conservation of the structures, with evidently some minor structural changes arising from time to time. Evolutionarily speaking, the small conserved domains would be homologous in all species (3). Another explanation can be proposed as well. Due to the very biased composition of MT, any motif similar to small structurally important domains could appear in any position of the MT sequences following very small structural changes. Such segments, that then would be shifted from one position to another depending on the mutation(s), should not be termed homologous, but rather analogous if they share a similar function (3). Our calculations show that it is impossible to choose between these two evolutionary possibilities. The conclusion is that no firm statement about evolution of MT can be proposed from direct protein comparisons, unless a very high level of isology is found as is the case for all vertebrate MT thus far sequenced. Additional information evidently is needed. The precise positioning of the intron in the Drosophila **Mtn** and **Mto** genes best illustrates this point.

ACKNOWLEDGMENTS

We are grateful to G. Maroni for providing the Drosophila Mtn clone and to J. Proust for her help with the in situ hybridization experiments. We thank J. Mesnard for excellent technical assistance. The manuscript was much improved by the criticisms of J. Deatrick. A.C.J. is a fellow of the "Ligue Nationale Française Contre le Cancer" and R.M.G. is a fellow of the "Fondation pour la Recherche Médicale".

REFERENCES

1. Maroni G, Otto E, Lastowski-Perry D (1986). Molecular and cytogenetic characterization of a metallothionein gene of Drosophila. Genetics 112:493.

2. Mokdad R, Debec A, Wegnez M (1987). Metallothionein genes in Drosophila melanogaster constitute a dual system. Proc Natl Acad Sci USA 84:2658.
3. Wegnez M (1987). Letter to the Editor. Cell 51:516.
4. Debec A, Mokdad R, Wegnez M (1985). Metallothioneins and resistance to cadmium poisoning in Drosophila cells. Biochem Biophys Res Commun 127:143.
5. Hamer DH (1986). Metallothionein. Ann Rev Biochem 55:913.
6. Maroni G, Wise J, Young JE, Otto E (1987). Metallothionein gene duplications and metal tolerance in natural populations of Drosophila melanogaster. Genetics 117:739.
7. Nemer M, Wilkinson DG, Travaglini EC, Sternberg EJ, Butt TR (1985). Sea urchin metallothionein sequence: Key to an evolutionary diversity. Proc Natl Acad Sci USA 82:4992.

Metal Ion Homeostasis: Molecular Biology and Chemistry, pages 101–108

PROTEIN-DNA INTERACTIONS AT THE YEAST CUP1 UPSTREAM ACTIVATOR SEQUENCES[1]

Cynthia R. Timblin[2] and Lawrence W. Bergman

Department of Chemistry, Clippinger Laboratories, Ohio University, Athens, Ohio 45701

ABSTRACT The yeast copper metallothionein gene (CUP1) is transcriptionally regulated in response to the presence of copper in the growth medium. Although the cis-acting elements responsible for the response are known, very little is known of the trans-acting factors involved in CUP1 expression. A promoter fusion between the CUP1 upstream activating sequences and the regulatory sequences of the yeast PHO5 gene was used to demonstrate that efficient transcriptional regulation in response to copper is mediated by DNA sequences contained on a 101 bp HinfI-HinfI CUP1 restriction fragment. DNA binding activity(s) specific for these sequences has been purified approximately 11,000-fold from yeast nuclear extracts and nucleoprotein electrophoresis reveals two discrete specific protein-DNA complexes. DNAase I footprinting detects two regions of protection, region I and II, which includes the repeated consensus sequence 5'-TCTTTTTGCT-3', which has been implicated in the transcriptional response to copper.

INTRODUCTION

In the yeast, *Saccharomyces cerevisiae*, metallothionein is encoded by the CUP1 locus and mediates resistance to elevated levels of copper (1-3). Interestingly although the yeast protein itself will bind a number of metal ions, the transcription of the CUP1 gene is induced only in response to the copper ion. The level of resistance to copper poisoning in yeast is directly porportional to the gene copy number. Copper resistant strains ($CUP1^r$) contain multiple, tandemly amplified copies of the structural gene and

---

1This work was supported by the National Institutes of Health (R01 GM 32319)

2Present Address: NCI-Naval Medical Oncology, National Medical Center, Bethesda, MD 20814

flanking regions, whereas strains sensitive to copper poisoning(CUP1^{s}) contain only a single copy of the copper metallothionein gene. In either CUP1^{r} or CUP1^{s} strains, CUP1 transcription is at a low constitutive basal level, in the absence of the inducer, but in the presence of copper transcription is induced approximately 30-fold. Taking advantage of the ability to do gene deletion experiments in yeast, strains deleted for the chromosomal CUP1 gene are viable but are hypersensitive to the presence of low levels of copper in the growth medium (4). In CUP1-deleted strains, transcription from a CUP1 promoter vector is at an elevated basal level in the absence of copper but still may be further induced (approximately 5-fold) in the presence of low levels of copper. This suggests that the CUP1 gene product functions in some manner to auto regulate its own expression (4). Both the detoxifying and autoregulatory functions can be restored by supplying either the yeast or monkey metallothionein proteins in _trans_ (5). Deletion analysis of the CUP1 promoter has identified two upstream activation sequences, denoted UAS_P and UAS_D, that are essential for efficient transcriptional regulation of CUP1 by copper ion (6). However, unlike other gene systems in yeast, the CUP1 system has not readily lent itself to the availability of yeast genetics. That is, there have been no genetically unlinked mutations isolated which affect the level of transcription of CUP1 or a CUP1 promoter vector. Thus, we have approached the protein-DNA interactions responsible for CUP1 transcription biochemically by fractionating yeast nuclei and assaying for the ability of vvarious protein fractions to specifically bind to a DNA fragment containing the CUP1 UAS

METHODS

Yeast Strains.

The following yeast strains were used in this study:

Y316	MAT α, pep4-3, trp1, CUP1^{r}
GG100-14D	MAT α, pho5, his3, ura3-52, trp1, CUP1^{r}
AB17	MAT a, arg4-8, leu2-3, leu2-112, his7-2, ade5, ura3-52, CUP1^{s}
AB17-DE1	same as AB17 except CUP1::LEU2

In regards to the transcription factor(s) which bind to the CUP1 UAS, all strains shown above proved to be identical.

Nuclear Extract Preparation.

Strain Y316 was grown at 30°C to an OD_{600}=1-1.5 in 18 liters of YEPDA plus 1 mM $CuSO_4$. Cells were collected by centrifugation (3,000 x g; 5 min.), washed once with sterile deionized water and spheroplasted as described (19). Spheroplasts were pelleted at 3,000 x g, washed once with 1.0 M sorbitol, 40 mM Tris-HCL pH7.5, 1.0 mM PMSF[1] and resuspended in 270 mls of lysis buffer containing 0.32 M sucrose, 10 mM NaC1, 0.5

mM $CaCl_2$, 1.0 mM PMSF, 20 mM Tris-HCL pH 8.0 plus 1 μg/ml Antipain and Leupeptin. Spheroplasts were lysed by the addition of 0.5% Triton-X100, vortexed and placed on ice for 10 min. Fifteen milliliter aliquots were layered onto 20 mls of ice-cold 1.7 M sucrose, 0.5 mM $CaCl_2$, 10 mM NaCl, 20 mM Tris-HCL pH 8.0, 1.0 mM PMSF, 1 μg/ml Antipain and Leupeptin and centrifuged at 7,600 x g for 45 min at 4°C. The nuclear pellet was resuspended in 0.375 M KCl in HGE (25 mM Hepes pH 7.6, 10% glycerol, 0.1 mM EDTA, 1 mM PMSF), placed on ice for 30 min., and spun at 5,000 x g for 15 min. The supernatant was collected and dialyzed against 0.05 M KCl in HGE containing 1.0 mM PMSF. Insoluble material was removed by centrifugation at 4,000 x g, 15 min and the supernatant was subjected to the chromatographic procedures described below.

Additional Procedures.

Chromatographic procedures, nucleoprotein gel and footprinting reactions were essentially as described in numerous published reports.

RESULTS

Sequences Responsible for CUP1 Regulation.

Our approach to identifying a binding activity, which specifically interacts with the CUP1 promoter, has been to utilize a 101 bp fragment from the CUP1 promoter to assay protein fractions extracted from yeast nuclei for specific binding to the fragment. To demonstrate directly that sequences in this fragment, which maps from -83 to -184 bp upstream of the most 5'-transcription start site, are sufficient for transcriptional regulation by copper (and therefore like to contain sequences recognized by specific transcription factors) we have constructed fusions which place the yeast phosphate-repressible acid phosphatase gene (PHO5) under control of the 101 bp CUP1 fragment.

TABLE 1
EXPRESSION OF CUP1-PHO5 FUSIONS

Plasmid	Induction Ratio[a]
p19	1.1
p19-101A	5.8
p19-101B	6.7

[a] The induction ratio is the ratio of acid phosphatase activities of strain GG100-14D transformed with the appropriate plasmid and grown in the presence of copper vs. the enzyme level in the same cells grown in the absence of copper.

In the control plasmid (p19), lacking the 101 bp HinfI-HinfI fragment, the expression of acid phosphatase was not affected by the addition of copper. However, in both CUP1-PHO5 plasmid (the difference lies in the orientation of the 101 bp fragment in the vector) there was an approximate 5-6 fold increase in response to the presence of copper. This suggests that the 101 bp fragment is sufficient to regulate expression in response to copper.

Isolation and Characterization of a CUP1-Specific Binding Activity.

Extracts prepared from yeast nuclei were fractionated by heparin agarose chromatography and monitored for CUP1-specific DNA binding activity by a gel retardation assay. As seen in Figure 1, three separate complexes, designated 2, 3 and 6 were seen which migrate slower than free, unbound DNA. These complexes were deteced in nuclear extracts from CUP1^{S}

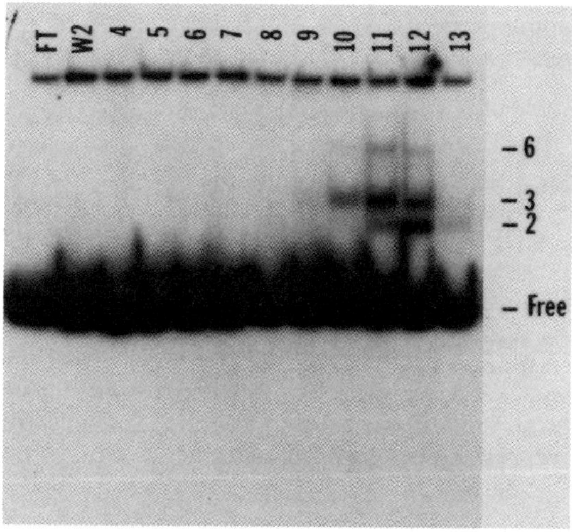

Figure 1: Assay for CUP1-specific binding acitivity in Heparin Agarose chromatographic fractions of a yeast nuclear extract. An aliquot (4.5 ng) of [^{32}P]-labeled 152 bp HindIII-EcoRI fragment containing the CUP1 UAS was incubated with 0.1 μg of each fraction for 1 minute at room temperature. Protein-DNA complexes were resolved by electrophoresis on an 8% polyacrylamide gel at 4° C for 2 hours at 200 volts. The protein concentration of each fraction was measured by the dye-binding method Bradford (Bio-Rad Laboratories) with bovine serum albumin as the standard. FT, 0.05 M KCl flow through fraction; W2, 0.05 M KCl fraction from the second column wash; fractions 4-6, 0.05 M KCL; fractions 7-13 0.32 M KCl; bands 2, 3, and 6, protein-DNA complexes; free, uncomplexed DNA.

and CUP1-deleted strains grown in the presence or absence of exogenous copper. To further purify the DNA-binding activities, the peak binding fractions from the heparin agarose column were pooled and applied to a CUP1-specific DNA cellulose column (the 101bp HinfI-HinfI fragment was covalently coupled to the cellulose). The column was eluted with a KCl step gradient and the CUP1 UAS binding activity localized by the gel retardation assay.The pattern of complex formation was identical to that seen with the heparin agarose fractions, however the band 2 activity eluted at 0.1 M KCl while the band 3 and band 6 activity eluted at 0.2 M KCl.

To specifically determine the sequences bound by the proteins eluted for the CUP1 DNA-celulose column at the higher KCl concentration, the <u>in vitro</u> binding reaction was subjected to DNAase I footprinting. As seen in Figure 2, the binding activity protects two regions from DNAase I digestion.

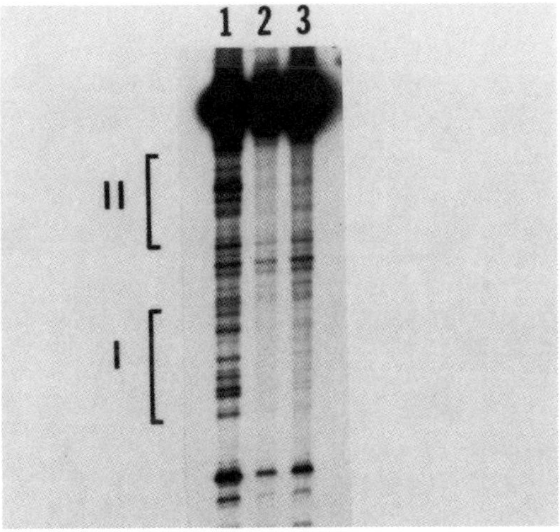

Figure 2: DNAase I footprint analysis of CUP1-specific DNA-binding activity. The 131 bp PstI-EcoRI CUP1 UAS fragment from p18-HF101 was [^{32}P] end-labeled on the noncoding strand (the EcoRI site) using the Klenow fragment of DNA Polymerase I. A 10.0 ul aliquot of CUP1-DNA-celluose fraction was incubated with 4.5 ng of the labeled fragment in the presence or absence of exogenously added copper for 1 min. at room temperature and subsequently digested with 50 ng DNAase I for 20 sec. The DNA was purified and run on a DNA sequencing gel Protected regions I and II are bracketed (see text) Region I includes sequences in both UAS_D and UAS_P and region II includes sequences in UAS_D. Lane 1, no protein added; lane 2, + 10.0 ul fraction 13; lane 3, + 10.0 ul fraction 13 and + 50 μM $CuSO_4$.

Region I includes sequences in both UAS$_D$ and UAS$_P$ and encompasses sequences -159 to -137, at a minimum. Region II includes nucleotides -181 to -168. Interestingly, there appears to be no qualitative differences if the footprinting reaction is done in the presence of copper. Preliminary evidence suggests that this footprint is identical to that seen in a fractionated band 6 complex, whereas the band 3 complex binds to a distinct subset of the region I sequences. The sequences protected from DNAase I digestion by the binding activity contain two complete copies and a portion of a third copy of the sequence 5'-GAAGCAAAA-3'/3'-CTTCGTTTT-5' as depicted in Figure 3. The binding activity specific for the CUP1 UAS has been purified approximately 11,000 fold for the yeast nucleus (see Table 2). Presently analysis of the protein components of the peak binding fraction from the CUP1 DNA-cellulose column reveals 5 major bands on SDS gel electrophoresis.

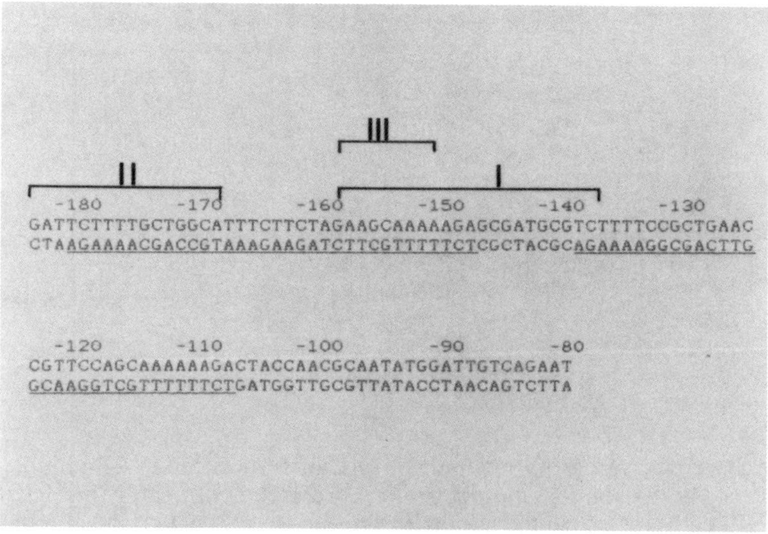

Figure 3: Summary of DNAase I footprint analysis. CUP1 UAS DNA sequences from -184 to -80 bp (relative to the most 5' mRNA start site) are shown. The two related elements shown by deletion analysis to be essential for regulation by copper (13), UAS$_D$ from -181 to -148 bp and UAS$_P$ from -139 to -108 bp, are underlined. Brackets indicate sequences protected from DNAase I digestion in band 3 (region III) and band 6 (regions I and II) complexes.

TABLE 2
PURIFICATION TABLE OF CUP1 TRANSCRIPTION FACTOR

Fraction	Total Protein	Yield	Binding Unit[a]	Overall Purification[b]
Nuclear Extract	63100 µg		100	
Heparinn Agarose	1024 µg	78%	95	50
CUP1 DNA-cellulose	4 µg	87%	87	11,125

[a] % of binding activity present in the peak fractions as assayed on nucleoprotein gel
[b] Taking the yield into account

DISCUSSION

The mechanism by which metal ions regulate transcription is not known. In yeast and mammalian systems, the DNA sequences have been defined that are necessary for the transcriptional response to heavy metal ions by the metallothionein genes. However, little is known about the *trans*-acting factors that interact with these sequencess. In human metallothionein, metal responsive sequence elements are interspersed with elements responsible for the maintainance of a basal level of expression. A factor (AP-1) has been shown to specifically interact with these basal level elements (7). In yeast, no *trans*-acting regulatory mutations have been isolated that specifically affect CUP1 gene expression. Therefore, we have utilized a biochemical approach to identify yeast nuclear factors that bind specifically to the upstream activator sequences of the CUP1 gene.

The CUP1 regulatory sequences have been mapped using DNA deletion mutants by Thiele and Hamer (6) and by promoter fusions (see Table 1). These sequences are located between -105 to -180 bp upstream from the start site of transcription (6). Two related elements within this region UAS_D (-148 to -181) and UAS_P (-108 to -139) are required for maximal induction in response to copper. Computer analyiss reveals that the consensus sequences 5'-TCTTTGCT-3' is present four times within these activator sequences. Using a promoter fusion, we have demonstrated that a 101 bp HinfI-HinfI fragment is sufficient to direct the heavy metal induced transcriptional response of a heterologous promoter system (see Table 1). This fragment functions in either orientation in the vector which is characteristic of a yeast UAS in general. Therefor, we have utilized this fragment as a probe to purify nuclear factors which specifically interact with the CUP1 UAS.

We have isolated from crude yeast nuclear extract a factor(s) that binds specifically to two regions of the CUP1 UAS. Region I encompasses sequences -137 to -159 and region II encompasses sequences -168 to -180. As shown in Figure 5, the protected sequences include two complete copies of the sequence 5'-TCTTTTGCT-3' and part of an additional third repeat of this sequence. This sequence has been shown to be an essential component of the CUP1 UAS by deletion analysis (6). Additionally, in fragment constructions which have deleted UAS_D, a consensus element centered a -134 is protect from DNAase I digestion (Timblin and Bergman, manuscript submitted). This suggests that UAS_D has a higher binding affinity for the specific factor.Identical protein-DNA complexes, as assayed by gel retardation and DNAase I footprinting, are observed with extracts prepared from $CUP1^r$, $CUP1^s$ or CUP1-deleted strains. This suggests that the copper metallothionein protein itself plays no role in the binding of the factor to the CUP1 UAS. Furthermore, at the present time, copper ion has no in vitro effect on the binding of the factor to the UAS region, either qualitative or quantitative. Therefore, we do not know for certain that the factor(s) we have partially purified are in fact copper responsive transcriptional factor. It is possible that these factors may represent the basal constitutive factor(s), analagous to AP-1 in the human metallothionein system (7), required for expression at a basal level. The observation of two specific complexes in the gel retardation assay (band 3 and band 6 in Figure 1) and the finding that the protected regions are both approximately 20 bp and separated by approximately 10 bp may suggest that we are truly working with multiple factors which interact with the CUP1 UAS. Purification to homogeneity of the factor or factors which bind to the CUP1 UAS and analysis of the role of copper ions in the binding will provide further insight on the mechanism of CUP1 gene expression.

REFERENCES

1. Fogel, S and Welch, JW (1982). Tandem Gene Amplification Mediates Copper Resistance in Yeast. Proc Natl Acad Sci USA 79: 5342-5346.
2. Karin, M, Najarian, R, Haslinger, A, Valenzuela, P, Welch, J and Fogel, S (1984) Primary Structure and Transcription of an Amplified Genetic Locus: the CUP1 locus of Yeast. Proc Natl Acad Sci USA 81: 337-341.
3. Butt, TR, Sternberg, EJ, Gorman, JA, Clark, P, Hamer, D, Rosenberg, M and Crooke, ST (1984) Copper Metallothionein of Yeast, Structure of the Gene, and Regulation of Expression. Proc Natl Acad Sci USA 81: 3332-3336.
4. Hamer, DH, Thiele, DJ and Lemontt, JE (1986). Function and Autoregulation of Yeast Copperthionein. Science 228: 685-690.
5. Thiele, DJ, Walling, MJ and Hamer, DH (1986). Mammalian Metallothionein is Functional in Yeast. Science 231: 854-856.
6. Thiele, DJ and Hamer, DH (1985). Tandemly Duplicated Upstream Control Sequences Mediate Copper-Induced Transcription of the Saccharomyces cerevisiae Copper-Metallotionein Gene. Mol Cell Biol 6: 1158-1163.
7. Lee, W, Haslinger, A, Karin, M and Tijan, R (1987). Activation of Transcription by Two Factors that Bind Promoter and Enhancer Sequences of the Human Metallothionein Gene and SV40. Nature 325: 368-372.

Metal Ion Homeostasis: Molecular Biology and Chemistry, pages 109–115

Post-Transcriptional Regulation of Ferritin and Transferrin Receptor Expression: The Role of Iron-Responsive Elements

Tracey A. Rouault, Matthias W. Hentze, S. Wright Caughman, John L. Casey, David M. Koeller, Joe B. Harford and Richard D. Klausner

Many of the enzymes and proteins of living cells require iron for function. These proteins include heme-containing molecules and ribonucleotide reductase, the rate-limiting enzyme in DNA synthesis. Though iron is abundant, it is found in the environment primarily as ferric iron which is highly insoluble. A variety of uptake systems specific for iron have evolved which assure that cells can acquire iron from the environment despite its relative inaccessibility. Once iron reaches the intracellular environment, its oxidative potential poses a potential threat to the integrity of proteins and nucleic acids. Therefore proteins which can store and detoxify iron have evolved.

The two major proteins which are involved in the regulation of iron uptake and detoxification in eukaryotes are, respectively, the transferrin receptor and ferritin. The transferrin receptor (TfR) is a 180 kD molecule which exists as a homodimer composed of two 90 kD subunits. The TfR is expressed on the cell surface where it binds and then mediates internalization of the serum iron-carrier, transferrin (Tf). After endocytosis of the TfR-Tf complex, iron is removed from Tf in an acidified endosome, and the complex is then recycled to the plasma membrane. Apo-Tf dissociates from the receptor at the cell surface, and the TfR is free to begin another cycle (1). Iron is distributed from the acidic endosomal compartment to proteins which require iron, and to ferritin. Ferritin is a highly conserved protein which is composed of two subunits, H and L, which assemble to form a 24 subunit spherical protein. Iron precipitates at the core of the ferritin molecule and is sequestered. The primary role of ferritin in cellular metabolism may be to absorb excess intracellular iron and protect the cell from the toxic effects of iron.

The biosynthetic rates of the TfR and ferritin are highly regulated. When cells are deprived of iron, the biosynthetic rate of the TfR increases and more iron is acquired from the environment. When intracellular iron levels are high, TfR biosynthesis decreases markedly, while ferritin biosynthetic rates increase dramatically (2,3). The increase in ferritin protein biosynthesis is accomplished without the alteration of ferritin mRNA levels. Ferritin mRNA shifts from a translationally inactive mRNP pool to polysomes where it becomes actively translated (4,5). In contrast, TfR mRNA levels decrease dramatically (approximately 20-fold) when intracellular iron levels increase. Very little of the change in mRNA levels can be accounted for by changes in the transcriptional rate of the TfR gene (the change in transcriptional rate is 2-4 fold) (6). Rather the changes in mRNA levels are most likely accounted for by regulated changes in the rate of TfR mRNA degradation. A summary of the function and regulation of TfR and ferritin is provided in Figure 1. The recent cloning of the human genes for the TfR (7) and ferritin subunits (8,9) has allowed the identification of mRNA sequences that are essential for post-transcriptional regulation of both genes.

Regulation of Ferritin Translation

The human ferritin H chain cDNA has been cloned and transfected into cultured mouse fibroblasts. Intracellular iron levels can be modified in tissue culture cells by treating cells either with the iron source, hemin (ferric protoporphyrin IX) or with the iron chelator, desferrioxamine. Biosynthetic rates of the transfected ferritin H chain subunit vary over a 50 to 100-fold range, the same range seen with the endogenous mouse ferritin, but the levels of ferritin mRNA do not change (3,4). When a 157 base segment of the 5' leader of the ferritin H chain mRNA is deleted, normal ferritin protein is produced, but regulation by iron is eliminated. When the 5' leader of ferritin is linked to the indicator genes bacterial chloramphenicol acetyltransferase or human growth hormone, the capacity for regulation by iron is transferred (10,11). Deletional analysis of the 5' leader has identified a region which is necessary for regulation and which is conserved among species and between ferritin H and L chains (11,12). Computer modeling suggests that this region has the capacity to assume a stem-loop structure (Figure 2). These stem-loops of ferritin mRNA have in common a 6-membered loop with the sequence CAGUGX and a base-paired stem with an unpaired cytidine located five bases 5' of the loop. When an oligonucleotide including these conserved features is synthesized with a lower stem provided by BAM HI linker sequences and subsequently cloned 5' to indicator genes so as to yield a ferritin stem-loop in the 5' UTR of chimeric mRNAs, the capacity for regulation of biosynthesis by iron is conferred (11). When the first cytidine of the loop is deleted, the capacity to confer regulation by iron is entirely eliminated (11,15). Chloramphenicol acetyl transferase (CAT) assays show a 6-fold range of regulation, and 15-fold levels of regulation are seen when the secretion of human growth hormone (hGH) is measured. When immunoprecipitations using antibodies to human growth hormone are performed on lysates of transiently transfected cells after biosynthetic labeling, treatment with either hemin or desferrioxamine reproduces the full range of regulation that is seen with the native ferritin mRNA (13). Regulation appears to be less than that seen for native ferritin biosynthetic rates with these constructs when CAT assays are performed because these assays measure steady-state enzyme levels rather than the instantaneous rates of biosynthesis. When the 3' untranslated region of ferritin H chain mRNA is deleted, no change in in vivo translational regulation by iron is observed (13). Thus, the information required for the translational regulation of ferritin mRNA is contained within a short 5' sequence which has the capacity to transfer full regulation to indicator genes. This sequence has been termed an iron-responsive element (IRE).

Regulation of Transferrin Receptor Expression

TfR expression is regulated by altered iron status partially at the transcriptional level where differing transcriptional rates produce a 2-4 fold range of regulation by iron (6). However, TfR mRNA levels often can vary by more than 20-fold in response to iron. Deletional analysis of the TfR mRNA has identified a 678 nucleotide fragment of the 3' UTR which is critical to iron regulation (6,14). When this region is cloned into the 3' UTR of indicator genes, it can confer iron-dependent control of mRNA levels. Computer-assisted structural analysis (14) of this region demonstrates that this region contains five potential stem-loop structures which resemble the ferritin IRE. All of the stem-loop structures contain an upper stem of 5 paired bases, an unpaired C residue separated from the loop by 5 paired bases, and a lower stem of variable length. Four of the five potential stem-loops contain a loop with the sequence CAGUGX and the fifth loop contains the sequence CAGAGX. The sequences of the stems of these stem-loop structures differ markedly from the sequence of the ferritin IRE. When synthetic oligonucleotides for two of the transferrin receptor stem-loops were cloned into the 5' untranslated region of indicator genes, mRNA levels were not regulated, but full translational regulation of the indicator gene was produced (14). Between the two synthetic IRE oligomers from the TfR that were tested, 22 of 32 bases differ from bases contained in the synthetic ferritin IRE. Despite these differences in sequence, IRE function is retained. These sequence changes are present in the stem of the IRE, and in instances in which a sequence change in the stem would prevent base-pairing, a complementary difference in the stem sequence exists such that the capacity for base-pairing is retained. These findings suggest that maintenance of structure of the stem of the IRE is important, but there is little sequence specificity in the stem region. The sequence of the loop is highly conserved in IRE's and the deletion of a single base in the loop can completely eliminate IRE function (11,15). These findings suggest a model for a "consensus IRE" (Figure 2). In this model, the CAGUGX sequence of the loop and the unpaired cytidine 5' of the loop are present in all IRE's, and intervening sequences may contain variations in sequence as long as base pairing is maintained.

Experiments in which TfR 3' UTR structures can produce translational regulation demonstrate that the context of an IRE within an mRNA is important in determining how regulation will be affected. If the IRE is located in the 5' UTR, mRNA levels are unchanged, but the rate of mRNA translation decreases when intracellular iron levels are low. In the native ferritin mRNA, this results in reduced synthesis of ferritin when iron is scarce. When the IRE is present in the 3' UTR, mRNA levels increase when iron is scarce, thus producing an increase in TfR biosynthesis proportional to the increase in mRNA levels. However, IRE's in the 3' UTR are not, in themselves, sufficient for regulation. Other elements in the 3' UTR of the TfR, in addition to the IRE's, are required for regulation of mRNA stability, but these sequences have not yet been fully identified.

A Regulatory Protein Interacts With the IRE

When ^{32}P-labeled RNA containing the IRE is incubated with cytosolic extracts, the mobility of the labeled RNA is altered in a non-denaturing gel. A gel-retardation complex is produced which is specific for the IRE and which requires protein (17,18). When the IRE is made non-functional in translational regulation by deletion of the first cytidine of the loop, formation of the gel retardation complex is also eliminated (18). These findings imply that a specific regulatory protein binds with the IRE and that its ability to bind correlates with its ability to produce translational regulation. When excess unlabeled mRNA containing the ferritin IRE is incubated in the gel-retardation assay along with labeled IRE-containing mRNA, the gel-retardation complex is eliminated. When excess unlabeled mRNA from the 3' UTR of the TfR containing the IRE's of the TfR is incubated in the assay, the gel-retardation complex is similarly eliminated (18) . Thus, it appears that the ferritin IRE and the TfR IRE's are capable of binding the same protein.

When cells are treated with desferrioxamine prior to making lysates, there is a 4- to 5-fold increase in the signal intensity of the IRE retardation complex (18). When lower desferrioxamine concentrations are used, intermediate increases in signal intensity are seen. Thus, desferrioxamine treatment of cells produces an increase in the amount and/or activity of the binding protein which is available for interaction with exogenous RNA probes. This information suggests that the IRE binding protein acts as a repressor, as intracellular iron depletion correlates with greater protein binding.

A repressor model suggests that when the IRE is located in the 5' UTR, the protein binds when iron is scarce, and prevents the translational apparatus from progressing to the coding portion of the mRNA. An increase in intracellular iron levels results in the release of binding such that translation of the coding region is permitted. Conversely, when the IRE is located in the 3' UTR, binding of the protein when iron is scarce may protect the mRNA from degradation, resulting in accumulation of mRNA and increases in the rate of biosynthesis. A single regulatory protein would thus have the capacity to simultaneously decrease ferritin biosynthesis and increase TfR biosynthesis (15). This simple and economical approach to regulation is the first example of a pleiotropic effect of an RNA regulatory element. Further studies on the molecular nature of these regulatory interactions will provide considerably more information about post-transcriptional control mechanisms.

References

1. Klausner, R.D., Ashwell, G., van Renswoude, J., Harford, J.B. and Bridges, K. (1983) Binding of apotransferrin to K562 cells: Explanation of the transferrin cycle. Proc. Natl. Acad. Sci. USA 80, 2263.

2. Mattia, E., Rao, K., Shapiro, D.S., Sussman, H. and Klausner, R.D. (1984) Biosynthetic regulation of the human transferrin receptor by desferrioxamine in K562 cells. J. Biol. Chem. 259, 2689.

3. Rouault, T.A., Hentze, M.W., Dancis, A., Caughman, S.W., Harford, J.B. and Klausner, R.D. (1987) Influence of altered transcription on the translational control of human ferritin expression. Proc. Natl. Acad. Sci. USA 84, 6335.

4. Aziz, N. and Munro, H.N. (1986) Both subunits of rat liver ferritin are regulated at a translational level by iron induction. Nucl. Acids Res. 14, 915.

5. Zahringer, J., Baliga, B.S. and Munro, H.N. (1976) Novel mechanism for translational control in regulation of ferritin synthesis by iron. Proc. Natl. Acad. Sci. USA 73, 857.

6. Casey, J.L., di Jeso, B., Rao, K., Klausner, R.D. and Harford, J.B. (1988) Two genetic loci participate in the regulation by iron of

the gene for the human transferrin receptor. Proc. Natl. Acad. Sci. USA 85, 1787.

7. McClelland, A., Kuhn, L. and Ruddle, F.H. (1984) The human transferrin receptor gene: genomic organization and the complete primary structure of the receptor deduced from a cDNA sequence. Cell 39, 267.

8. Hentze, M.W., Keim, S., Papadopoulos, P., O'Brien, S., Modi, W., Drysdale, J., Leonard, W.J., Harford, J.B. and Klausner, R.D. (1986) Cloning, characterization, expression and chromosomal localization of a human ferritin heavy-chain gene. Proc. Natl. Acad. Sci. USA 83, 7226.

9. Costanzo, F., Colombo, M., Staempfli, S., Santoro, C., Marone, M., Frank, R., Deluis, H. and Cortese, R. (1986) Structure of gene and pseudogenes of human apoferritin H. Nucl. Acids Res. 14, 721.

10. Hentze, M.W., Rouault, T.A., Caughman, S.W., Dancis, A., Harford, J.B. and Klausner, R.D. (1987) A cis-acting element is necessary and sufficient for translational regulation of human ferritin expression in response to iron. Proc. Natl. Acad. Sci. USA 84, 6730.

11. Hentze, M.W., Caughman, S.W., Rouault, T.A., Barriocanal, J.G., Dancis, A., Harford, J.B. and Klausner, R.D. (1987) Identification of the iron-responsive element for the translational regulation of human ferritin mRNA. Science 238, 1570.

12. Murray, M.T., White, K. and Munro, H.N. (1987) Conservation of ferritin heavy subunit gene structure: Implications for the regulation of ferritin gene expression. Proc. Natl. Acad. Sci. USA 84, 7438.

13. Caughman, S.W., Hentze, M.W., Rouault, T.A., Harford, J.B. and Klausner, R.D. (1988) Manuscript in preparation.

14. Casey, J.L., Hentze, M.W., Koeller, D.M., Caughman, S.W., Rouault, T.A., Klausner, R.D. and Harford, J.B. (1988) Iron-responsive elements: Regulatory RNA sequences that control mRNA levels and translation. Science 240, 924.

15. Hentze, M.W., Caughman, S.W., Casey, J.L., Koeller, D.M., Rouault, T.A., Harford, J.B. and Klausner R.D. (1988) A model for

the structure and functions of iron-responsive elements. Gene, in press.

16. Zuker, M. and Stiegler, P. (1981) Optimal computer folding of large RNA sequences using thermodynamics and auxiliary information. Nucl. Acids Res. 9, 133.

17. Leibold, E. and Munro, H. (1988) A cytoplasmic protein binds in vitro to a highly conserved sequence in the 5'-untranslated region of ferritin H and L subunit mRNAs. Proc. Natl. Acad. Sci. USA 85, 2171.

18. Rouault, T.A., Hentze, M.W., Caughman, S.W., Harford, J.B. and Klausner, R.D. (1988) Binding of cytosolic proteins to the iron-responsive element of human ferritin mRNA. Manuscript submitted.

Metal Ion Homeostasis: Molecular Biology and Chemistry, pages 117–126

CHARACTERIZATION OF A REPRESSOR OF FERRITIN mRNA TRANSLATION FROM RABBIT RETICULOCYTES[1]

W.E. Walden*, L. Gaffield*, P.H. Brown+, S. Daniels-McQueen+, D. Bielser+, and R.E. Thach+

*Department of Microbiology and Immunology
College of Medicine
The University of Illinois at Chicago
Chicago, IL, 60612, USA

+Department of Biology
Washington University
St. Louis, MO, 63130, USA

ABSTRACT A specific repressor of ferritin mRNA translation has been partially purified from rabbit reticulocytes by differential ultracentrifugation, ammonium sulfate fractionation, and chromatography on phosphocellulose, DEAE cellulose, and Sephacryl S300. The elution profile from the latter suggests an aggregate molecular weight of approximately 180,000 Da for the repressor. The inhibitory activity of this repressor against native ferritin mRNA can be relieved by adding in vitro transcripts of ferritin light chain RNAs that contain the first 92 nucleotides of the 5′ untranslated region. No other sequences appear to be necessary for this effect.

INTRODUCTION

Synthesis of ferritin, the major iron storage protein in eukaryotes, is regulated in coordination with iron levels (1-8). Early evidence suggested that the primary site of this regulation is translational (4-9). The identification

---

1. This work was supported by a UIC, CRB institutional award to W.E.W. and by grants from the National Science Foundation (DMB-8518050) and the United States Public Health Service (A120484) to R.E.T.

of a region of ferritin mRNA responsible for iron regulated translation has confirmed this (10-12). This region, called the "iron responsive element" (IRE), is a highly conserved sequence of 28 nucleotides, located in the 5′ untranslated region (5′ UTR) of all ferritin mRNAs for which sequences are known (13). The presence of this sequence in ferritin mRNA is absolutely required for proper regulation of translation, since deletion of the IRE renders the mRNA unresponsive to iron (10). Moreover, appending the IRE to the 5′ end of a heterologous mRNA confers iron responsiveness to it (10-12).

It seems likely that iron does not interact directly with the IRE, but rather that its effects are mediated by one or more "repressors". These inhibitors would be expected to bind specifically to the IRE sequence within the mRNA on the one hand, and be sensitive to intracellular iron pools on the other. Support for such a model has recently been reported (14). Short RNA fragments, corresponding to the 5′ UTR of ferritin mRNAs (containing the IRE sequence), specifically bind to protein factors in a rat liver cytoplasmic extract in an iron sensitive fashion.

Repression of ferritin mRNA translation has recently been observed _in vitro_ (15, Walden _et al_., in preparation). Translation of ferritin mRNAs is repressed in rabbit reticulocyte lysates. In contrast, ferritin mRNAs are translated very efficiently in wheat germ lysates (7,15). The repression seen in reticulocyte lysate can be transferred to wheat germ extracts by the addition of small aliquots of a 100,000 x g supernatant fraction (S100) from reticulocyte (Walden _et al_., in preparation). This repression is highly specific and it suggest that reticulocytes contain a trans-acting repressor of ferritin mRNA translation.

In this communication we describe the partial purification and initial characterization of a repressor from rabbit reticulocytes. Its distinctive feature is that it inhibits translation of both light (L) and heavy (H) chain ferritin mRNAs in a wheat germ lysate with a high degree of specificity. Moreover, this inhibition can be outcompeted by the addition of short RNA transcripts containing the IRE sequence. No other region of the light chain ferritin mRNA sequence is required for this effect, suggesting that the repressor recognizes specifically the IRE.

METHODS

In Vitro Translation: _In vitro_ translations in wheat

germ lysates (Promega Biotec) were performed following the manufacturer's recommendations, except that 10 µL reactions were used. ^{35}S methionine (Amersham; $\geq$800 Ci/mmol) was used as label in all cases. Analysis of translation products by immune precipitation and SDS-PAGE was as described previously (16), except that 12% polyacrylamide gels were used.

Repressor Purification: All procedures were conducted at 4°C using Buffer A, which consists of 20 mM Tris-HCl, pH 7.4, 1 mM DTT, 0.1 mM EDTA and 10% glycerol. Buffer A-100, A-150, A-500 and A-1000 are Buffer A containing 100 mM, 150 mM, 500 mM and 1 M KCl, respectively. To purify repressor from rabbit reticulocytes, solid ammonium sulfate was added to the S100 (generously provided by Dr. W.C. Merrick) to 55% saturation. The precipitate which forms was dissolved, dialyzed exhaustively against Buffer A-150, and loaded onto a phosphocellulose column in the same buffer. Protein was eluted from the column with a step gradient consisting of Buffer A-150 followed by Buffer A-500 and Buffer A-1000; one column volume of each. The material eluting at each KCl concentration was pooled yielding the PCI, PCII, and PCIII fractions, for material eluting at 150 mM, 500 mM and 1 M KCl, respectively. Ferritin specific repressor activity elutes in the PCI fraction. PCI was concentrated by ammonium sulfate precipitation at 70% saturation, dissolved and dialyzed against Buffer A. It was loaded onto a DEAE cellulose column in Buffer A and eluted with a linear KCl gradient (0 mM KCl to 100 mM KCl) in the same buffer. The peak of ferritin translational repressor activity elutes between 0.5 mM KCl and 1 mM KCl. Repressor from DEAE (DEPI) was concentrated, dialyzed against Buffer A-100 and loaded onto a Sephacryl S300 column in the same buffer. Repressor activity eluted from this column was designated S3PI.

RESULTS

A repressor of ferritin mRNA translation was partially purified from reticulocyte S100 by a combination of ammonium sulfate precipitation, ion exchange chromatography, and gel filtration chromatography as described in the Methods section. The 55% ammonium sulfate precipitate contains all of the ferritin specific repressor found in S100. When this fraction was loaded onto a phosphocellulose column in 150 mM KCl at pH 7.4, ferritin repressor activity did not bind and was eluted in the starting buffer. Material which eluted in higher ionic strength buffer washes showed no specific repressor activity. However, there was a low level of generally inhibitory activity present in some of these washes (data not shown). The PCI fraction was further fractionated on DEAE cellulose and Sephacryl S300 chromatography. Ferritin repressor activity eluted from DEAE cellulose at very low ionic strength in pH 7.4 buffer, eluting at between 0.5 mM and 1.0 mM KCl. When this peak material (called DEPI) was further fractionated on a Sephacryl S300 column in 100 mM KCl, it eluted with an apparent native molecular weight of approximately 180,000 Da (this peak activity is called S3PI).

The repressor isolated from reticulocyte S100 is highly specific towards ferritin mRNA. This is evident from the data shown in Fig 1. Here, aliquots of PCI, DEPI, and S3PI were tested for repressor activity against total mouse poly A^+ RNA in wheat germ extracts. It is clear in panel A that the repressor in each preparations has no effect on the translation of the majority of mRNAs. In contrast, ferritin mRNA translation is very sensitive to this repressor(Fig 1B).

It was of interest to determine whether particular sequences within the ferritin mRNAs are recognized by the repressor. To this end, a competition assay was devised in which various transcripts of L ferritin cDNAs were included in the basic translation assay, to see whether their presence would relieve the repression of translation of natural ferritin mRNAs. The various transcripts that were employed are shown in Fig 2. The inclusion of small amounts of most of these transcripts relieved the repression caused by 2 μl of PCI (Figure 3A-C). This is most clearly evident with the M1 transcript, which encodes no protein, yet restores translation of both rabbit H and L ferritin (Fig 3A & 3B; compare lane 3, which shows fully repressed translation, with lane 5 which contains, in addition to repressor, 0.06 pmoles of M1 transcript). Similar results are seen when the M2 transcript is present (lane 4, Fig 3B), although only the relief of H ferritin repression is an accurate indicator, since the M2 transcript itself encodes L ferritin. In contrast, the SX9 transcript (which also encodes L ferritin) did not significantly relieve repression of H ferritin synthesis (lane 6 and 9 in Fig 3B). We suspect that this is due to the fact that it lacks the 5′ UTR region which contains the IRE. No clear-cut requirement for the 3′ UTR was obtained, inasmuch as the D3 transcript relieved repression of H ferritin synthesis (Fig 3C, lane 5). However, in our hands 3′ UTR deleted transcripts tend to be inefficiently translated themselves, which may suggest an involvement of the 3′ UTR in translational regulation (15).

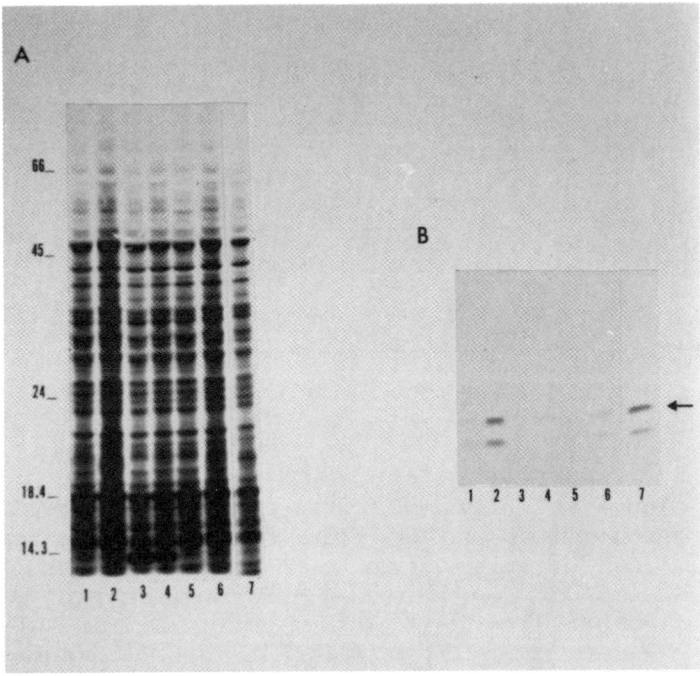

Figure 1: Demonstration of specificity of repressor activity. Mouse poly A$^+$ RNA was translated in a wheat germ lysate in the presence of various amounts of PCI (lanes 3 and 4), DEPI (lanes 5 and 6) or S3PI (lanes 1 and 2) repressor fractions. Translation products were analyzed by SDS-PAGE followed by fluorography. Panel A: A fluorogram showing the complete pattern of protein synthesized. Panel B: A fluorogram showing the product immune precipitated with a mouse ferritin specific antiserum. Control, lane 7. S3PI, lane 1 (1 µL); lane 2 (0.125 µL). PCI, lane 3 (1 µL); lane 4 (0.125 µL). DEPI, lane 5 (1 µL); lane 6 (0.125 µL). The arrow indicates the position of mouse ferritin. Ferritin subunits are not readily detectable in the total protein synthesis pattern.

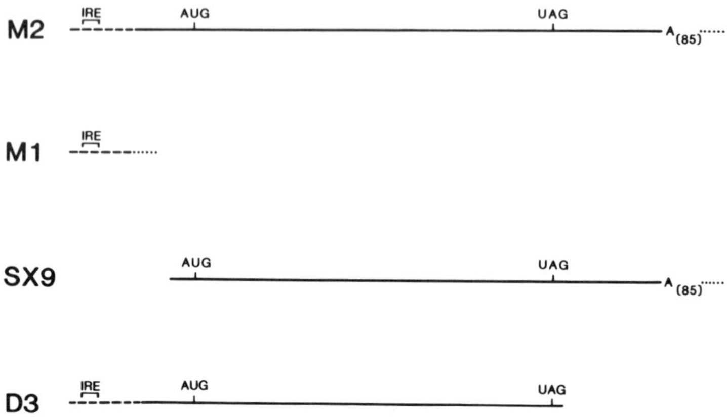

Figure 2: Diagrammatic representation of RNAs used in repression relief assays. The four RNAs diagrammed were synthesized _in vitro_ from a T7 promoter using T7 RNA polymerase. M2 is a chimeric mRNA consisting of the first 92 nucleotides from the 5′ UTR of human ferritin light chain mRNA (dashed line; 17) appended by a 4 nucleotide linker region to a rabbit ferritin mRNA (solid line; 18). The rabbit ferritin part of the mRNA has 101 nucleotides of 5′ UTR, making the chimeric 5′ UTR 197 nucleotides long. It also contains the complete ORF for light chain synthesis (528 nucleotides), a complete 3′ UTR (128 nucleotides) and a poly A sequence. M1 is a RNA fragment consisting of the first 92 nucleotides from the 5′UTR of human ferritin light chain mRNA (dashed line). SX9 is a rabbit ferritin light chain mRNA that has only 30 nucleotides of 5′ UTR. It contains a complete ORF for light chain synthesis and a complete 3′ UTR, but lacks the IRE. D3 is the same as M2 except that it lacks all but 3 nucleotides of 3′ UTR. Extra nucleotides encoded by the DNA vector that are appended during transcription to the 5′ and 3′ termini of each inserted sequence are, respectively: 8 and 42 for M2; 8 and 33 for M1; 3 and 36 for SX9; 8 and 0 for D3. "IRE" over the bracket indicates the approximate location of the 5′ iron responsive element.

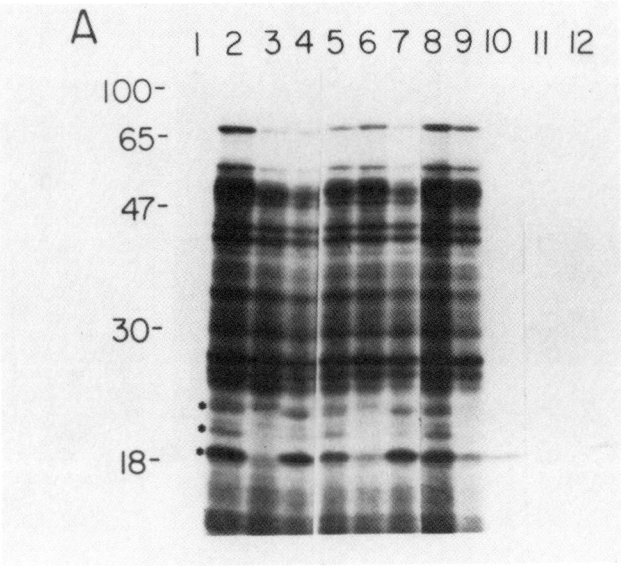

Figure 3: Determination of regions of the ferritin mRNA important to in vitro repression. Each of the RNAs diagrammed in Fig 2 were tested for the ability to relieve repression in vitro. Translations were done in wheat germ lysates using total rabbit liver poly A^+ mRNA and 2 µL of the PCI fraction repressor. Panel A: Total translation pattern. Additions were: lane 1, none; lane 2, mRNA; lane 3, mRNA and PCI; lane 4, mRNA, PCI and 0.06 pmol M2; lane 5, mRNA, PCI and 0.06 pmol M1; lane 6, mRNA, PCI and 0.06 pmol SX9; lane 7, mRNA, PCI and 0.09 pmol M2; lane 8, mRNA, PCI and 0.09 pmol M1; lane 9, mRNA, PCI and 0.09 pmol SX9; lane 10, 0.09 pmol M2; lane 11, 0.09 pmol M1; lane 12, 0.09 pmol SX9. Panel B: Products immune precipitated by mouse anti-rabbit ferritin serum from the translations shown in panel A. Lane assignments are the same as in panel A. Panel C: Products immune precipitated by mouse anti-rabbit ferritin serum. Additions were: lane 1, none; lane 2, mRNA; lane 3, mRNA and PCI; lane 4, mRNA, PCI and 0.09 pmol M2; lane 5, mRNA, PCI and 0.09 pmol D3; lane 6, mRNA, PCI and 0.09 pmol M1; lane 7, mRNA, PCI and 0.09 pmol SX9. The asterisks (*) mark the position of peptides synthesized from rabbit ferritin mRNA. Lower band, light chain. Middle band, heavy chain.

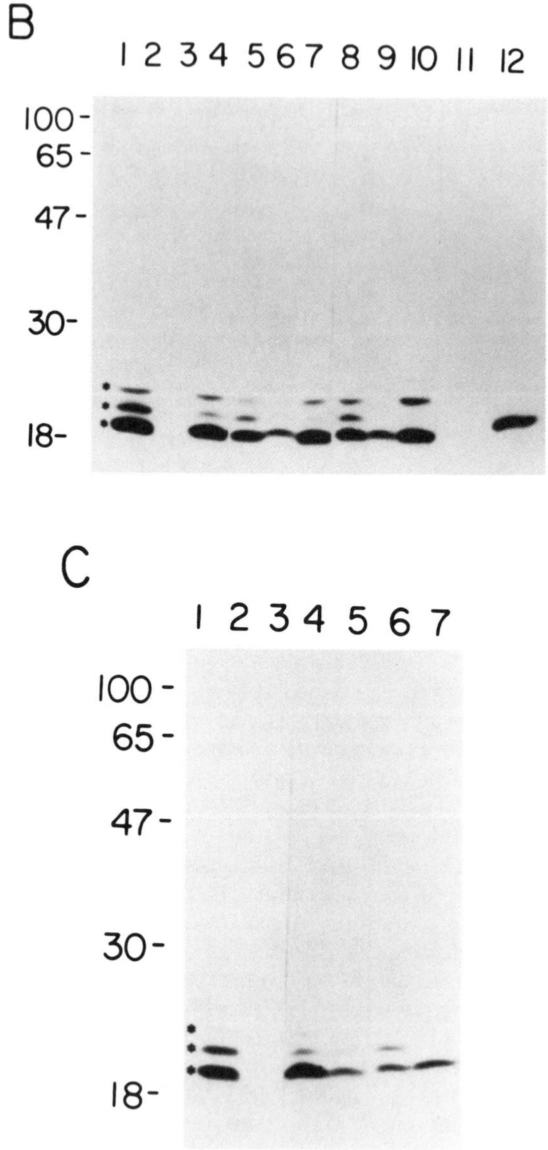

Figure 3 continued.

DISCUSSION

The results described in this communication show that a repressor that selectively inhibits translation of L and H ferritin mRNAs can be partially purified from rabbit reticulocytes. The specificity of this inhibition is extremely high: no effect of the repressor on the translation of any other component of mouse poly A(+) RNA could be detected (Fig 1). A likely basis for this specificity was revealed by competition studies, where a transcript of the first 92 nucleotides of the 5′ UTR of the L ferritin mRNA was sufficient to reverse the inhibitory effect of repressor. Since this region contains the IRE sequence that has been shown to regulate translation *in vitro* (10-12), we feel that it is highly likely that repressor binds specifically to this sequence and thereby inhibits translation.

Gel filtration studies suggest that the ferritin repressor from reticulocytes has a native molecular weight of approximately 180,000 Da, assuming it to be a globular protein. Interestingly, the protein in rat liver extracts which can be crosslinked to the 5′ UTR of ferritin mRNAs has an estimated molecular weight of 87,000 Da (14). This is not inconsistent with the findings reported here since it may well be that the ferritin translational repressor is an aggregate of smaller polypeptide chains, as well as other substances (such as RNA). In any case, its sensitivity to phenol, and its high A_{280}/A_{260} ratio, suggest that the repressor is primarily comprised of protein.

Experiments to determine the relationship of the reticulocyte repressor to the "iron sensitive" repressor of ferritin mRNA translation are in progress. The fact that the reticulocyte repressor recognizes a region of ferritin mRNA which contains the IRE is very good evidence that it is closely related to the iron sensitive repressor. The mechanism by which iron addition relieves repression is also of interest. Iron could modify repressor activity by simply binding to it, or by a more complicated mechanism, such as by inducing a covalent modification of the repressor. In any event, an obvious step in this de-repression process could be the removal of the repressor from the IRE.

It is likely that other repressors, with specificities for other mRNAs, may be isolatable from eukaryotic cell lysates by the approach used here. This deduction follows from the fact that the ferritin mRNAs are not unique; rather, they are members of a significant subclass of mRNAs which share kinetic properties that would seem to be diagnostic for translational repression (9).

REFERENCES

1. Munro HN, Linder MC (1978). Physiol Rev 58:317.
2. Aisen P, Listowsky I (1980). Annu Rev Biochem 49:3571.
3. Theil EC (1987). Annu Rev Biochem 56:289.
4. Drysdale JW, Munro HN (1966). J Biol Chem 241:3630.
5. Chu LLH, Fineberg RA (1969). J Biol Chem 244:3847.
6. Zahringer J, Baliga BS, Munro HN (1976). Proc Natl Acad Sci USA 73:857.
7. Shull GE, Theil EC (1982). J Biol Chem 257:14187.
8. Aziz N, Munro HN (1986). Nucleic Acids Res 14:915.
9. Walden WE, Thach RE (1986). Biochemistry 25:2033.
10. Hentze MW, Rouault TA, Caughman SW, Dancis A, Harford JB, Klausner RD (1987). Proc Natl Sci USA 84:6730.
11. Hentze MW, Caughman SW, Rouault TA, Barriocanal JG, Dancis A, Harford JE, Klausner RD (1987). Science 238: 1570.
12. Aziz N, Munro H (1987). Proc Natl Acad Sci USA 84:8478.
13. Murray MT, White K, Munro HN (1987). Proc Natl Acad Sci USA 84:7438.
14. Liebold EA, Munro HN (1988). Proc Natl Acad Sci USA 85:2171.
15. Dickey LF, Wang Y-H, Shull GE, Wortman IA, Theil EC (1988). J Biol Chem 263:3071.
16. Walden WE, Daniels-McQueen S, Smith LL, Thach RE (1987). Gene 61:317.
17. Santoro C, Marone M, Ferrone M, Costanzo F, Colombo M, Minganti C, Cortese R, Silengo L (1986). Nucleic Acids Res 14:2863.
18. Daniels-McQueen S, Ray A, Ray BK, Walden WE, Thach RE (1988). Nucleic Acids Res, in press.

Metal Ion Homeostasis: Molecular Biology and Chemistry, pages 127–136

THE REGULATION OF FERRITIN EXPRESSION: A PARADIGM FOR TRANSLATIONAL CONTROL[1]

Yuh-Hwa Wang, Steven R. Sczekan, and Elizabeth C. Theil

Department of Biochemistry, North Carolina State University Raleigh, North Carolina 27695-7622

ABSTRACT The induction of ferritin synthesis by iron has provided a number of novel insights into the translational control of gene expression, which are now finding applications in understanding the control of expression of a variety of other genes induced by agents such as stress and light. The historical aspects of the regulation of ferritin synthesis are reviewed in terms of ferritin mRNA derepression and recruitment by iron, translational competition by ferritin mRNA, and coordinate regulation of the translation of mRNA for the multiple ferritin subunit types. Recent results related to the repression of ferritin mRNA in cell-free systems, to the role of regions of conserved structure in the 3′-untranslated region of ferritin mRNA, and to mRNA folding for translational control are presented and evaluated in terms of the conserved sequences in the 5′-untranslated regions that respond to iron and in terms of ferritin mRNA binding proteins.

OVERVIEW OF TRANSLATIONAL CONTROL OF FERRITIN SYNTHESIS

Iron has been known to regulate ferritin gene expression for more than 40 years; Granick first observed the accumulation of ferritin with excess iron (1). More contemporary studies of the regulation of ferritin synthesis began with the work of Munro and colleagues, who showed that increases in ferritin synthesis in the presence of iron

[1]This work was supported in part by NIH grant DK20251.

depended in part on the recruitment of stored ferritin mRNA (2–4). Our studies then showed that recruitment of stored ferritin mRNA involved derepression of the message (5,6). In addition, we showed that in some cell types, e.g., those with reduced capacities for protein synthesis, full induction of ferritin synthesis required not only derepression of the ferritin mRNA but also competition for translation initiation factors (6–8). Now, the molecular basis for translational control of ferritin synthesis is being characterized with a variety of novel observations, e.g., conserved sequences in the 5′-untranslated regions of the ferritin message are required for the response to Fe (9–11); specific proteins bind to ferritin mRNA (12); specific proteins alter ferritin mRNA translation (13); and the 3′-untranslated region of ferritin mRNA contains sequences which influence translation (14). In this volume alone are three papers about the translational control of ferritin mRNA: Klausner and coworkers describe their fascinating results on the role of the iron-responsive element (IRE) in the 5′-untranslated region (UTR) of ferritin mRNA and in the 3′-UTR of the transferrin receptor where, by affecting mRNA stability, it appears to permit Fe to coordinately increase ferritin expression and decrease transferrin receptor expression; Walden and Thach describe their results on the isolation of a protein that specifically represses ferritin mRNA translation; and we describe our results on a short sequence in the 3′-UTR of ferritin mRNA that, apparently by specific folding, regulates translation of the message.

The role of ferritin is to provide a reservoir of iron for proteins, such as those involved in DNA synthesis, respiration, and photosynthesis; accordingly, ferritin is found in animals, plants, and microorganisms. In addition, ferritin sequesters excess amounts of iron in a safe form, thus playing a role in detoxification during iron overload. Ferritin consists of a multisubunit protein which forms a coat around an iron core of hydrous ferric oxide that can consist of as many as 4500 Fe atoms with variable amounts of phosphate. The general structure is conserved among all organisms, and the primary sequence of the subunits is conserved among plants (15) and among animals (16). However, the relationship between the plant and animal proteins remains to be determined (reviewed in ref. 16).

Ferritin can be formed from heterologous types of protein subunits which maintain large blocks of conserved sequences interspersed among specific sequences (16). The distribution of specific sequences is sufficiently frequent that the subunits appear to represent distinct genes; separate genes have already been characterized for two of the subunits in humans (17,18). As many as three subunits have been found to be expressed in a single cell type in amphibia (19).

The ratio of each ferritin protein subunit in a specific cell type reflects the concentration of each subunit mRNA, which appears to be determined, in general, by alterations in mRNA concentration (transcription?) at the time of differentiation (20-22). The mRNA for each subunit type is stored with translational regulation by Fe appearing to be coordinate for each subunit type (19,22). An exception occurs in the case of Fe stress, in a cell overloaded with Fe by experimental manipulation, when the relative concentration of mRNA and protein for a minor subunit increased threefold in a highly differentiated cell (19).

Derepression of ferritin mRNA by Fe was indicated by comparing the functional and structural concentration of ferritin mRNA in animals treated with excess Fe to those with normal levels of Fe (6,7,23). Ferritin mRNA levels were unchanged by Fe, although both the rates of synthesis and the accumulation of the protein increased 20-40x (6,7). Moreover, the amount of ferritin mRNA was high in the absence of Fe. Thus, constitutive rates of ferritin synthesis reflected repression of the high concentrations of ferritin mRNA in the cell, and induced rates of synthesis were the result, in part, of derepression of the stored ferritin mRNA.

Competitive translation of ferritin mRNA has been shown to occur in cells with relatively low capacities for protein synthesis, such as reticulocytes (7,8) and hepatocytes from starved animals (8). In cells with high capacities for protein synthesis, such as fibroblasts or normal hepatocytes, induced rates of ferritin synthesis appear to be mainly achieved by derepression of stored ferritin mRNA (7,23). Competitive translation of ferritin mRNA was indicated in red blood cells by the sensitivity of Fe-induced ferritin synthesis to inhibitors of elongation such as cycloheximide and emetine (7). Concentrations of the inhibitors which reduced the relative rate of ferritin synthesis 50% had no effect on total protein synthesis; ten times the concentration of the inhibitors were required to reduce the synthesis of non-ferritin proteins 50%. Normally, initiation is the rate-limiting step in the translation of mRNA. When elongation inhibitors are used, mRNAs which are relatively poor initiators (the non-ferritin messages in this case) have the opportunity to "catch up" with the mRNAs that are efficient initiators; the rate of synthesis of proteins such as ferritin, which are encoded by efficiently translated mRNAs, thus appears to decrease relative to the synthesis of the other proteins. The same phenomenon can be demonstrated in wheat germ extracts, by measuring the synthesis of ferritin and other proteins under the direction of poly(A+) RNA from erythroid cells (5,6). The concentration of ribosomes and initiation factors (soluble proteins) was held constant while the concentration of poly(A+) RNA was increased; synthesis of ferritin and total protein synthesis were monitored. As the concentration of poly(A+) RNA was increased, the rate of total protein synthesis became constant, but in the range of RNA concentrations used, rates of ferritin synthesis continued to increase linearly; the synthesis of ferritin relative to the other proteins appeared to increase. Thus when translation factors became limiting, ferritin mRNA formed initiation complexes with a disproportionately high frequency. The competition between ferritin mRNA and other messages could be relieved by adding a crude preparation of initiation factor 4B (kindly provided by J. Ravel) to the translation mixture; the rate of ferritin synthesis was unchanged, but the rate of synthesis of non-ferritin proteins increased and the relative rate of ferritin synthesis decreased (M. Shull and E. C. Theil, unpublished results). Note that the competitive translation of ferritin mRNA appears to be independent of Fe.

REPRESSION OF FERRITIN mRNA IN CELL-FREE SYSTEMS

Analysis of the molecular basis for ferritin mRNA repression and competition can be studied to advantage in cell-free systems, because the contribution to the results of the effects of transcription and mRNA stability do not need to be separated from those involving translational control, in contrast to results obtained during transient or permanent expression of ferritin mRNA sequences in transfected cells. Recently, we have been able to find a cell-free system from rabbit reticulocytes that specifically represses ferritin mRNA encoded in red cell poly(A+) RNA from amphibia by 4-20 fold (14).

Repression of ferritin mRNA in the cell-free system was independent of any form of Fe which we have so far tested [heme, Fe(III)ATP, Fe(III)transferrin, Fe(II)]. Degradation did not cause the apparent repression of the ferritin mRNA because the ferritin mRNA could be recovered after translation and translated at normal rates in the cell-free system from wheat germ; moreover, the electrophoretic mobility of the ferritin mRNA was unchanged by translation. However, the location of ferritin mRNA was different after translation of the red cell poly(A+) RNA in wheat germ or rabbit reticulocyte lysates (Fig. 1). Although ferritin and non-ferritin protein (mainly globin) mRNAs were able to form

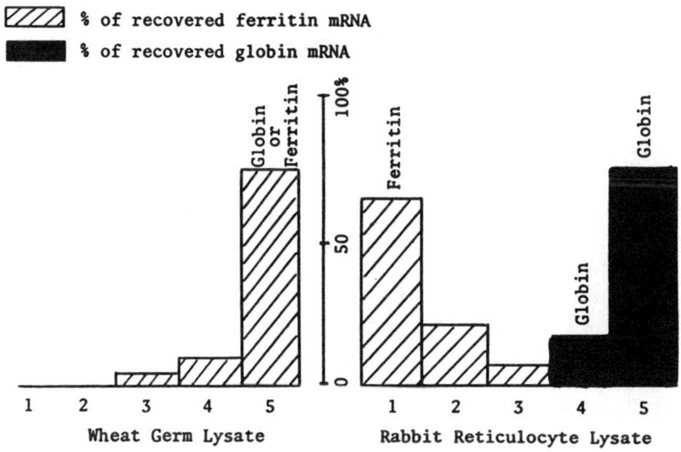

FIGURE 1. Location of ferritin mRNA after translation in permissive (wheat germ) or repressive (rabbit reticulocyte) cell-free extracts. Poly(A+) RNA from reticulocytes of the embryonic (tadpole) cell line (a rich source of ferritin and globin mRNAs) was translated in both cell-free systems, which were then fractionated by sedimentation through gradients of sucrose. The recovered RNA was fractionated by electrophoresis through agarose gels and located by hybridization to radioactively labeled cDNA probes for ferritin or globin after transfer to nylon membranes. Note the presence of ferritin and globin mRNA in the polysome fraction of the wheat germ extracts and globin mRNA in the polysomes of the rabbit reticulocyte lysates. However, ferritin mRNA remains in the monosome or oligosome fraction of the rabbit reticulocyte lysate. The data are taken from ref. 14.

polysomes in the wheat germ extract, and globin mRNA was able to form polysomes in the rabbit reticulocyte lysate and the wheat germ lysate, ferritin mRNA could only form monosomes or oligosomes in the rabbit reticulocyte lysate. Thus the translation of ferritin mRNA was repressed because the mRNA was a poor initiator in the rabbit reticulocyte lysate. Addition of purified reticulocyte initiation factors (kindly provided by W. Merrick) did not relieve the repression. Coordinate repression for all three ferritin subunits encoded in the poly(A+) RNA used was observed as shown by hybridization under high stringency with subunit mRNA-specific probes in the Northern blot analysis and by examining the relative synthesis of the subunits using electrophoresis in SDS gels and autoradiography (14).

THE ROLE OF THE 3′-UTR IN FERRITIN mRNA REPRESSION

We decided to determine which part of the ferritin mRNA was required for repression and whether or not factors present in the poly(A+) RNA itself were required (14). Two types of transcripts were prepared: RNA transcribed from pBF-8 contained 24 nucleotides in the 3′-UTR (a deletion of 151 nucleotides), and RNA transcribed from pBF-15 contained 94 nucleotides in the 3′-UTR (a deletion of 81 nucleotides). pBF-8 was not repressible and was translated equally well in either wheat germ or rabbit reticulocyte lysates; when capped, pBF-8 directed the synthesis of protein at the same rate as poly(A+) RNA in either cell-free extract. Moreover, there was no change in the rate of translation of RNA from pBF-8 when poly(A+) RNA was added, nor was there any evidence for antisense ferritin RNA in the poly(A+) RNA. Thus, repression of ferritin mRNA in rabbit reticulocytes appeared to be a function of sequences contained in the deleted portion of the 3′-UTR.

In order to locate the sequences required for repression of ferritin mRNA, the translation of RNA transcribed from pBF-15 was compared to pBF-8. Translation of the mRNA with the longer 3′-UTR was reduced 3x in the rabbit reticulocyte lysate compared to pBF-8, showing that repression was essentially restored by the additional 70 nucleotides in the 3′-UTR. Comparison of the sequences in the 3′-UTR of eight ferritin mRNAs from mammals, chickens, or bullfrogs (17-25) shows no conservation of sequence (Table 1) nor complementarity to the 5′-UTR. However, prediction of secondary structure in the 94 bases of the 3′-UTR required for

TABLE 1
PERCENT HOMOLOGY IN UNTRANSLATED REGIONS
OF FERRITIN SUBUNIT mRNAs[a]

	5′-UTR	3′-UTR
Human H	100	100
Human L	49	44
Rat H	60	78
Rat L	54	27
Chicken H	51	53
Tad H	38	48
Tad M	39	46
Tad L	--	39
Tad pseudogene	37	40

[a]Sequences were used from references 17-20 and 24,25 were screened using search programs provided by BIONET™. The number of base pairs matched ranged from 66-208.

repression showed a strong conservation of STRUCTURE (14). A common, stable, stem/loop structure was present in the ferritin mRNAs (-ΔG ranged from 18-40 kcal/mol). The predicted structures contain a series of 3 stems involving 8-11 base pairs interrupted by small bulges or loops. mRNA transcribed from pBF-8, with only 24 nucleotides in the 3′-UTR, cannot form the predicted stems and loops.

The linear distance between the 3′-UTR sequences and the sites of initiation in the 5′-UTR is so great that the effect of the 3′-UTR is unlikely to be exerted along the mRNA chain. More likely, the effect of the 3′-UTR is exerted through complex folding of the mRNA in three dimensions, which influences or perhaps even contacts the 5′-IRE (9-12). The folded structure may contain the actual features recognized by ferritin mRNA-specific proteins (12,13). In some cell types, the appropriate cytoplasmic factors may respond only to the IRE, permitting quantitative reproduction of translation rates in constructs with only the IRE. Whatever the final features turn out to be, studies of ferritin mRNA have, for yet another time, provided new insights into the details of translational control. The role of predicted 3′-UTR structures in regulating rates of translation (14) has never been reported before in a eukaryotic mRNA, to our knowledge. The importance of mRNA folding for function, emphasized by the data with ferritin mRNA, has also been little considered. Clearly, studies of the regulation of ferritin mRNA expression and translational control are important not only for understanding Fe metabolism and metal homeostasis in particular, but also for understanding general aspects of biochemistry and molecular biology.

ACKNOWLEDGMENTS

The computer resources used were provided by the BIONET™ National Computer Resource for Molecular Biology, part of whose funding is provided by the Biomedical Research Technology Program, Division of Research Resources, National Institutes of Health Grant 1 U41 RR-01685-02.

REFERENCES

1. Granick S (1946). Ferritin: its properties and significance for iron metabolism. Chem Rev 38:379.
2. Drysdale JW, Olafsdottir E, Munro HN (1968). Effect of ribonucleic acid depletion on ferritin induction in rat liver. J Biol Chem 243:552.
3. Zahringer J, Baliga BS, Munro HN (1976). Novel mechanism for translational control in regulation of ferritin synthesis by iron. Proc Natl Acad Sci USA 73:857.
4. Aziz N, Munro HN (1986). Both subunits of rat liver ferritin are regulated at a translational level by iron induction. Nucleic Acids Res 14:915.
5. Shull GE, Theil EC (1982). Translational control of ferritin synthesis by iron in embryonic reticulocytes of the bullfrog. J Biol Chem 257:14187.
6. Shull GE, Theil EC (1983). Regulation of ferritin mRNA: a possible gene-sparing phenomenon. J Biol Chem 258:7921.
7. Schaefer FV, Theil EC (1981). The effect of iron on the synthesis and amount of ferritin in red blood cells during ontogeny. J Biol Chem 256:1711.
8. Theil EC (1987). Storage and translation of ferritin messenger RNA. In Ilan J (ed): "Translational Regulation of Gene Expression," New York: Plenum Press, p 141.

9. Aziz N, Munro HN (1987). Iron regulates ferritin mRNA translation through a segment of its 5′ untranslated region. Proc Natl Acad Sci USA 84:8478.
10. Hentze MW, Rouault TA, Caughman SW, Dancis A, Harford JB, Klausner RD (1987). A cis-acting element is necessary and sufficient for translational regulation of human ferritin expression in response to iron. Proc Natl Acad Sci USA 84:6730.
11. Hentze MW, Caughman SW, Rouault TA, Barriocanal JG, Dancis A, Harford JB, Klausner RD (1987). Identification of the iron-responsive element for the translational regulation of human ferritin mRNA. Science 238:1570.
12. Leibold EA, Munro HN (1988). Cytoplasmic protein binds in vitro to a highly conserved sequence in the 5′ untranslated region of ferritin heavy- and light-subunit mRNAs. Proc Natl Acad Sci USA 85:2171.
13. Walden WE, Gaffield L, Brown PH, McQueen SD, Bielser D, Thach RE (1988). Characterization of a repressor of ferritin mRNA translation from rabbit reticulocytes. Fed. Proc. 2(4):A778.
14. Dickey LF, Wang Y-H, Shull GE, Wortman IA III, Theil EC (1988). The importance of the 3′-untranslated region in the translational control of ferritin mRNA. J Biol Chem 263:3071.
15. Briat JF, personal communications.
16. Theil EC (1987). Ferritin: structure, gene regulation, and cellular function in animals, plants, and microorganisms. Ann Rev Biochem 56:289.
17. Santoro C, Marone M, Ferrone M, Costanzo F, Colombo M, Minganti C, Cortese R, Silengo L (1986). Cloning of the gene coding for human L apoferritin. Nucleic Acids Res 14:2863.
18. Costanzo F, Colombo M, Staempfli S, Santoro C, Marone M, Frank R, Delius H, Cortese R (1986). Structure of gene and pseudogenes of human apoferritin H. Nucleic Acids Res 14:721.
19. Dickey LF, Sreedharan S, Theil EC, Didsbury JR, Wang Y-H, Kaufman RE (1987). Differences in the regulation of messenger RNA for housekeeping and specialized-cell ferritin. J Biol Chem 262:7901.
20. Chou C-C, Gatti RA, Fuller ML, Concannon P, Wong A, Chada S, Davis RC, Salser WA (1986). Structure and expression of ferritin genes in a human promyelocytic cell line that differentiates in vitro. Mol Cell Biol 6:566.

21. Beaumont C, Jain SK, Bogard M, Nordmann Y, Drysdale J (1987). Ferritin synthesis in differentiating friend erythroleukemic cells. J Biol Chem 262:10619.
22. Didsbury JR, Theil EC, Kaufman RE, Dickey LF (1986). Multiple red cell ferritin mRNAs, which code for an abundant protein in the embryonic cell type, analyzed by cDNA sequence and by primer extension of the 5′-untranslated regions. J Biol Chem 261:949.
23. Walden WE, Thach RE (1986). Translational control of gene expression in a normal fibroblast. Characterization of a subclass of mRNAs with unusual kinetic properties. Biochemistry 25:2033.
24. Stevens PW, Dodgson JB, Engel JD (1987). Structure and expression of the chicken ferritin H-subunit gene. Mol Cell Biol 7:1751.
25. Leibold EA, Munro HN (1987). Characterization and evolution of the expressed rat ferritin light subunit gene and its pseudogene family. J Biol Chem 262:7335.

Metal Ion Homeostasis: Molecular Biology and Chemistry, pages 137–148

STRUCTURE, FUNCTION AND SYNTHESIS OF PLANT FERRITINS

J.F. Briat, J.P. Laulhère, A.M. Labouré, D. Proudhon, and A.M. Lescure

Laboratoire de Biologie Moléculaire Végétale, CNRS UA 1178, Université Joseph Fourier, BP 53X, F-38041 Grenoble Cédex

ABSTRACT Ferritins from seeds of maize, pea and soyabean were purified. They are built up from two polypeptides of respectively 28 kDa and 26.5 kDa. The molecular weight of the native pea seed ferritin has been estimated to be 540 000. The 26.5 kDa subunit is produced by a specific cleavage of the 28 kDa subunit. Pea and maize seed ferritins appear very similar when compared by reverse phase HPLC chromatography, amino acids composition and 2-D gel electrophoresis.

Rabbit antibodies were produced in response to pea seed ferritin. It was shown by Western blot analysis that ferritins of the three plants analysed shared immunological determinants. However, horse spleen ferritin was not recognized by the plant ferritin antibodies.

Iron release by chelation using desferrioxamine and by reduction using ascorbate has been investigated. Reduction appeared to be a necessary step for the release as well as for the uptake of iron into ferritin molecules.

Antibodies were used to investigate ferritin synthesis. It was demonstrated that ferritins were not uniformly distributed within different pea organs of 30 day old iron unloaded plants ant that ferritin synthesis was induced by iron in soyabean cell suspensions at the transcriptional or mRNA stability level.

INTRODUCTION

Ferritins are a class of proteins present in all living organisms. They are known to sequester and thus detoxify iron that is taken up by cells but not utilized for metabolic requirements. On the other hand, under conditions of iron need, ferritin iron can be released for cellular utilization. Therefore, ferritins are key proteins acting as a buffer for iron, protecting cells from a harmful overdose of free iron and tuning properly their immediate need (for review see 1, 2).

Animal ferritins have been extensively studied and a wealth of information is now available concerning their structure, synthesis and function (2). However, as mentioned in a recent review (2), very little is known about plant ferritins. They have been purified from a few dicotyledons (3, 4, 5) and one monocotyledon (6). As in mammals, isoforms could exist (7). Bean ferritin has been shown to be nuclear encoded and plastids located (8). The control of iron induction of their synthesis in bean leaves could occur at the transcriptional level (9).

We have initiated a study of the structure, synthesis and function of plant ferritins. In this paper, we present the main results obtained in our laboratory concerning maize, pea, and soyabean.

PURIFICATION AND CHARACTERIZATION OF SEED FERRITINS FROM MAIZE, PEA, AND SOYABEAN

Purification of Ferritins from Plant Seeds

Ferritins were isolated from roots of sorghum (6), and from seeds of soyabean (5), bean (4), pea and lentil (3). For these four legumes, only one polypeptide was observed on SDS-polyacrylamide gel and the molecular weight of this ferritin subunit was variable: 28 kDa for soyabean (5), 26.5 kDa for bean (4), 21.4 kDa for lentil and 20.3 kDa for pea (3) ferritins.

This discrepancy provoked further investigations. We purified the pea seed ferritin by two different procedures and the soyabean and maize seed ferritins were also partially purified.

By SDS-polyacrylamide gel electrophoresis it was shown that ferritins from the seeds of the three plants analysed contained two polypeptides of molecular weight respectively 28 kDa and 26.5 kDa (Fig. 1).

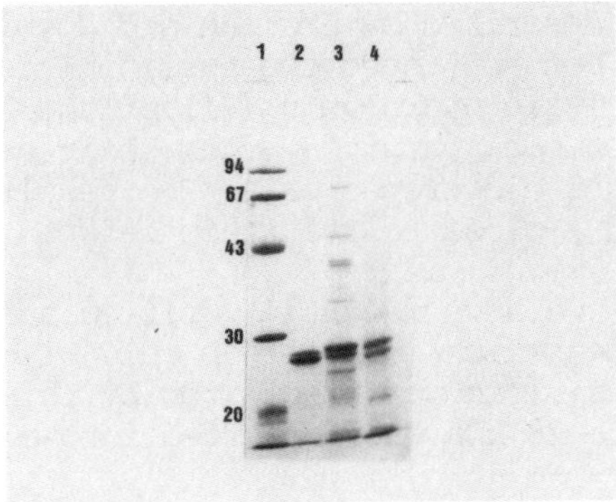

FIGURE 1. Subunit composition of ferritins isolated from seeds of: pea (lane 2), soyabean (lane 3), and maize (lane 4). Lane 1 contains 4 µg of marker. 5 µg of each phytoferritin fraction were loaded on a 12.5 % polyacrylamide-SDS gel (lanes 2 to 4).

Gel filtration at 0.5 M NaCl on a Biorad 1.5 M column of the pea seed ferritins gives one sharp peak of protein corresponding to an iron peak. These gel filtration fractions contained the two polypeptides of 28 kDa and 26.5 kDa which therefore belongs to the same oligomer.

Characterization of Ferritins from Plant Seeds.

Relationship between the 28 kDa and the 26.5 kDa subunits. Scekman and Joshi (5) have recently purified a soyabean ferritin which has a molecular mass of 572 kDa and which is built up from a unique 28 kDa polypeptide. There are two possibilities to explain the presence of a 26.5 kDa polypeptide in our preparation. The first postulates that isoforms of phytoferritins could exist in plants. This hypothesis would involve different genes for phytoferritin subunit synthesis as is the case for the H and L subunits in animal ferritins (10). The second postulates that a specific cleavage of some of the 28 kDa polypeptides could occur leading to hidden breaks within the native oligomer. The product of these hidden breaks could be the 26.5 kDa polypeptide. If this was the case, the 28/26.5-ferritin we have characterized would be a functional intermediate in the pathway leading from the 28-ferritin to the 22-ferritin described by Scekman and Joshi (5). In order to discriminate between these two possibilities we performed a partial proteolytic hydrolysis (11) of the 28 kDa and of the 26.5 kDa from pea seed ferritin after their separation by SDS-polyacrylamide gel electrophoresis. The result of this experiment indicated a close relationship between these two polypeptides.

Furthermore we have recently purified a ferritin from dried seeds of pea which is built up only with the 28 kDa subunit. After *in vitro* uptake of iron by this ferritin, we observed that the 28 kDa was converted into a 26.5 kDa polypeptide.

These results indicate that the 26.5 kDa arises from a specific cleavage of the 28 kDa polypeptide in relation to the iron loading of the oligomer.

Determination of the molecular weight of the ferritin oligomer of pea seeds. The molecular weight of pea seed ferritin was estimated to be 540 kDa. This value is higher than the value of 463.7 kDa calculated from neutron scattering experiments by Crichton *et al.* (3) but confirms that phytoferritins have a higher molecular weight than mammal ferritins.

Comparison of the pea and maize seed ferritins. Ferritins of these two plants elute at 48 % acetonitile on reverse phase HPLC column indicating a strong hydrophobicity.

Four isoforms of the pea seed 28 kDa subunit have a pHi ranging from 5.94 to 6.17. No isoforms have been detected for the maize 28 kDa subunit and for the 26.5 kDa subunit of both plants.

The amino acid compositions of the ferritin of pea and maize seeds are very similar. The percentage of non-polar amino acids for pea and maize is of 40.9 and 40.2 respectively. Basic amino acids represent 15.1 % of the pea seed ferritin and 15.7 % of the maize one. Acidic amino acids represent 26.4 % of the pea seed ferritin and 28 % of the maize one.

Immunological relationship of ferritins from different sources. Western blot analysis of purified ferritins of pea, soyabean and maize seeds and of horse spleen (Fig. 2) shows that antibodies raised against the pea seed ferritin recognise not only the two polypeptides of its own antigen (Fig. 2B, lane 2) but also the two polypeptides of the soyabean (Fig. 2B, lane 3) and maize (Fig. 2B, lane 4) seed ferritins. The horse spleen ferritin is not recognized (Fig. 2B, lane 5).

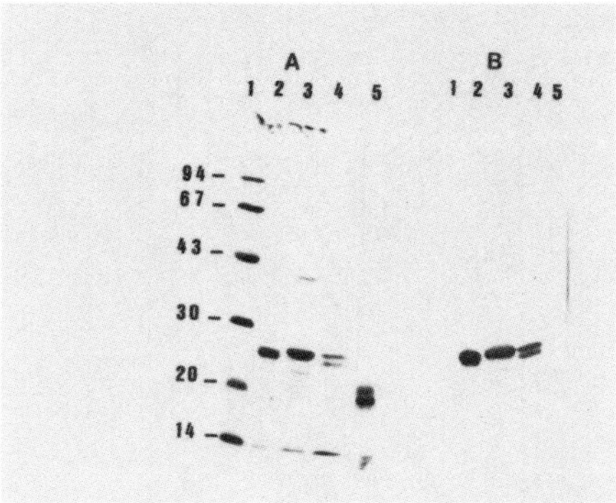

FIGURE 2. Characterization of antibodies raised against the pea seed ferritins purified by reverse phase HPLC. A-Coomassie staining of a 12.5 % polyacrylamide-SDS gel loaded with: 4 µg of marker (lane 1), 5 µg of pea seed ferritins (lane 2), soyabean seed ferritins (lane 3), maize seed ferritins (lane 4), and horse spleen ferritins (lane 5). B-Western blot analysis of a duplicate of the gel shown in A using pea seed ferritin antibodies diluted at 1/500 and anti-rabbit IgG sheep immunoglobulins coupled with peroxydase diluted at 1/800.

FUNCTION OF PEA SEED FERRITIN: IRON RELEASE AND UPTAKE

Iron Release by Chelation.

Strong chelates of various stabilities are able to remove ferric iron from ferritin. Using desferrioxamine we observed a slow release of ferric iron from pea seed ferritin. These results are comparable to those obtained with horse spleen ferritin (12).

Iron Release by Reduction.

Ascorbate was used as a reducer in studies about mobilization of iron from horse spleen ferritin (13, 14). We chose similar conditions using pea seed ferritins. The release of Fe^{2+} was monitored by the appearance of a Fe^{2+} O-Phenantrolin complex, using the same mixture without ascorbate as control.

In the presence of 50 mM ascorbate and oxygen more than

50 % of iron is released in one hour (Fig. 3, curve A).

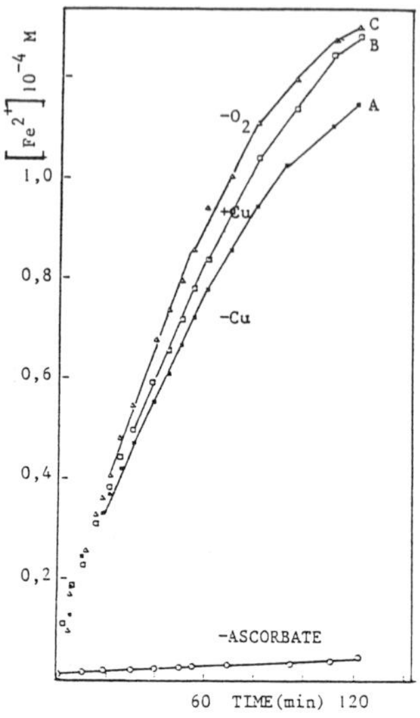

FIGURE 3. Reduction of phytoferritin iron in the presence or in the absence of 10^{-5} M Cu^{2+} in air or under nitrogen.

This rate is fast but it is not noticeably accelerated by the presence of copper (Fig. 3, curve B). The release was repeatedly found slightly faster in the absence of oxygen (Fig. 3, curve C).

The active release of Fe^{2+} in the absence of oxygen or of Cu^{2+} is noticeably different from that found by Bienfait *et al.* (13) using horse spleen ferritin. The reaction does not show a lag phase and so does not require the accumulation of putative intermediate products to take place. The slope of the curve obtained in the control without ascorbate increases slightly, suggesting either a slow endogenous reduction or a Fe^{2+} leak from ferritins.

Influence of Ascorbate on Iron Uptake.

Various amounts of ascorbate were added to labeled $^{59}Fe^{3+}$ and pea seed ferritin. The ascorbate added varied from 0 to amounts which could reduce all the Fe^{3+} (i.e. both the added

$^{59}Fe^{3+}$ and the internal ferritin Fe^{3+}).

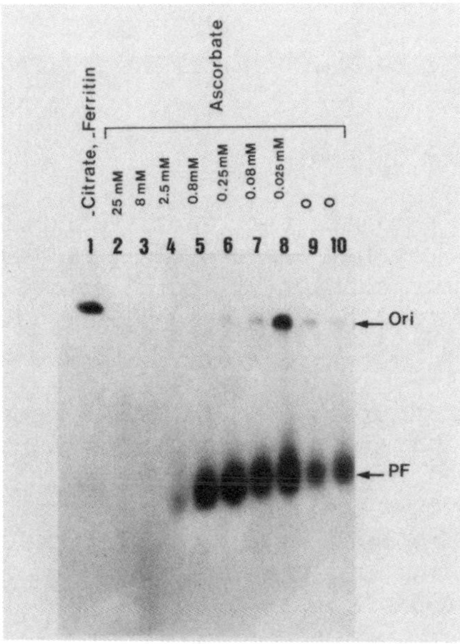

FIGURE 4. Influence of ascorbate concentrations on the incorporation of ^{59}Fe into phytoferritin. $^{59}Fe^{3+}$-citrate was added to phytoferritin in the presence of ascorbate and incubated at 25° C for 17 hours. Phytoferritin was then separated by agarose gel electrophoresis. The gel was dried and autoradiographed. Lane 1: $^{59}Fe^{3+}$ without citrate nor phytoferritin; lanes 2 to 8: $^{59}Fe^{3+}$-citrate with phytoferritin and various amounts of ascorbate; lanes 9 and 10: $^{59}Fe^{3+}$-citrate without ascorbate.

After several hours of contact between external iron and ferritin, the protein was separated by electrophoresis in agarose gel. The ferritin iron was identified by autoradiography of the dried gels.

Figure 4 shows the autoradiographs obtained after an exchange of 17 hours. It appears that the amount of reducer triggers the direction of the overall uptake or release flow. When ascorbate concentrations range from 0 to 2.5 mM an important amount of iron is incorporated into ferritins while with ascorbate concentrations ranging from 2.5 to 25 mM no uptake of Fe^{2+} by ferritin can be observed. The control without ascorbate shows an incorporation of iron, so at least a partial uptake accompanies the reduction.

FERRITIN SYNTHESIS IN DIFFERENT PLANT SYSTEMS.

Organ Distribution in Pea.

In order to estimate the relative abundance of ferritin in different pea organs, Western blot analysis was performed using total protein extract of roots, leaves, flowers and fruits from 30 days old normal pea plants (i.e. not iron loaded) and rabbit antibodies raised against pea seed ferritins.

As shown in figure 5, the relative amount of ferritin is variable in the different organs analyzed. No ferritins are detected in normal green leaves (Fig. 5, lane 3). The maximum quantity of ferritin is detected in flowers (Fig. 5, lane 4). Less ferritin is present in fruits (Fig. 5, lane 5) than in flowers. The amount detected in roots (Fig. 5, lane 2) is itself lower than in fruits.

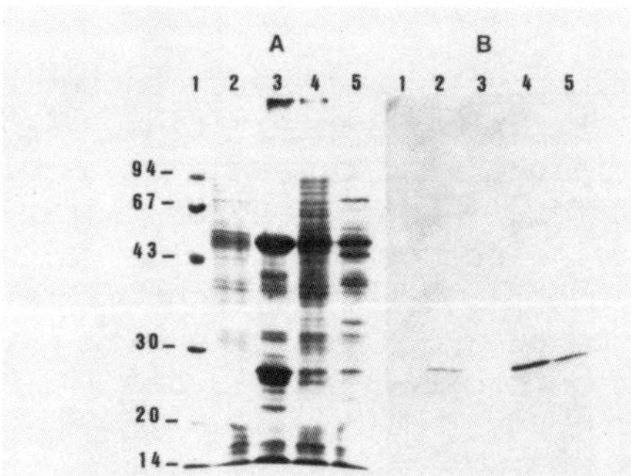

FIGURE 5. Distribution of ferritins in various organs of pea. A-Coomassie staining of a 12.5 % polyacrylamide-SDS gel loaded with: 4 µg of marker (lane 1), 15 µg of total proteins extracted from roots (lane 2), leaves (lane 3), flowers (lane 4) and fruits (lane 5). B-Western blot analysis of a duplicate of the gel shown in A using pea seed ferritin antibodies diluted at 1/500 and anti-rabbit IgG sheep immunoglobulins coupled with peroxydase diluted at 1/800.

It has to be noticed that in fruits a faint second polypeptide is present below the predominant one. In flowers (Fig. 5, lane 4), two faint polypeptides are located below the major one.

The difference in the relative amount of ferritin observed in different organs of pea raises the question of the nature of the molecular mechanisms controlling these amounts. Ferritins were not detected in pea leaves. However, immunoprecipitation of *in vitro* translation products from poly A$^+$ RNA of bean leaves with bean seed ferritin antibodies indicates that translatable ferritin mRNAs are present in this organ (9). We have performed the same kind of experiments with *in vitro* translation products of RNAs isolated from leaves, flowers and fruits of pea. Preliminary results indicate that translatable ferritin mRNAs are present in similar amounts in the three organs of pea we have analysed. Work is in progress to further investigate this problem, but it seems that the variations observed in the amount of ferritin in different organs of normal plants (i.e. not iron loaded) could be post-transcriptionally regulated.

Iron Induction in Soyabean Cell Suspensions.

Iron induction of ferritin synthesis in animal systems is essentially controlled at the translational level (2). In order to address the question of iron induction of plant ferritin synthesis soyabean cell suspensions were used. Cells were grown under different iron-citrate concentrations and after 72 hours of culture, total proteins and RNAs were extracted.

Western blot analysis of protein extracts shows clearly that synthesis of ferritin is correlated to the amount of iron added in the culture medium (Fig. 6A).

No ferritins were detected in absence of iron and more ferritins were synthesized in presence of 500 µM Fe-citrate than in presence of 100 µM Fe-citrate.

RNAs from cells grown 72 hours without iron or with 500 µM Fe-citrate were translated in a reticulocyte system. Translation products were immunoprecipitated with antibodies raised against pea seed ferritin. This experiment indicates clearly that no translatable ferritin mRNA were present when iron was absent in the cell cultures. On the other hand, translatable ferritin mRNAs were present when cells were grown in presence of iron (Fig. 6B). It has to be noticed that the molecular weight of the ferritin precursor detected is of 32 kDa, confirming the results obtained with bean leaves (8).

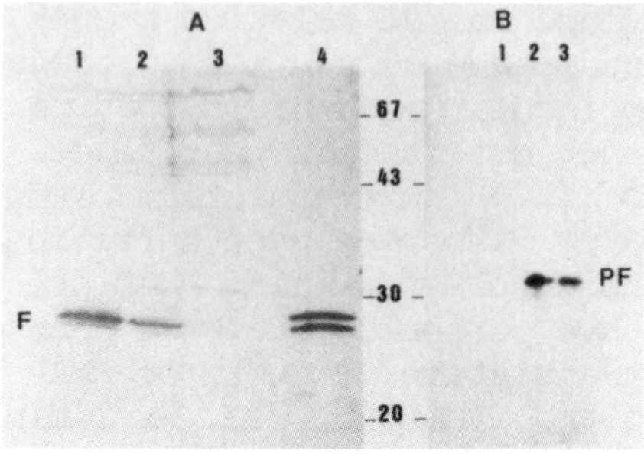

FIGURE 6. Iron induction of ferritin synthesis in soyabean cell suspensions. A-Western blot analysis of protein extracts of cells grown for 72 hours in presence of 500 μM (lane 1), 100 μM (lane 2) and no Fe-citrate (lane 3). Each lane contains 40 μg of proteins. Lane 4 contains 400 ng of purified seed soyabean ferritin. Antibodies used were raised against pea seed ferritin. B-Immunoprecipitation of *in vitro* translation products with pea seed ferritin antibodies: lane 1: rabbit reticulocyte + 16 μg of RNAs extracted from soyabean cells grown 72 hours without iron; lane 2: +16 μg of RNAs extracted from soyabean cells grown 72 hours with 500 μM iron; lane 3: +1 μg of poly A^+ RNAs extracted from soyabean cells grown 72 hours with 500 μM iron. F = ferritin; P.F. = precursor ferritin

These results led us to conclude that iron induction of ferritin synthesis in soyabean cell suspensions is not controlled at the translational level as in animal cells but at the transcriptional or mRNA stability level. Work is in progress in our laboratory to discriminate between these two possibilities.

CONCLUSION AND PROSPECTIVE.

Plant ferritins constitute an attractive system and the following points will deserve further investigations:

- what are the mechanisms responsible for the specific cleavage of the 28 kDa subunit leading to smaller subunits ?

- What are the mechanisms controlling iron release and uptake ?

- What is the nature of the mechanism controlling, at the molecular level, the distribution of ferritin in different plant organs ?

- What is the mechanism of iron induction which is different from animal systems ?

- What is the origin of plant ferritin which is nuclear encoded and plastid located ?

REFERENCES

1. Aisen P, Listowsky I (1980). Iron transport and storage protein. Ann Rev Biochem 49:357.
2. Theil EC (1987). Ferritin: structure, gene regulation, and cellular function in animals, plants and microorganisms. Ann Rev Biochem 56:289.
3. Crichton RR, Poze-Ortiz Y, Koch MHJ, Parfait R, Stuhrmann HB (1978). Isolation and characterization of phytoferritin from Pea and Lentil. Biochem J 171:349.
4. van der Mark F, de Lange T, Bienfait HF (1981). The role of ferritin in developing primary bean leaves under various light conditions. Planta 153:338.
5. Sczekman SR, Joshi JG (1987). Isolation and characterization of ferritin from soyabeans. J Biol Chem 262:13780.
6. Laulhère JP, Lambert J, Berducou J (1972). Les ferritines du sorgho, caractérisation et modalités d'échange du fer. C R Acad Sc Paris 275:759.
7. van der Mark F, van den Briel W (1985). Purification and partial characterization of ferritin from normal and iron-loaded leaves of Phaseolus vulgaris. Plant Science 39:55.
8. van der Mark F, Klerk H, Bienfait F (1983). Two higher molecular weight form related to ferritin are synthesized in vitro by RNA from developing cotyledons and leaves of Phaseolus vulgaris. Plant Mol Biol 2:311.
9. van der Mark F, Bienfait F, van den Hende H (1983). Variable amounts of ferritin translatable ferritin mRNA in bean leaves with various iron content. Biochem Biophys Res Comm 115:463.
10. Arioso P, Adelman TG, Drysdale JW (1978). On ferritin heterogeneity. J Biol Chem 253:4451.
11. Cleveland DW (1983). Peptide mapping in one dimension by limited proteolysis of sodium dodecyl sulfate solubilized protein. Methods in Enzymol 96:22.

12. Treffry A, Harrison PM (1979). The binding of ferric iron by ferritin. Biochem J 181:709.
13. Bienfait HF, van den Briel ML (1980). Rapid mobilization of ferritin iron by ascorbate in the presence of oxigen. Biochem Biophys Acta 631:507.
14. Dognin J, Girardet JL, Chapron Y (1973). Etude polarographique de la mobilisation du fer de la ferritine. Biochim Biophys Acta 297:276.

Metal Ion Homeostasis: Molecular Biology and Chemistry, pages 149–158

RELATIONSHIP OF THE SUPEROXIDE DISMUTASE GENES, *sodA* AND *sodB*, TO THE IRON UPTAKE (*fur*) REGULON IN *ESCHERICHIA COLI* K-12[1]

Eric C. Niederhoffer, Cleo M. Naranjo, and James A. Fee

Isotope and Structural Chemistry Group, INC-4 C345, Los Alamos National Laboratory, Los Alamos, NM 87545

ABSTRACT Expression of *sodA*, as indicated by MnSod activity is normal in *fur* mutants. This suggests that *sodA* is not a member of the *fur* regulon and that the putative Fe-binding, regulatory protein of *sodA*, suggested by Moody and Hassan (1984 J Biol Chem 259:12821), is not the Fur protein. By contrast, expression of *sodB*, as indicated by FeSod activity, is completely blocked in *fur* mutants and the effect is restored by transformation with a plasmid having a normal *fur* locus. The observations suggest that Fur, either directly or indirectly, controls SodB biosynthesis. Additional observations are described which indicate that SodB and Fur act together in a complicated fashion to control the biosynthesis of enterobactin.

INTRODUCTION

Metalloproteins that catalyze the disproportionation of superoxide anion are denoted superoxide dismutases (EC 1.15.1.1) (1-3), which function by the following catalytic cycle.

---

[1]This work was supported by NIH GM21519.

$$E_{ox} + O_2^- \rightarrow E_{red} + O_2$$
$$E_{red} + O_2^- + 2H^+ \rightarrow E_{ox} + H_2O_2$$

There are three forms: an iron-requiring protein (FeSod) found in bacteria and plants, a manganese-containing protein (MnSod) present in bacteria and mitochondria, and a copper-zinc protein (CuZnSod) found in the cytosol of eukaryotic cells. In *Escherichia coli*, the 38.7-kDa FeSod and 45.8-kDa MnSod are dimeric structures with one iron or manganese atom bound to each subunit; the bacterial enzymes have remarkably similar structures (4). Their common physical structures and catalytic properties have suggested a common biological function, namely protection against oxygen toxicity through superoxide dismutation. This idea has been criticized by Fee (5).

The regulation of Sod activities in *E. coli* is remarkably complex. Expression of the *sodA* gene, which codes for the Mn-protein, is apparently coupled to respiration. Thus, *sodA* is expressed during respiratory growth with dioxygen, nitrate, trimethylamine oxide, dimethyl sulfoxide, and ferricyanide acting as electron acceptors (6-10); during either aerobic or anaerobic respiration, MnSod is induced by added methyl-viologen (11). Contrary to previous claims (11), *sodA* expression is not dependent on O_2^- (12). The work of Moody and Hassan (13) suggested that *sodA* expression was enhanced by removing Fe^{2+} from the culture medium. By contrast, expression of sodB appears to be enhanced with increased levels of Fe^{2+} in the medium (14, and references in 15). Thus, both genes may be responsive to Fe in the environment.

In this work we present evidence that *sodB* is closely related to the iron uptake systems of *E. coli*. Iron is required for growth of the cell and is available through several genetically unlinked uptake systems (16-18). These include the siderophores; enterobactin, aerobactin, and ferrichrome; the ferric citrate transport system; and a low-affinity, iron transport apparatus. The ferric ion uptake regulatory locus (*fur*) negatively controls expression of these systems through the Fur protein (19). *In vivo*, the Fur protein binds, as a Fe^{2+}-containing dimer, to a specific sequence of DNA (20), denoted the "iron box" (GATAATGATAATCATTACT), which is in the promoter regions of genes coding for iron uptake. An apparent "iron box" can be found in the *sodA* promoter region (21) suggesting the possibility that *sodA* may interact with Fur. By contrast, the *sodB* promoter region does not contain this iron-regulatory element (15).

Recognition of the "iron box" in *sodA* prompted us to examine *sodA* and *sodB* regulation in *fur* mutants. We predicted that *sodA* would be constitutively expressed in *fur* mutants while *sodB* would be unchanged (wild-type phenotype) in *fur* mutants. Our predictions were completely wrong.

METHODS

Some of the *E. coli* K-12 strains and the multicopy plasmids used in this work are listed in Table 1. Cells were grown in LB medium supplemented with glucose (0.2%) and the appropriate antibiotics. Cultures were induced with 100 μM methylviologen. Anaerobic cultures were grown in sealed flasks. Minimal medium consisted of Chelex-treated M9, glucose, thiamine, amino acids, $MgCl_2$, $CaCl_2$, $FeCl_2$, and the appropriate antibiotics. Catechol production was assayed by the Arnow method (22).

Saturated cultures were harvested, washed twice with 50 mM KPi, 0.5 mM EDTA, 1 mM $MnCl_2$, pH 8.0 buffer, lysed with lysozyme, treated with p-toluenesulfonyl fluoride, sonicated at 4°C for 90 s, and centrifuged at 10 000 x *g* for 10 min to remove cell debris. Protein concentrations were determined by the BCA method (23); bovine gamma globulin served as the calibration standard. Proteins were analyzed on 10% polyacrylamide gels (24). The gels were stained for protein with 26% isopropanol, 11% glacial acetic acid, and 0.1% Coomassie blue R, and destained with 10%

isopropanol and 10% glacial acetic acid, or stained for Sod activity with nitro blue tetrazolium (25).

TABLE 1
Strains And Plasmids

Strains	Genotype	Source or Reference
GC4468	F^- Δ*lac*4169 *rps L*	D. Touati
QC774	GC4468Φ(*sodA-lacZ*)49 Φ(*sodB*-Km)1-Δ_2 Cm^R Km^R	D. Touati
AB1157	*his arg thr*	G. Walker
BN402	P1 *lac*ΔU169xAB1157 *galK*$^-$	J. Neilands
BN407	P1 *lac*ΔU169xAB1157 *galK*$^-$ pColVA3:Tn10 Tc^R	J. Neilands
BN4020	P1 *lac*ΔU169xAB1157 *galK*$^-$ *fur*:Tn5 Km^R	J. Neilands
BN4023	P1 *lac*ΔU169xAB1157 *galK*$^-$ *fur*:Tn5 Km^R pColVA3:Tn10 Tc^R	J. Neilands
EN3	P1 QC774xBN407 Km^R	J. A. Fee
EN5	P1 QC774xBN402 Km^R	J. A. Fee
EN9	P1 QC774xBN402 Cm^R	J. A. Fee
EN13	P1 QC774xBN4020 Cm^R	J. A. Fee
Plasmids		
pHS1-4	*sodB*$^+$ Tc^R	D. Touati
pJM19	*sodB*$^+$ Am^R	J. A. Fee
pABN203	*fur*$^+$ Tc^R	J. Neilands

The xanthine-xanthine oxidase coupled reduction of cytochrome *c* was used to quantify soluble Sod activity (26).

RESULTS

As determined by PAGE using the assay of Beauchamp and Fridovich (24) (not shown), strains BN402 (wild-type) and BN 4020 (*fur*$^-$) showed the normal phenotype for MnSod activity. No MnSod was produced under anaerobic growth, but O_2 and methylviologen induced MnSod activity. The FeSod activity was present in BN402, but FeSod activity was not detected in BN4020. Transformation of BN4020 with pABN203 (*fur*$^+$) restores the wild-type levels of FeSod activity. The plasmid pHS1-4 (*sodB*$^+$) produced large amounts of FeSod activity in BN402 but barely detectable levels of FeSod activity in BN4020.

Both MnSod and FeSod contribute to the total activity in cell extracts. In *sodA* mutants, the cytosolic activity is due only to FeSod; specific activities for *sodA* mutants appear in Table 2. The values directly parallel the observation made in the PAGE analysis and provide a quantitative base for our conclusions (see below).

TABLE 2
Specific Iron Superoxide Dismutase Activities[a] For *sodA* Mutants

	fur$^+$			*fur*$^-$		
	BN402	pABN203	pHS1-4[b]	BN4020[c]	pABN203	pHS1-4
No O_2	15	40	1700	<2	38	36
O_2	29	170	1800	<2	29	36
O_2 + PQ^{2+}	56	41	1700	<7	12	49

[a] Sod activity measured by the xanthine-xanthine oxidase coupled cytochrome *c* method (26); units/mg protein; 5-28 μg protein/mL in assay
[b] 0.2 μg protein/mL in assay
[c] 50-250 μg protein/mL in assay

Enterobactin production assayed by the method of Arnow (22) indicates total catechol content of the sample. Selective strains were tested and the results appear in Table 3. As expected, *fur*$^+$ strains did not produce catechol in iron sufficient conditions while *fur* mutant strains excreted catechol independently of iron in the medium. Unexpectedly, interuption of the *sodB* gene led to an apparent *fur* phenotype, and this was not restored by transformation of *sodB* strains with pJM19 *sodB*$^+$. Indeed, multiple copies of either *fur*$^+$ (pABN203) or *sodB*$^+$ (pJM19) led to the apparently complete suppression of enterobactin biosynthesis in all strains tested.

TABLE 3
Production of Catechols[a] by Selected Strains

Strain Genotype	-Fe	+Fe	Phenotype
fur$^+$	+	-	wt
fur$^-$	+	++	*fur*
fur$^-$ pABN203 *fur*$^+$	-	-	
fur$^+$*sodB*$^-$	+	++	*fur*
fur$^+$*sodB*$^-$ pJM19 *sodB*$^+$	-	-	
fur$^+$ pJM19 *sodB*$^+$	-	-	

[a] Ref. (22); + indicates a wild-type level of catechol production, ++ indicates an obvious increase in catechol production, - indicates that no catechol was present.

DISCUSSION

Three general conclusions can be drawn from our observations: First, expression of *sodA* is not affected by mutation in the *fur* locus. Second, *sodB* expression is directly or indirectly controlled by the *fur* locus. Finally, *sodB* (or downstream cistrons) is somehow controlling expression of enzymes involved with enterobactin biosynthesis.

Based on the observation that *E. coli* expresses MnSod under anaerobic conditions in iron-depleted media, Moody and Hassan (13) postulated the existence of an Fe-binding protein that acts as a repressor of *sodA*. To account for the apparent "oxidative" induction of MnSod activity, it was further speculated that oxidation of Fe^{2+} to Fe^{3+} relieved repression of *sodA* (8). The recognition of a nearly perfect "iron box" in the -35 region of *sodA* prompted us to test the idea that *sodA* was part of the *fur* regulon. The data indicate that *sodA* is not regulated by *fur;* it is possible that another iron-binding protein is involved.

In contrast to *sodA*, *sodB* expression is dramatically influenced by the *fur* locus. Thus, the *sodB* gene product, FeSod, is absent in *fur* mutants and is restored when supplemented with pABN203 fur^{+}. The return of FeSod activity parallels the appearance of protein (not shown) indicating that *sodB* regulation by fur involves synthesis of new enzyme rather than activation of incompetent enzyme. Assuming regulation of transcription, our observations indicate that *fur* through its product Fur, is acting in trans to control *sodB* expression. The question of whether this is a direct activation of *sodB* by Fur or an indirect activation through some other loci, is presently under experimental consideration.

The *ent* operon, responsible for the biosynthesis of enterobactin, is central to *E. coli*'s strategy for obtaining Fe from its environment (18). It has been known for some time that expression of *ent* is controlled by the *fur* locus whose gene product, in combination with Fe^{2+}, acts as a repressor of *ent*. As shown in Table 3, our observations confirm this basic hypothesis and extend it to include a role for *sodB* (or downstream cistrons). Excess *fur* in the cell, as obtains in the presence of the multicopy plasmid, pABN203 fur^{+}, suppresses enterobactin biosynthesis even in iron poor medium. Similarly, the multicopy plasmid, pJM19, bearing *sodB*$^+$ also suppresses enterobactin biosynthesis. Broadly speaking, cells lacking either Fur or SodB appear to sense iron starvation and the *ent* operon is turned on, while cells having excess Fur or SodB sense iron sufficiency and expression of the *ent* operon is shut down. We are currently exploring the genetic and molecular basis of *sodB* in these regulatory processes.

ACKNOWLEDGMENTS

We thank J. B. Neilands and his group for providing a preprint of their review and for generously donating the *fur* strains and plasmids used in our work. Generous thanks go to Danielle Touati whose pioneering effort in isolating *sodA* and *sodB* strains made this work possible. We are also indebted to the many individuals at this Symposium who helped focus our efforts.

REFERENCES

1. Rotilio G (1986). "Superoxide and Superoxide Dismutases in Chemistry, Biology and Medicine." Amsterdam: Elsevier/North Holland Biomedical Press.
2. Fridovich I (1986). Superoxide dismutases. Adv Enzymol 58:61.
3. Fee JA (1981). The copper/zinc superoxide dismutase. In Sigel H (ed): "Metal Ions in Biological Systems," New York: Dekker, p 259.
4. Stallings WC, Pattridge KA, Strong RK, Ludwig ML (1984). Manganese and iron superoxide dismutases are structural homologs. J Biol Chem 259:10695.
5. Fee JA (1982). Is superoxide important in oxygen poisoning? TIBS 7:84.
6. Smith MW, Neidhardt F (1983). Proteins induced by aerobiosis in *Escherichia coli*. J Bacteriol 154:344.
7. Miyake K (1986). Effect of nitrate on the level of superoxide dismutase in anaerobically grown *Escherichia coli*. J Gen Appl Microbiol 32:527.

8. Hassan HM, Moody CS (1987). Regulation of manganese-containing superoxide dismutase in *Escherichia coli*: anaerobic induction by nitrate. J Biol Chem 262:17173.
9. Schiavone JR, Hassan HM (1988). The role of redox in the regulation of manganese-containing superoxide dismutase biosynthesis in *Escherichia coli*. J Biol Chem 263:4269.
10. Privalle CT, Fridovich I (1988). Inductions of superoxide dismutases in *Escherichia coli* under anaerobic conditions: accumulation of an inactive form of the manganese enzyme. J Biol Chem 263:4274.
11. Hassan HM, Fridovich I (1977). Regulation of the synthesis of superoxide dismutase in *Escherichia coli*: induction by methyl viologen. J Biol Chem 252:7667.
12. Nettleton CJ, Bull C, Baldwin TO, Fee JA (1984). Isolation of the *Escherichia coli* iron superoxide dismutase gene: evidence that intracellular superoxide concentration does not regulate oxygen-dependent synthesis of the manganese superoxide dismutase. Proc Natl Acad Sci USA 81:4970.
13. Moody CS, Hassan HM (1984). Anaerobic biosynthesis of the manganese-containing superoxide dismutase in *Escherichia coli*. J Biol Chem 259:12821.
14. Simons RS, Jackett PS, Carroll MEW, Lowrie DB (1976). Superoxide independence of paraquat toxicity in *Escherichia coli*. Toxicol Appl Pharmacol 37:271.
15. Carlioz A, Ludwig ML, Stallings WC, Fee JA, Steinman HM, Touati D (1988). Iron superoxide dismutase: nucleotide sequence of the gene from *Escherichia coli* K12 and correlations with crystal structures. J Biol Chem 263:1555.
16. Neilands JB (1982). Microbial envelope proteins related to iron. Ann Rev Microbiol 36:285.
17. Crichton RR, Charloteaux-Wauters M (1987). Iron transport and storage. Eur J Biochem 164:485.
18. Winkelmann G, vander Helm D, Neilands JB (eds) (1987): "Iron Transport in Microbes, Plants, and Animals," Weinheim, Federal Republic of Germany: VCH Publishers.

19. Bagg A, Neilands JB (1987). Molecular mechanism of regulation of siderophore-mediated iron assimilation. Microbiol Rev 51:509.
20. Bagg A, Neilands JB (1987). Ferric uptake regulation protein acts as a repressor, employing iron(II) as a cofactor to bind the operator of an iron transport operon in *Escherichia coli*. Biochemistry 26:5471.
21. Takeda Y, Avila H (1986). Structure and gene expression of the *E. coli* Mn-superoxide dismutase gene. Nucl Acids Res 14:4577.
22. Arnow LE (1937). Colorimetric determination of the compounds of 3,4-dihydroxyphenylalanine-tyrosine mixtures. J Biol Chem 118:531.
23. Smith PK, Krohn RI, Hermanson GT, Mallia AK, Gartner FH, Provenzano MD, Fujimoto EK, Goeke NM, Olson BJ, Klenk DC (1985). Measurement of protein using bicinchoninic acid. Anal Biochem 150:76.
24. Hames BD (1987). An introduction to polyacrylamide gel electrophoresis. In Hames BD, Rickwood D (eds): "Gel Electrophoresis of Proteins: A Practical Approach," Oxford: IRL Press, p 1.
25. Beauchamp C, Fridovich I (1971). Superoxide dismutase: improved assays and an assay applicable to acrylamide gels. Anal Biochem 44:276.
26. McCord JM, Fridovich I (1969). Superoxide dismutase: an enzymatic function for erythrocuprein (hemocuprein). J Biol Chem 244:6049.

II. STRUCTURE AND CHEMISTRY OF METALLOPROTEINS

Metal Ion Homeostasis: Molecular Biology and Chemistry, pages 161–168

THE STRUCTURE OF MIXED-LIGAND CYANO COMPLEXES OF TRANSFERRIN[1]

N.D. Chasteen*, P.A. Snetsinger*, S.K. Swope*, H. van Willigen**, J.B. Cornelius***, and J. McCracken***

*Department of Chemistry, University of New Hampshire, Durham, New Hampshire 03824

**Department of Chemistry, University of Massachusetts at Boston, Boston, Massachusetts 02125

***Biotechnology Resource in Pulsed EPR Spectroscopy, Albert Einstein College of Medicine, Bronx, New York 10461

ABSTRACT The binding of cyanide to the Fe(III) centers of human serum transferrin was investigated by ultraviolet-visible, electron paramagnetic resonance and electron-nuclear double resonance spectroscopies, and by electron spin echo envelope modulation measurements. A low spin Fe(III) cyanide adduct is formed at the C-terminal iron binding site of transferrin. There is evidence for cyanide binding to the N-terminal site as well but the Fe(III) in this site remains high-spin. Spectroscopic and binding data suggest the coordination of two tyrosines, a histidine, and three cyanides to the metal in the low spin complex. The complex is unusual in that cyanide appears to function in place of (bi)carbonate in the structure of the metal site.

---

[1] This work was supported by NIH grants GM20194 and RR02583.

INTRODUCTION

The transferrins are a class of iron binding proteins widely distributed in the physiological fluids of vertebrates. Serum transferrin, whose principal function is iron transport, plays a major role in the homeostasis of metabolic iron. The protein binds two Fe(III) ions strongly, but reversibly, at sites located in separate, and largely independent, globular domains of the protein (1).

The mechanism by which serum transferrin exchanges iron(III) with its environment is poorly understood. Physiological chelators of iron are most likely involved in the process, however. Chelators containing phosphate groups are especially effective at delivering iron to transferrin or removing it from the protein. It has long been speculated that (bi)carbonate binding, which is required for iron to bind to the protein, may be disrupted by chelators during the iron removal reaction.

In recent years Bates and coworkers (2-3) have obtained visible spectroscopic evidence for the formation of mixed-ligand complexes during iron exchange reactions between transferrin (Tf) and the chelators (Chel) citrate, pyrophosphate, and acetohydroxamate, viz.

$$\text{Fe-Tf-CO}_3 + \text{Chel} \longrightarrow \underset{\text{(intermediate)}}{\text{Chel-Fe-Tf-CO}_3} \longrightarrow \text{Chel-Fe} + \text{Tf} + \text{CO}_3^{=}.$$

Because of their short half-lives however, such mixed-ligand intermediates have proven difficult to study in detail. It is not known which, if any, protein ligands are displaced by the incoming chelator upon formation of the intermediate and whether (bi)carbonate remains bound in the complex as shown in the reaction above.

We have undertaken a study of cyano complexes of transferrin to gain insight into the nature of these mixed ligand intermediates and to develop methods for probing the structure of Fe(III) sites in metalloproteins in general. Cyanide was chosen for this work because its ligand exchange reactions are slow, thus permitting careful studies of the cyano intermediates that are formed. Furthermore, cyano complexes of iron(III) lend themselves to study by a variety of spectroscopic methods.

RESULTS AND DISCUSSION

Formation of a Low Spin Adduct.

The addition of cyanide to a solution of diferric transferrin produces the EPR spectrum shown in Figure 1A. The details of sample preparation have been reported elsewhere (4). Signals from both high and low-spin Fe(III) are observed (Fig. 1). The rhombic low-spin signal around 300 mT has g-factors g_x=2.34, g_y=2.15 and g_z=1.92, and arises from the formation of a cyanide adduct. The high-spin signal at g′=4.3 (150 mT)

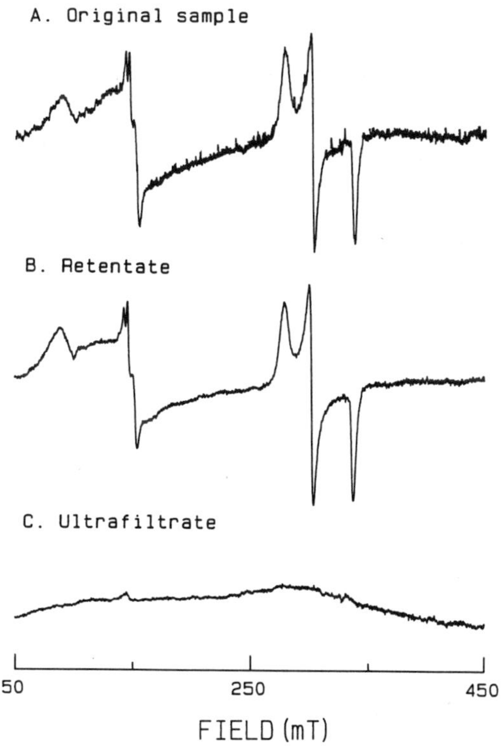

FIGURE 1. Ultrafiltration experiment. EPR spectrum of transferrin plus cyanide (A) before and (B) after ultrafiltration and of (C) the ultrafiltrate. Conditions: 0.4 M NaCN, pH 10.4, 0.1 mM protein.

characteristic of the native protein also shows perturbations from cyanide. In particular the broad shoulder to low field is normally absent in the native protein.

The ultrafiltration experiment summarized in Figure 1 demonstrates that both the high and low-spin EPR signals arise from protein bound species. Double integrations indicate that all of the iron in the sample is accounted for by the EPR spectrum. The iron is distributed about equally between the high and low spin forms (4). EPR measurements on monoferric transferrin samples, where the iron is in one site or the other, indicate that only Fe(III) in the C-terminal site forms a low-spin complex. Iron(III) in the N-terminal site, while perturbed by the presence of cyanide, remains high-spin and gives the g′=4.3 signal in Figure 1.

The kinetics of Fe(III) removal from the C-terminal site by cyanide to produce ferricyanide and apotransferrin was measured by EPR spectroscopy (Figure 2). The kinetic profile shows that the low-spin adduct is formed as an

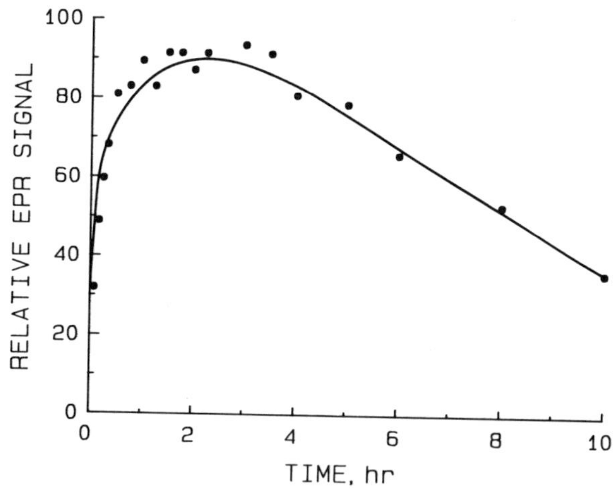

FIGURE 2. Kinetics of adduct formation and decay as measured by EPR spectroscopy. Conditions: 0.5M NaCN, 0.5 mM protein, pH 10.3, 20°C.

intermediate in the reaction. Therefore the properties of the adduct were investigated further.

Binding studies have shown that the (bi)carbonate present in the iron site of the native protein is displaced by cyanide according to the following reaction (4).

$$\text{Fe-Tf-CO}_3 \; + \; 3\text{CN}^- \longrightarrow \text{Fe-TF-(CN)}_3 \; + \; \text{CO}_3^{=}$$

Here only the iron in the C-terminal site is depicted.

UV-Visible Spectroscopy.

Ultraviolet difference spectroscopy (5) was employed to establish the number of coordinating tyrosine residues in the adduct. The native protein has two per iron binding site (5,6). The UV difference spectrum of the C-terminal monoferric adduct measured against the native monoferrin protein at pH 7.1 showed cancellation of the tyrosine ligand bands at 245 and 295 nm. This result indicates that the same number of tyrosines (namely two) also binds to the iron in the adduct. Moreover, the adduct shows a tyrosine-to-iron charge transfer band at 420 nm in the visible spectrum that is typical of Fe(III) phenolate proteins. The band is blue-shifted from the value of 465 nm observed for the native protein but has approximately the same intensity (4).

Electron-Nuclear Double Resonance (ENDOR).

ENDOR spectroscopy was used to probe the environment of the iron. ENDOR is used to measure electron-nuclear couplings that are too small to be resolved within the EPR linewidth. Measurements were made on a number of samples prepared with H_2O and D_2O, 57-Fe, and 15-N and 13-C isotopically enriched cyanide.

Comparisons of the proton ENDOR of samples prepared in H_2O <u>vs.</u> D_2O have been revealing. The matrix proton peak is little affected by hydrogen-deuterium exchange, indicating that the metal site is buried below the surface of the protein, a finding consistent with the recent x-ray structure of lactoferrin (6).

Upon replacement of H_2O with D_2O, some protons with 1 MHz couplings are exchanged, however. They probably arise from second coordination sphere water or exchangeable protons on the protein. No proton couplings in the 5 to 15 MHz range were observed, a result suggesting that water or hydroxide is probably absent in the first coordination sphere of the adduct.

A 13-C coupling from 13-C cyanide was observed in the ENDOR of the isotopically enriched sample. The anisotropy of this coupling was measured as a function of field. The dipolar contribution to the coupling can be accounted for by assuming that a cyano group(s) is located on the g_z axis at an iron-carbon distance of 2 Å. This result places the group(s) perpendicular to the d_{xy} ground state orbital as derived from an analysis of the g-factors using a crystal field model for spin-orbit coupling within the t_{2g} set of orbitals. The ENDOR signals from additional cyano group(s) have not been located thus far.

Electron-Spin Echo Envelope Modulation (ESEEM).

ESEEM is generally used to measure smaller couplings that can be observed by ENDOR. Thus the two techniques provide complementary information. ESEEM measurements on 15-N cyanide enriched samples show couplings from 14-N as well as 15-N, demonstrating that a nitrogen containing ligand from the protein is also present. This ligand is presumably the imidazole group of a histidine residue present in the native structure of the metal site (see below). Neither 13-C or 57-Fe couplings were observed, presumably because of their large values.

Structure of the Adduct.

The x-ray crystal structure of lactoferrin to 3.2 Å resolution shows that the protein provides two tyrosines, one histidine, and one aspartate residue as ligands to the iron(III) in each site (6). The other two coordination positions are presumably occupied by (bi)carbonate and water (or hydroxide).

We propose that the cyanide adduct has the structure shown in Figure 3. The presence of three cyano groups is consistent with the stoichiometry of binding as noted above. We suggest that one cyano group substitutes for (bi)carbonate, forming a salt bridge to a nearby Arg residue as frequently proposed for the Fe-TF-CO_3 complex.

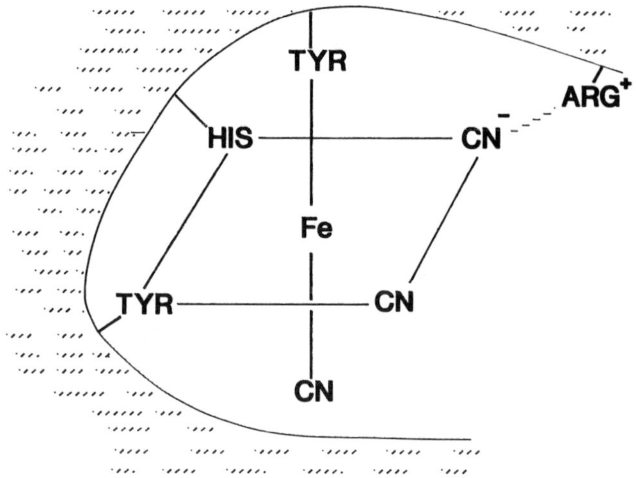

FIGURE 3. Proposed structure of the cyanide adduct.

The ENDOR data implies the absence of first coordination sphere water in the adduct as shown. The coordination of two tyrosine residues and a histidine residue is consistent with the UV-visible spectral and the ESEEM data.

We have no direct evidence for or against the presence of the Asp residue in the complex. However, if the iron remains six coordinate in the adduct, then one expects the Asp residue to be displaced. Displacement of Asp by cyanide may be the first step in the unfolding of transferrin ultimately leading to the loss of its iron. This residue is located far from the other ligands in the amino acid sequence and appears to play a major role in the folding of the protein about the metal site. A more detailed analysis of ENDOR and ESEEM data should help to clarify the structure of the metal site and lead to a better understanding of the molecular events that lead to the release of iron by transferrin.

ACKNOWLEDGMENTS

The authors thank Mr. John K. Grady for preparing the figures and some of the samples.

REFERENCES

1. Chasteen ND (1983). Transferrin. A perspective. In Theil EC, Eichorn GL, Marzilli LG (eds): "Advances in Inorganic Biochemistry,", Vol 5, New York, Elsevier, p 201.
2. Cowart RE, Swope S, Loh TT, Chasteen ND, Bates GW (1986). The exchange of Fe(III) between pyrophosphate and Transferrin. J Biol Chem 261:4607.
3. Cowart RE, Kojima N, Bates G (1982). The exchange of iron between acetohydroxamic acid and transferrin. J Biol Chem 257:7560.
4. Swope SK, Chasteen ND, Weber KE, Harris DC (1988). A low spin cyanide adduct of transferrin. J Am Chem Soc 110: in press.
5. Pecoraro VL, Harris WR, Carrano CJ, Raymond KN (1981). Siderophilin metal coordination. Difference ultraviolet spectroscopy of di-, tri-, and tetravalent metal ions with ethylenebis-[o-hydroxyphenyl)glycine] Biochemistry 20:7033.
6. Anderson BF, Baker HM, Dodson EJ, Norris GE, Rumball SV, Waters JM, and Baker EN, (1987). Structure of human lactoferrin at 3.2 Å resolution. Proc Natl Acad Sci USA 84:1769.

Metal Ion Homeostasis: Molecular Biology
and Chemistry, pages 169–178

IRON DOES NOT AFFECT THE SECONDARY STRUCTURE OF OVOTRANSFERRIN OR ITS HALF-MOLECULE DOMAINS[1]

Anne B. Mason and Stephen A. Brown

Department of Biochemistry, University
of Vermont College of Medicine
Burlington, Vermont 05405

ABSTRACT Circular dichroism studies on holo-ovotransferrin in 10 mM sodium phosphate buffer show no significant differences in secondary structure attributable to either the presence of iron or to a change in pH from 7.4 to 5.0. Significant differences in the secondary structure of the isolated half-molecule domains of ovotransferrin do exist. In contrast to studies with the isolated lactoferrin half-molecule domains, however, there does not appear to be additional secondary structure due to the association of the two ovotransferrin half-molecule domains in solution.

INTRODUCTION

Past work from our laboratory has been concerned with defining the structural features of the transferrin (TF) molecule which are important in binding to the specific TF receptor found on the surface of most actively dividing cells. TF is a single chain 80-kDa glycoprotein roughly divided into two globular domains each of which binds iron (1-4). The TF receptor is a disulfide-linked glycoprotein made up of two identical 95-kDa monomers each of which apparently binds one molecule of TF (5,6). Studies have been performed which examine the binding of chicken ovotransferrin (OTF) and the two proteolytically derived half-molecule iron-binding domains of OTF to the specific TF receptors on chick embryo red blood cells. This work has shown that: 1) both domains of OTF must be present,

---

[1]This work was supported by USPHS Grant DK/ML-31729

associated and iron loaded to obtain levels of binding to receptor which are comparable to holo-OTF (7,8), 2) the receptor recognition sites on the two half-molecule domains are different as demonstrated by domain specific monoclonal antibodies (9), 3) there appears to be a conservation of sequence involved in receptor recognition and binding which occurs in both domains as shown by binding studies with quail OTF (10) and 4) there is a tyrosine residue in or near the receptor recognition site of the C-terminal half-molecule domain of OTF (11).

Diferric TF binds with high affinity to the TF receptor at pH 7.4. Apo-TF binds with low affinity to receptor at pH 7.4 (12,13). The reverse situation occurs at pH 5.0. These observations are cornerstones of the current view of the TF-TF receptor cell cycle (6). But what is the exact nature of the changes in the TF molecule and/or the receptor at the two pH's which contribute to the phenomenon? Little can be said about the conformation of the receptor. In contrast, several studies have examined the effect of iron (and in one case pH) on the secondary structure of human serum TF (14-17), of OTF (15,18-20) and of lactoferrin (LTF)(16,21) Conflicting reports on the influence of iron on secondary structure have resulted. In the present report we have examined OTF and the half-molecule domains of OTF by circular dichroism. We compare our findings with those of other laboratories and discuss the implications with regard to receptor interaction.

METHODS

OTF and the amino- and carboxyl-terminal half-molecule domains (designated OTF/2N and OTF/2C) were prepared as described in detail previously (7-10).

Circular dichroism (CD) spectra were obtained at room temperature using a Jobin-Yvon Dichrograph Mark V (Instuments S./A.). A solution of isoandrosterone in dioxane was used to calibrate the machine. The CD readings at 1.0 nm intervals between 260 and 185 were recorded on line in an Apple IIe computer. The protein solutions at a concentration of 0.3-0.5 mg/ml were filtered through 0.2-micron filters and placed into a cylindrical cuvette with a 0.1 cm pathlength. For each run the average of 16 scans was saved, a buffer baseline was subtracted and the data was converted to ellipticity expressed as mean residue molar ellipticity [Θ] in degrees cm^2 $dmol^{-1}$. Mean residue weights of 113.7, 113.0, 114.4, were used for holo-OTF, N-, and C-terminal OTF, respectively. Concentrations were determined from the absorbance of the samples at 280 nm using the molar extinction coefficients which we have reported previously (8). Buffers at pH 7.4 and 5.0 were prepared from 10 mM sodium phosphate, treated with Chelex 100 (Bio-Rad) to eliminate iron and filtered through 0.22-micron filter units. Apo-OTF was also run in 10 mM citrate buffer which was treated as above. All of the measurements were repeated multiple times with different preparations of protein. Aliquots of all isolated proteins were subjected to sodium dodecyl sulfate and urea-polyacrylamide gel electrophoresis (7) to confirm the purity and the iron status, respectively.

Prediction of secondary structure was performed using the computer program of Chang _et al_ (22) which compares the spectrum of the unknown protein to 15 reference proteins whose secondary structure has been determined by x-ray diffraction.

RESULTS AND DISCUSSION

Representative spectra of diferric OTF at pH 7.4 in 10 mM sodium phosphate buffer and of apo-OTF at pH 7.4 and pH 5.0 in phosphate buffer and at pH 5.0 in 10 mM sodium citrate buffer are shown in Figure 1. Analysis and averaging of 4 to 9 runs of 16 scans each resulted in the estimates of secondary structure which are presented in Table 1. Large differences in secondary structure occur only in the case of apo-OTF run in sodium citrate buffer. Statistically insignificant differences in secondary structure exist between diferric OTF at pH 7.4 and apo-OTF at pH 7.4 and pH 5.0 in sodium phosphate buffer.

These results conflict with those presented by Teeters _et al_ (17) for human serum TF. This group reports that human apo-TF at pH 7.0 is significantly different from diferric TF at the same pH. Differences include: 1) the shapes of the two spectra, 2) the magnitudes of the spectra (apo-TF is smaller by 10% than diferric TF) and 3) a decrease in the amount of β-sheet in apo-TF at pH 7.0. Apo-TF at pH 4.5 in citrate buffer has a spectrum which is similar to the diferric spectrum at pH 7.0.

Our studies do not show differences in the shape of the spectra of diferric OTF and apo-OTF at pH 7.4. The magnitude of the two spectra appears to be the same within

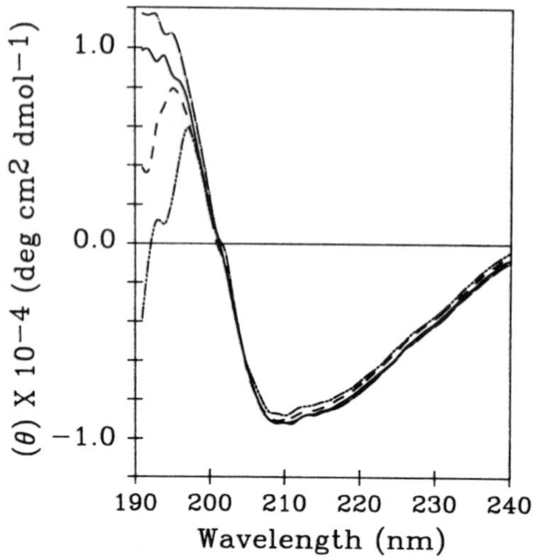

FIGURE 1. Representative CD spectra of diferric OTF (·—·—) and apo-OTF (——) in 10 mM sodium phosphate buffer pH 7.4. Also shown are the spectra of apo-OTF (----) at pH 5.0 in 10 mM sodium phosphate and of apo-OTF (—··—·) at pH 5.0 in 10 mM sodium citrate buffer.

statistical limits. At pH 7.4, diferric OTF has an average minima at 209 nm of -9204 (±558) and apo-OTF has an average minima at 209 nm of -9144 (±477). In OTF there is a reduction in the content of β-sheet in the apo-OTF in sodium citrate buffer. In fact the only significant differences in the secondary structure of OTF occur when the apo-OTF is analyzed in citrate buffer. In particular, the content of β-turn is approximately doubled. Maxima at 224 and 202 and a strong minima at 190 nm are associated with β-turn (23). In the study by Teeters _et al_ the analysis does not appear to include points below 195 nm.

Additional explanations for the observed differences include possible species variation and differences in the method used to estimate the secondary structure in the two studies. A review by Yang _et al_ (23) details the complexities of determining secondary structure by CD. A survey of the literature for CD studies of TF shows scattered estimates of secondary structure (see Table 2). Much of the variation can be attributed to the method used to calculate the parameters. Nevertheless the majority of studies to date show that the secondary structure does not

TABLE 1
PERCENTAGES (±SD) OF SECONDARY STRUCTURE OF IRON AND APO OVOTRANSFERRIN AND HALF-MOLECULE DOMAINS OF OVOTRANSFERRIN[a]

Sample	pH	α-helix	β-sheet	β-turn	Random	n
Fe_2OTF	7.4	22.1 ±1.4	45.3 ±3.8	8.7 ±2.4	23.9 ±1.3	9
ApoOTF	7.4	21.9 ±1.1	39.3 ±7.6	12.9 ±5.4	25.9 ±1.5	8
ApoOTF	5.0	20.8 ±0.9	40.8 ±0.5	13.0 ±0.9	25.5 ±0.0	4
ApoOTF	5.0[b]	25.3 ±1.3	19.0 ±2.0	26.9 ±0.8	28.9 ±1.8	4
FeOTF/2N	7.4	15.1 ±1.4	64.2 ±3.1	1.5 ±1.3	19.3 ±1.2	8
ApoOTF/2N	7.4	15.9 ±1.7	58.6 ±6.0	5.0 ±4.2	20.5 ±0.9	7
FeOTF/2C	7.4	25.4 ±2.4	34.6 ±7.7	12.7 ±3.8	27.2 ±1.9	7
ApoOTF/2C	7.4	25.3 ±1.9	31.8 ±6.0	14.0 ±3.0	29.0 ±1.7	10
(OTF/2N + OTF/2C)[c]	7.4	20.0 ±1.3	50.3 ±0.6	7.5 ±0.5	22.2 ±0.8	3
OTF/2N + OTF/2C[d]	7.4	20.2 ±1.0	49.0 ±1.8	8.0 ±0.9	22.8 ±1.3	3

[a]The secondary structure was determined using the program of Chang et al (22). Each **n** unit equals the average of 16 scans collected at 1 nm intervals between 240 and 191 nm. Abbreviations are: OTF, ovotransferrin; OTF/2N, N-terminal half-molecule domain; OTF/2C, C-terminal half-molecule domain.
[b]This sample was run in 10 mM sodium citrate buffer, pH 5.0.
[c]The CD spectrum of equimolar amounts of OTF/2N and OTF/2C in solution together was analyzed to give these results.
[d]The CD spectra of individually run OTF/2N and OTF/2C were added together by computer, averaged and analyzed to give these results.

TABLE 2
PERCENTAGES OF SECONDARY STRUCTURE OF THE TRANSFERRINS[a]

Protein[b]	α-helix	β-sheet	β-turn	Random	Reference
ApoTF	21.0	-[c]	-	-	(15)
	17.0	68	-	15	(16)
	24.0	32	17	28	(17)
Fe_2TF	17.5	-	-	-	(14)
	23.0	62	-	15	(16)
	20.0	40	13	28	(17)
ApoOTF	19.0[d]	32	-	40	(18)
	16.0	-	-	-	(15)
	11.0	32	-	57	(19)
Fe_2OTF	19.0[d]	32	-	40	(18)
	7.0	46	-	47	(20)
	14.0	21	-	65	(20)
ApoLTF	28.0	64	-	16	(16)
Fe_2LTF	26.0	57	-	17	(16)

[a]The secondary structure was determined from circular dichroism spectra analyzed by a variety of fitting programs as indicated in each reference. The samples were run at pH's ranging from 7 to 8.9 in a variety of buffers as indicated in the references.
[b]Abbreviations are TF, human serum transferrin; OTF, chicken ovotransferrin; LTF, human lactoferrin.
[c]Percentage not estimated.
[d]Recalculated by Tomimatsu and Vickery (15) using a mean residue weight of 112 instead of the 131 originally used by Tan (18). Using 131, Tan estimated 28% α-helix.

appear to change much upon binding of iron. It is clear that additional studies on both human serum TF and OTF and/or a comparison of results from different methods of analysis are necessary to resolve the differences.

Representative spectra for the N- and C-terminal domains of OTF are shown in Figure 2. Also given is the spectrum of the domains present in equimolar amounts. This spectrum is compared to that obtained by adding the spectra for the individual domains together and averaging the results. All of the samples were run in 10 mM sodium phosphate buffer, pH 7.4. As indicated by the estimate of secondary structure (Table 1) there is no significant difference attributable to the presence of iron in either domain. There are however large differences in the secondary structure of the individual domains. The C-terminal domain appears to have a degree of organization which is somewhat higher than that found for holo-OTF. The N-terminal domain has a degree of organization which is much lower than holo-OTF. The C-terminal domain has a minimum approximately 6% larger and the N-terminal domain has a minimum which is approximately 20% smaller than holo-OTF. The equimolar combined domains yield a spectrum which closely approximates the spectrum generated by computer addition and averaging of the two separate domains. This implies that there is no additional secondary structure which is attributable to the association of the two domains in solution. From our previous work measuring the association of the half-molecule domains in solution an apparent dissociation constant for the interaction was determined (8). In the present study we estimate that the domains are approximately half associated at the concentration examined (0.4 mg/ml).

The results which we obtained for the CD of the individual half-molecules are grossly consistent with previous work with OTF domains (20) and with LTF domains (21). In all cases the N-terminal domain has considerably less organized structure than the C-terminal domain. In our study and in the work of Keung and Azari (20) the C-terminal domain of OTF has more α-helix than holo-OTF. In LTF, however, the C-terminal domain shows somewhat less α-helix than LTF (21). The x-ray structure of LTF predicts 32% α-helix and 22% β-sheet (3) vs. 26% α-helix and 57% β-structure by the CD studies (16,21). Possible differences in the secondary structure of the two domains of LTF were not discussed in the report of the x-ray structure although clearly such differences do exist.

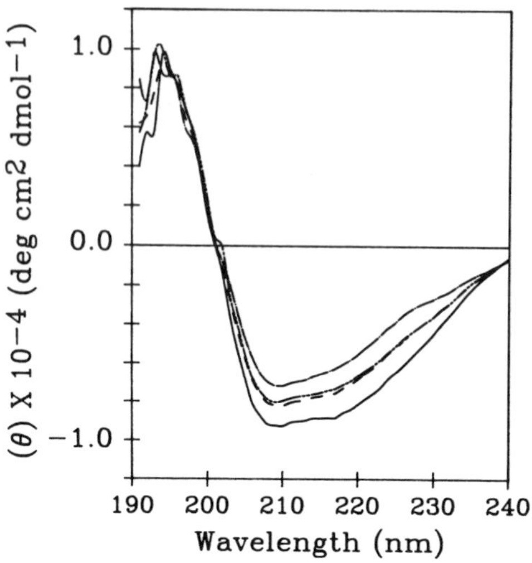

FIGURE 2. Representative CD spectra of the OTF/2N alone (-—·—); of OTF/2C alone (———); of equimolar OTF/2N and OTF/2C combined and analyzed together (··——··) and of individually run half-molecule domains added together by computer and averaged (– – –).

Analysis of the two domains of LTF by CD shows that there is an apparent 30% reduction in the sum of α-helical content compared to holo-LTF (21). When the domains are allowed to reassociate and are examined by CD the amount of α-helix is comparable to that found in holo-LTF. These findings indicate that, in LTF, the association of the two domains induces a significant conformational change not seen in OTF. Some of the difference may be due to the presence of a 3-turn helical bridge joining the two domains of LTF. Cleavage of diferric LTF by trypsin occurs at Arg 281 such that the C-terminal domain includes the helical bridge and is 50-kDa vs. the 30-kDa N-terminal domain. In OTF cleavage occurs at Arg 341 giving rise to two domains of approximately 40-kDa each. In addition the N-terminal domain of LTF appears to become extremely disordered in the absence of the C-terminal domain.

In conclusion our studies with OTF indicate no influence of iron on secondary structure and no increase in secondary structure attributable to the association of the two domains in solution. The higher affinity of the TF receptor for diferric TF at pH 7.4 implies that the binding of iron exposes regions of the sequence involved in receptor recognition. A small angle x-ray study suggests that binding of iron appears to alter the relative position of the two lobes rather than the structure per se (24). Twisting of the two domains would explain how diferric TF could change without changes in the secondary structure.

REFERENCES

1. Aisen P, Listowsky I (1980). Iron transport and storage proteins. Ann Rev of Biochem 49: 357.
2. Chasteen ND (1983). Transferrin: A perspective. Adv Inorg Biochem 5: 201.
3. Anderson BF, Baker HM, Dodson EJ, Norris GE, Rumball SV, Waters JM, Baker EN (1987). Structure of human lactoferrin at 3.2-A resolution. Proc Natn Acad Sci USA 84: 1769.
4. Metz-Boutigue M-H, Jolles J, Mazurier J, Schoentgen F, Legrand D, Spik G, Montreuil J, Jolles P (1984). Human lactotransferrin: amino acid sequence and structural comparisons with other transferrins. Eur J Biochem 145: 659.
5. Trowbridge IS, Newman RA, Domingo DL, Sauvage C (1984). Transferrin receptors: structure and function. Biochemical Pharmacology 33: 925.
6. Ward JH (1987). The structure, function, and regulation of transferrin receptors. Invest Radiol 22:74.
7. Brown-Mason A, Woodworth RC (1984). Physiological levels of binding and iron donation by complementary half-molecules of ovotransferrin to transferrin receptors on chick reticulocytes. J Biol Chem 259: 1866.
8. Mason AB, Brown SA, Butcher ND, Woodworth RC (1987). Reversible association of half-molecules of ovotransferrin in solution: Basis of cooperative binding to reticulocytes. Biochem J 245: 103.
9. Mason AB, Brown SA, Church WR (1987). Monoclonal antibodies to either domain of ovotransferrin block binding to transferrin receptors on chick reticulocytes. J Biol Chem 262: 9011.
10. Mason AB, Brown SA, Church WR (1988). Domain specific antibodies to ovotransferrin indicate conservation of determinants involved in avian transferrin receptor recognition. Comp Biochem Physiol In press.
11. Mason AB, Brown SA (1987). Differential effect of iodination of ovotransferrin and its half-molecule

domains on binding to transferrin receptors on chick embryo red blood cells. Biochem J 247: 417.

12. Huebers H, Csiba E, Josephson B, Huebers E, Finch CA (1983). Interaction of human diferric transferrin with reticulocytes. Proc Natn Acad Sci USA 78: 621.
13. Young SP, Bomford A, Williams R (1984). The effect of the iron saturation of transferrin on its binding and uptake by rabbit reticulocytes. Biochem J 219: 505.
14. Nagy B, Lehrer SS (1972). Circular dichroism of iron, copper, and zinc complexes of transferrin. Arch Biochem Biophys 148: 27.
15. Tomimatsu Y, Vickery LE (1972). Circular dichroism studies of human serum transferrin and chicken ovotransferrin and their copper complexes. Biochim Biophys Acta 285: 72.
16. Mazurier J, Aubert J-P, Louchoux-Lefebvre M-H, Spik G (1976). Comparative circular dichroism studies of iron-free and iron-saturated forms of human sero-transferrin and lactotransferrin. FEBS Lett 66: 238.
17. Teeters CL, Lodish HF, Ciechanover A, Wallace BA (1986). Transferrin and apotransferrin: pH-dependent conformational changes associated with receptor-mediated uptake. Ann NY Acad Sci 463: 403.
18. Tan A-T (1971). Circular dichroism properties of conalbumin and iron and copper complexes. Can J Biochem 49: 1071.
19. Yeh Y, Iwai S, Feeney RE (1979). Conformations of denatured and renatured ovotransferrin. Biochemistry 18: 882.
20. Keung W-M, Azari P (1982). Structure and function of ovotransferrin. J Biol Chem 257: 1177.
21. Legrand D, Mazurier J, Aubert J-P, Louchoux-Lefebvre M-H, Montreuil J, Spik G (1986). Evidence for interactions between the 30 kDa N- and 50 kDa C-terminal tryptic fragments of human lactotransferrin. Biochem J 236: 839.
22. Chang TC, Wu C-SC, Yang JT (1978). Circular dichroic analysis of protein conformation: inclusion of the b-turns. Anal Biochem 91: 13.
23. Yang JT, Wu C-SC, Martinez HM (1986). Calculation of protein conformation from circular dichroism. In "Methods in Enzymology," Orlando, Academic Press, p 208.
24. Kilar F, Simon I (1985). The effect of iron binding on the conformation of transferrin. Biophys J 48: 799.

Metal Ion Homeostasis: Molecular Biology and Chemistry, pages 179–188

SPECTROSCOPIC STUDIES OF HAFNIUM-TRANSFERRIN-COMPLEXES: DIFFERENCES BETWEEN THE TWO BINDING SITES[1]

D.M. Taylor and W.-G. Thies

Institut für Genetik und für Toxikologie von Spaltstoffen, Kernforschungszentrum Karlsruhe,

H. Appel, M. Neu-Müller, F. Schwab, and H.W. Weber

Institut für Experimentelle Kernphysik, Universität Karlsruhe

Postfach 3640, D-7500 Karlsruhe 1, Federal Republic of Germany

ABSTRACT Hafnium-transferrin complexes were prepared in serum and in different buffer systems. Differences between the two metal-binding sites of ^{181}Hf-labeled transferrin were studied using the Time Differential Perturbed Angular Correlation (TDPAC-) Technique. This method is highly sensitive to the local environment of the metal ions. The distribution of the hafnium ions between the two binding configurations proved to be dependent on the pH of the samples, the temperature and the components of the buffer system. UV difference spectroscopy was used to obtain complementary information.

INTRODUCTION

Serum transferrin (TF) is the essential transport protein for metabolic iron in vertebrates. Besides iron, TF also binds a variety of other metals. After their uptake

[1] Support of this work by the Deutsche Forschungsgemeinschaft, Bonn-Bad Godesberg, is gratefully acknowledged

these metal ions are bound to TF until their intracellular absorption, their incorporation into metal-storage proteins or their eventual deposition onto bone surfaces.

There are two metal-binding sites per TF molecule, situated in the C- and N-terminals of the protein, respectively. Differences between the two binding sites were demonstrated by a wide variety of thermodynamic, kinetic, and spectroscopic studies. There is evidence for conformational changes within the protein when metal ions are bound [1], [2].

The details of the metal binding to TF and the distribution between the two binding sites is dependent on the parameters

- pH,
- temperature,
- salt concentration, and
- components of the buffer system.

It appears unique in biochemistry that the simultaneous binding of a synergistic anion (e.g. (bi)carbonate) is necessary for the formation of a stable metal-TF complex.

A considerable number of studies were performed to characterize the two metal-binding configurations in TF. Spectroscopic methods like UV difference spectroscopy, ESR and NMR were applied predominantly. In the work presented here, the Perturbed Angular Correlation (PAC-) technique was used [3]. It is based on the observation of an anisotropic correlation between the emission directions of two successive γ-rays while the nucleus of the metal ion is exposed to an electromagnetic perturbation. This perturbation reflects the electron configuration characteristic for the metal-binding site. Thus the method is highly sensitive and specific to details of the metal binding. As will be demonstrated, the observed parameters are different though complementary to those obtained from UV difference spectroscopy studies. The PAC technique also provides insight into the dynamic behavior of the molecule.

It is an important advantage of the method that proteins can be studied under nearly physiological conditions, i.e. as liquid samples with low protein concentrations of the order nmol/ml.

THE PERTURBED ANGULAR CORRELATION TECHNIQUE

It has been pointed out previously [4] that hafnium (Hf) can be considered to be a convenient analogue of plutonium for metabolic studies in mammals. The isotope ^{181}Hf is particularly suitable for the PAC-method. It decays by β-emission to the excited I = 1/2 level in ^{181}Ta. The γγ-correlation of the following 133 keV - 482 keV cascade is observed using a fast coincidence system. It consists of four BaF_2- detectors which allow a time resolution of σ = 600 ps. During the lifetime of the intermediate I = 5/2 state ($t_{1/2}$ = 10.8 ns) the nuclear quadrupole moment of the probe atom interacts with the electric field gradient (EFG) arising at the position of the nucleus due to the chemical environment of the metal ion.

The coincidence spectra N(Θ,t) were taken as a function of the delay time t for fixed angles Θ = 90° and Θ = 180°. It is usual to calculate the ratio

$$R(t) = 2\,\frac{N(180°,t) - N(90°,t)}{N(180°,t) + 2N(90°,t)} = A_{22}^{eff}G_{22}(t)$$

where A_{22}^{eff} is the anisotropy coefficient of the γγ-cascade.

The R(t) ratio is proportional to the perturbation function $G_{22}(t)$ which can be interpreted in terms of the following parameters:

ν_Q = electric quadrupole coupling constant
= eQV_{zz}/h [MHz]
= $(10/3\pi)\omega_0$ for I = 5/2.

The observed frequencies are ω_k [Mrad] = $n_k\omega_0$ with k = 1,2,3 for the given I = 5/2 intermediate level in ^{181}Ta. For an axially symmetric EFG (η = 0) the frequencies ω_2 and ω_3 are harmonics of the fundamental frequency ω_0, i.e. the n_k are integers. This is no longer true for finite η-values, where the coefficients n_k have to be evaluated numerically.

Q = nuclear quadrupole moment of the nucleus
V_{zz} = z-component as largest component of the diagonalized EFG tensor due to the electron configuration of the metal ion.

η = the asymmetry parameter

= $(V_{xx} - V_{yy})/V_{zz}$ (with $0 \leq \eta \leq 1$).

V_{xx} and V_{yy} are the x- and y-components, respectively, of the diagonalized EFG tensor, η reflects the deviation of the binding configuration from axial symmetry.

δ = the width of the EFG distribution for which a Lorentzian or Gaussian shape may be assumed. This parameter should be small (δ < 5%) for well-defined binding configurations.

λ = the relaxation constant which provides information on the dynamical properties of the molecule.

HUMAN SERUM DATA

A series of PAC measurements have been performed with human serum samples [5]. Two characteristic and well-defined binding states were observed. The parameters taken from the PAC spectra are listed in table 1.

TABLE 1
QUADRUPOLE INTERACTION PARAMETERS FOR THE BINDING OF ^{181}Hf TO HUMAN SERUM TRANSFERRIN

Configuration	ν_Q[MHz]	η
f_1 (low pH)	869 (2)	0.95 (1)
f_2 (high pH)	1435 (1)	0.577 (1)

The distribution between these two configurations was dependent on:

- the pH of the samples. The R(t) spectra of human serum samples at pH 7.8 and 9.3, respectively, are plotted in fig. 1. The Fourier transforms show the characteristic resonance frequencies ω_k that correspond to the two binding configurations with their respective asymmetry parameters η_1 and η_2.
- temperature. Higher temperatures seem to favour the occupation of the second binding configuration f_2 [6].

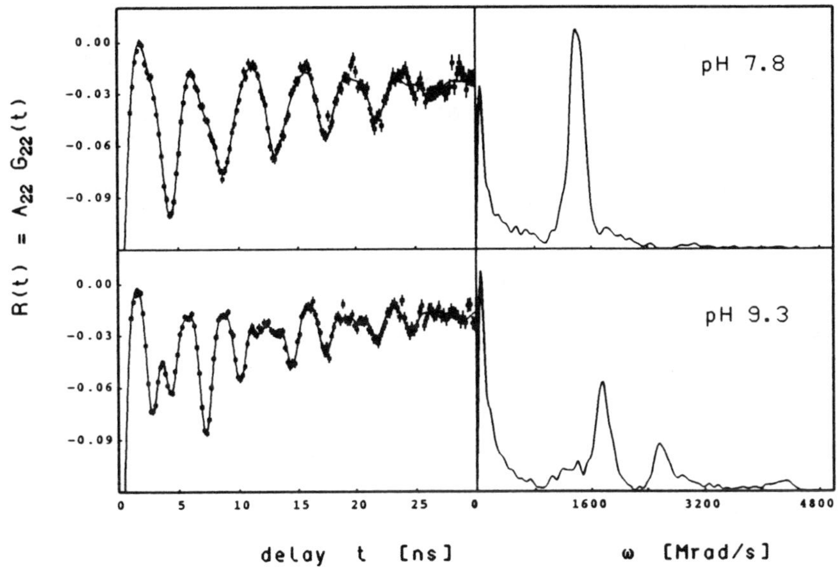

FIGURE 1. R(t) spectra of [181]Hf-labeled human serum TF (left hand side). The Fourier transforms (right hand side) show the characteristic angular frequencies corresponding to two different and well-defined binding configurations. The spectra were taken at pH 7.8 and pH 9.3, respectively. For the evaluated parameters ν_Q and η see table 1.

PAC STUDIES ON THE INFLUENCE OF THE BUFFER SYSTEM

Human apotransferrin, purchased from Serva was dissolved in different buffer systems (25 nmol TF/ml, sample volume 0.5 ml). Before addition to TF the ^{181}Hf-NTA solution (5 nmol Hf, 20 nmol NTA) was titrated with NaOH to just below the required pH and then adjusted to the final value by adding small amounts of sodium-bicarbonate.

The results of a series of PAC measurements indicate a possible influence of the buffer medium on the Hf-TF binding. For several samples, prepared in HEPES, MOPS, MES/HEPPS or TRIS buffers (50 mM, containing 1 mM $NaHCO_3$) new binding configurations were observed: they could not be described by the parameters of the two configurations in serum.

However, a phosphate buffered sample (PBS) led to the low- and high pH configuration previously observed in serum. This was a strong indication for the influence of phosphate on the Hf-binding in TF [7].

More systematic studies showed the following results:

- The low pH serum configuration can only be reproduced in buffers containing phosphate, e.g. HEPES or MES/HEPPS with approximately 8mM phosphate ($H_2PO_4^-$/ HPO_4^{2-}).
- Above pH 8.5 the high pH serum configuration could also be observed in the absence of phosphate (TRIS or MES/HEPPS).

As an example the results of PAC measurements with MES/ HEPPS buffered ^{181}Hf-TF samples are shown in figures 2 and 3.

We conclude that the difference between the two specific binding configurations in serum is very probably due to phosphate which is involved in the metal binding at physiological pH or its release at higher pH values.

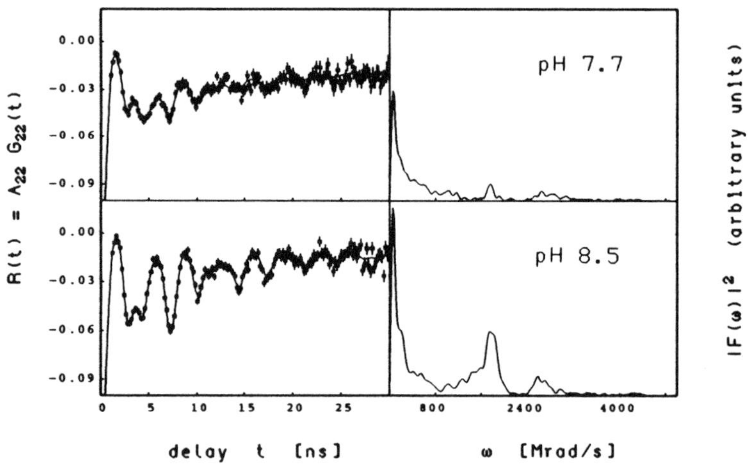

FIGURE 2. R(t) spectra of ^{181}Hf-labeled apotransferrin in MES/HEPPS buffer. The results from phosphate-free samples are characterized by the f_2 configuration at pH 8.5, whereas at pH 7.7 the amplitude of this signal is significantly reduced.

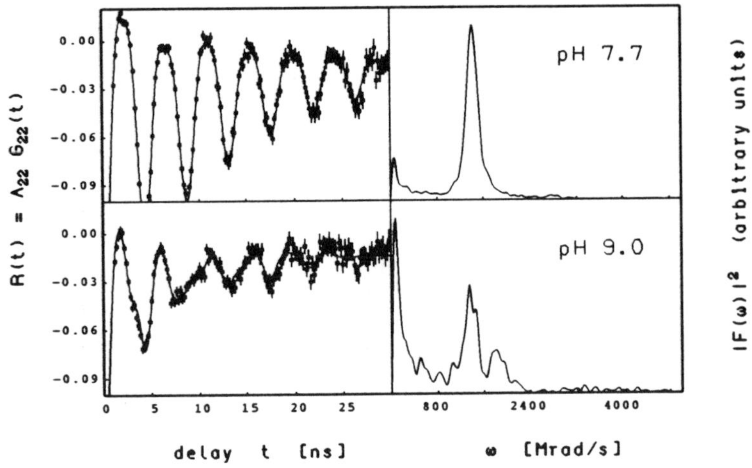

FIGURE 3. R(t) spectra of ^{181}Hf-labeled apo-transferrin in MES/HEPPS buffer with phosphate. In the presence of phosphate the spectra are closely similar to the human serum data with configuration f_1 at pH 7.7 and the appearance of the configuration f_2 at higher pH values.

PAC SATURATION STUDIES

In order to study the possible non-equivalence of the two metal-binding sites Hf-TF samples with different metal saturations were measured. The results of a series of PAC studies at various pH values were rather surprising:

The amplitudes of the well-defined and sharp signals decrease markedly for low and high pH configurations with a higher metal saturation. Figure 4 shows a series of measurements with ratios r = [Hf]/[TF] ranging from 0.4 to 1.2.

The missing signal amplitude may be due to a broad distribution of slightly different configurations for a large fraction of the bound Hf-ions.

Thus the following hypothesis may be considered:

- The observed well-defined configurations arise from single-loaded TF. The probability of their observation decreases with higher metal concentration.

- Conformational changes which occur with double-loaded TF molecules lead to metal-binding sites that are either perturbed by relaxation effects and/or characterized by a non-uniform, slightly distorted EFG distribution.

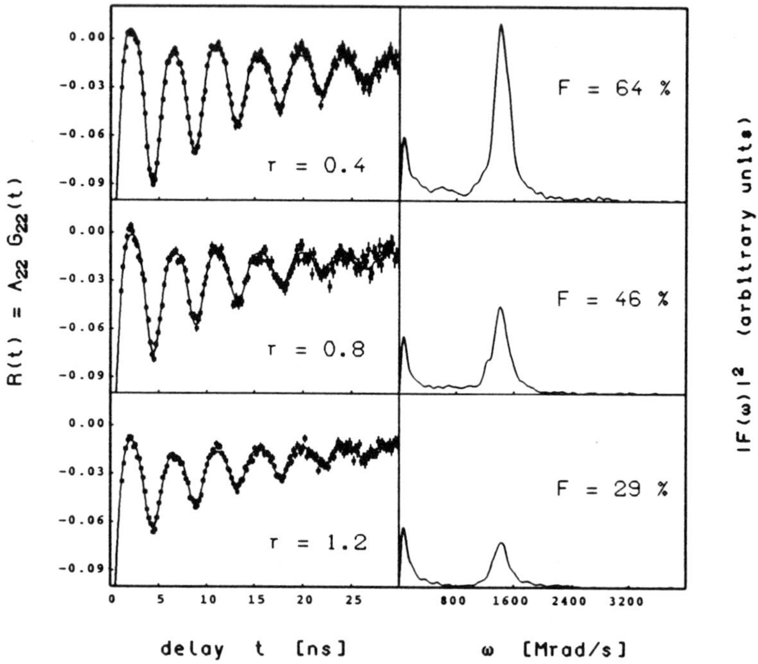

FIGURE 4. R(t) spectra and Fourier transforms for ^{181}Hf-TF samples (MES/HEPPS buffer + phosphate, pH 7.8) with Hf-saturation ratios r = [Hf]/[TF] between 0.4 and 1.2. The amplitude F [%] of the f_1-signal, characteristic for low pH Hf-binding, is strongly reduced for higher saturation values.

DIFFERENCE UV SPECTROSCOPY

The UV difference spectra of the Hf-TF complex versus apo-transferrin were recorded at room temperature using a Beckman DU 65 spectrometer. The samples were prepared in

MES/ HEPPS buffer at pH values between pH 7.5 and pH 9. Some of the measurements were performed on samples which had been prepared in phosphate-free buffers.

No dependence on phosphate was observed in the difference UV spectra. Figure 5 shows the titration of the Hf-TF complex (MES/HEPPS + phosphate at pH 9). The absorption at 240 nm plotted versus the ratio r = [Hf]/[TF] indicates a saturation at about r = 2 and thus the binding of two Hf-ions per molecule, as expected.

However, the curve below r = 2 is not exactly linear. This might be due to a different number of tyrosyl residues coordinated to the metal at the N-terminal site compared to the C-terminal site.

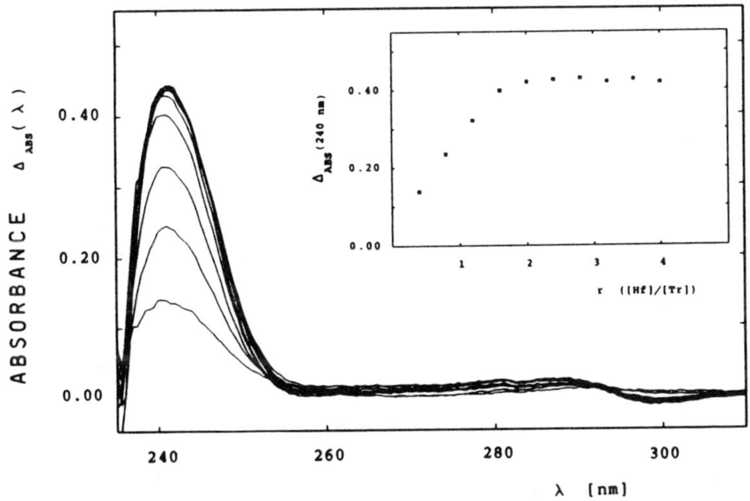

FIGURE 5. Difference UV spectroscopy of Hf-TF (pH 9, MES/HEPPS + phosphate). The inserted titration curve shows the absorption at 240 nm for different ratios r =[Hf]/[TF].

CONCLUSIONS

The UV difference spectra clearly demonstrate the stoichiometric binding of Hf to TF. The data appear to indicate that a different number of tyrosyl residues are involved in the metal-binding at the two sites.

The PAC spectra prove the existence of two characteristic and chemically different binding configurations. Whether these can be attributed to the C- and N-terminal binding site, respectively, or whether the metal binding at both sites changes as a function of pH, remains open. This ambiguity may be cleared up by PAC measurements on either C- or N-terminal monoferric, ^{181}Hf-labeled TF. Such studies are in preparation.

An essential result is that the low-pH configuration observed in human serum TF can only be reproduced in the presence of phosphate.

The data also show that for metal saturations above 40% a substantial proportion of the Hf ions do not bind in 'specific' configurations characterized by sharp electric field gradients. The well-defined binding states may be characteristic for single-loaded Hf-TF only.

REFERENCES

1. Chasteen ND (1983). Transferrin: A Perspective. Adv Inorg Biochem 5:201.
2. Baldwin DA, Egan TJ (1987). An inorganic perspective of human serum transferrin. South Afr J Science 83:22.
3. Bauer R (1985). Perturbed angular correlation spectroscopy and its application to metal sites in proteins: possibilities and limitations. Quarterly Rev Biophys 18:1.
4. Taylor DM, Lehmann M, Planas-Bohne F, Seidel A (1983). The Metabolism of Radiohafnium in Rats and Hamsters: A possible Analog of Plutonium for Metabolic Studies. Radiat Res 95:339.
5. Appel H, Duffield J, Taylor DM, Then GM, Thies WG (1987). TDPAC Studies of the Metal-Binding Sites in Serum Transferrin: Comparison between ^{181}Hf-labeled Human and Rat Transferrin. J Inorg Biochem 31:229.
6. Appel H, Duffield J, Taylor DM, Then GM, Thies WG (1987). ^{181}Hf-labeled Rat Serum Transferrin: Influence of Temperature on the Metal-Binding Configurations. Hyperfine Interactions 35:953.
7. Penner MH, Osuga DT, Meares CF, Feeney RE (1987). The Interactions of Anions with Native and Phenylglyoxal-Modified Human Serum Transferrin. Arch Biochem Biophys Vol 252, 1:7.

Metal Ion Homeostasis: Molecular Biology and Chemistry, pages 189–196

SPECTROSCOPIC PROPERTIES OF THE FUR (FERRIC UPTAKE REGULATION) PROTEIN OF *Escherichia coli* K-12[1]

R. J. P. Williams and M. R. Wormald
Inorganic Chemistry Laboratory, University of Oxford, South Parks Road, Oxford, OX1 3QR

R. C. Hider
Department of Pharmacy, King's College, University of London, Manresa Road, London, SW3 6LX

J. B. Neilands
Biochemistry Department, University of California, Berkeley, CA 94720

ABSTRACT The Fur (ferric uptake regulation) protein of *Escherichia coli* K-12 is known to require, under *in vitro* conditions, a divalent heavy metal ion as co-repressor. *In vivo*, the main active metal ion is assumed to be Fe(II). Isolation of Fur from bacterial cells cultured in a medium relatively low in iron yields protein containing detectable amounts of Fe(III) thus suggesting that the repressor may also have a certain affinity for ferric iron. The proton NMR spectra of Fur were recorded at 300, 500 and 600 MHz. From these measurements it is possible to conclude, among other structural details, that the Fur protein exhibits definite tertiary interactions in solution and that it is largely α-helical. The latter conclusion is reinforced by structure predictions derived from the CD spectrum of the protein.

---

[1]This work was supported in part by Grants AI04156, PCM78-12198 and CRCR-1-1633 from the NIH, NSF and USDA, respectively.

INTRODUCTION

A mutation designated *fur* (ferric uptake regulation) results in constitutive expression of iron uptake systems in *Salmonella typhimurium* (1) and *Escherichia coli* (2). The mutation is apparently pleiotropic since *fur* mutants of *E. coli* fail to grow on certain small carbon substrates (3,4), the cells are resistant to Mn (4) and there is some connection, as yet unexplained, to expression of the *sod*B gene coding for the iron form of superoxide dismutase (5). The product of the *fur* gene of *E. coli* K12, a 17 kDa protein rich in histidine (6), has been obtained in high yield from an expression vector and purified by affinity chromatography over zinc iminodiacetate agarose (7). The availability of the pure protein has enabled *in vitro* studies of its mode of action, from which it was concluded that at least one function is that of a classical repressor (8). An *in vitro* footprinting analysis of the promoter regions of the aerobactin operator (9) of pColV-K30 and of the *fur* gene (10) itself demonstrated that the repressor, when activated by certain divalent heavy metal ions, binds to an "iron box" operator with the consensus sequence 5'GAT AAT GAT A ATC ATT ATC. A synthetic nucleotide with this sequence inserted in a construct bearing the *omp*F promoter fused to *lac*Z was found to impart *fur* gene dependent iron regulation of expression of β-galactosidase (11).

In the present report we describe some of the physical properties of Fur, namely, the electronic, NMR and CD spectra of the protein.

METHODS

Unless otherwise noted, the Fur protein was prepared as previously described (7) and dialyzed vs 10 mM phosphate buffer pH 7.2. The protein solution was then lyophilized and stored in sealed serum vials over N_2. Inter-laboratory shipment, when required, was generally by regular airmail.

It should be noted that in the construction of pMON2064 (7), the expression vector for Fur, a *Nco*I restriction site was introduced immediately after the initiator methionine codon. This required a change from an A to a G in the second codon in the *fur* gene and resulted in replacement of a threonine by an alanine residue. Following post-translational modification the alanine became N-terminal.

Visible and ultraviolet absorption spectra were recorded with a Beckman 25 and with a Bausch and Lomb 2000 spectrophotometer.

The following methodology was used for the NMR experiments. Samples were prepared by dissolving between 10 and 20 mg of the protein (lyophilized from a solution of potassium phosphate in water) in 0.55 ml of either deuterated water (D_2O sample) or 90% water, 10% deuterated water (H_2O sample) giving a final phosphate concentration of about 25 to 50 mM. The pH was then adjusted to between 6.8 and 7.0 before the sample was transferred to a 5 mm NMR tube.

Nuclear magnetic resonance spectra were recorded at 300 MHz on a Bruker WH300 spectrometer, at 500 MHz on a Bruker AM500 spectrometer and on a home built spectrometer with an Oxford Instruments magnet and a Nicolet 1280 computer, and at 600 MHz on a Bruker AM600 spectrometer. All spectra were acquired at a temperature of 300 K.

One dimensional (1D) spectra, with solvent suppression, were recorded at 300, 500 and 600 MHz on the samples in H_2O and D_2O. Two dimensional (2D) COSY (J-correlated), RELAY (relayed J-correlated) and NOESY (nuclear Overhauser) spectra (12) were recorded at 500 MHz on the samples in D_2O. Mixing times of 20 to 30 ms were used for the RELAY experiments, and mixing times of 200 to 300 ms (varied randomly by ± 5% during the experiment) used for the NOESY experiments. Attempts were made to measure the 2D spectra in H_2O, but the protein concentration proved to be too low to obtain reasonable signal to noise ratios in 24 hours.

The circular dichroism spectra were recorded with a JASCO Model J40CS instrument and the data were analyzed by the method of Dufton and Hider (13).

RESULTS

Electronic Absorption Spectra

As noted previously (7), preparations of Fur containing of the order of a few milligrams per ml may display a distinct amber color. This color persisted when the protein was isolated from *E. coli* JRB (pMON2064) grown on M9 minimal medium supplemented with casamino acids and glucose, but without added iron, rather than on the usual iron-rich LB medium. The cells proved to be much lighter in color when cultured on the low iron medium.

A 7 g quantity of packed cells obtained from 2.5 l of M9-casamino acids-glucose medium was used to investigate the nature of the amber color, which appeared to be coincident with the concentration of Fur in histidine elution profiles from the zinc iminodiacetate agarose column. The absorption spectrum of the solution in the peak tube from the histidine elution step, shown by Bradford measurement to contain 1.64 mg protein per ml and confirmed by SDS-PAGE analysis as Fur, was examined from 500 to 200 nm. Weak absorption through the visible was evident with a shoulder at 460 nm and a broad peak centered at 410 nm with absorbancy of 0.09. In the ultraviolet the absorbancy reached a maximum of 0.7 at 276 nm and thereafter fell to a minimum value at 254 nm, where the absorbancy was 0.4. Upon addition of a knife-point of solid $Na_2S_2O_4$ the absorption in the visible was almost completely abolished. When excess 1,10-phenanthroline was then added a significant level of pink coloration, characteristic of the Fe(II)-phenanthroline complex, appeared. This indicates that the Fur protein may have an appreciable affinity for Fe(III).

Nuclear Magnetic Resonance Spectra

The 1D NMR spectrum of Fur shows considerable dispersion in the aromatic (9 to 6 ppm.) and methyl (1 to -1 ppm.) regions from typical 'random coil' shifts (12), indicative of considerable tertiary interactions in the protein. However, very few resonances are seen in the 'α-CH β-region' (6 to 5 ppm) of the spectrum (14) showing very little β-structure (<10%).

No residual backbone amide (-NH) resonances are seen in the D_2O spectra, thus exchange between the backbone -NHs and solvent must be fast, suggesting that the structure is highly mobile allowing easy solvent access to the center of the protein.

At pH 6.8, separate resonances for all twelve histidine C-2 protons can be seen (8.5 to 7.5 ppm.) On raising the pH, all these resonances shift upfield and the degree of dispersion dramatically decreases. It has not been possible to obtain protein solutions at low enough pH to determine accurate dissociation constants, pKas, for the histidine residues, but they all titrate close to the expected value, 6.6, and thus must be exposed to the solvent. The fact that all resonances are separated at pH 6.8 but not at 9.0 implies that the histidines must have slightly different pKas and thus be in slightly different chemical environments, again indicative of tertiary structure.

The molecular mass of the protein (about 17 kDa) is high by two-dimensional NMR standards, but we have been able to record COSY, RELAY and NOESY spectra in D_2O.

The COSY and RELAY spectra, showing correlations between J-coupled (through bond) nuclei, give a large number of cross-peaks. These are most useful for assigning resonances to amino-acid type. There is considerable spectral overlap in several regions of the spectrum (often found in mostly helical proteins), which will make the task of resonance assignment more difficult. However, many of the aromatic spin systems can be identified, as can about 16 aliphatic systems in the upfield methyl region.

The NOESY spectra, showing correlations between dipole-coupled (through space) nuclei, did not give as many cross-peaks as would be expected. In particular, very few cross-peaks between the upfield shifted methyl resonances and the aromatic resonances were found. This is unusual, as large upfield shifts on methyl resonances are usually due to neighboring aromatic residues. However, in other regions of the spectra very intense cross-peaks were seen. The observation of an NOE (nuclear Overhauser effect) is very dependent on the correlation and relaxation times of the nuclei in the protein (15). This total absence of expected NOEs and the strong presence of others suggests that different regions of the protein have different relaxation and correlation times, probably due to differing degrees of mobility in different regions of the protein. One or two α-CH to α-CH NOEs are seen, typical of β-structure, but these may well be from β-turns rather than long β-sheets. From all the 2D measurements it is apparent that there are a very large number of α-protons with resonances typical for α-helical structures (14).

Circular Dichroism Spectra

From its CD spectrum and the structural prediction analysis of Dufton and Hider (13), it is apparent that the protein contains about 50% α-helix. The helices are distributed predominantly in four long sequences representing residues 15-36, 80-96, 105-123 and 130-147. The region 52-73 appears to represent a section of anti-parallel β-sheet.

DISCUSSION

A previous study (8) examined the capacity of different metal ions to act as co-factors for Fur, thus enabling the protein to both negatively regulate *in vitro* transcription-translation of a gene for a reporter enzyme fused downstream from an iron-sensing promoter and protect a *Hinf*1 site in the operator. From these data it was concluded that Fur is an Fe(II) binding protein. In the present work the detection of a certain quantity of iron in purified preparations of Fur suggests that the protein may have an appreciable affinity for Fe(III). This raises the possibility that the Fe(II) form of the protein recognizes the iron box array in the operator while the Fe(III) form is either inactive or possessed of another function.

Contamination of the Fur protein with trace amounts of iron would appear to account for the varying quality of NMR spectra found for preparations which were mono-disperse by SDS-PAGE analysis.

From the NMR analysis, various conclusions can be drawn about FUR in solution. The protein shows clear tertiary interactions in solution. It contains a little, but not much, β-structure suggesting a basically α-helical structure. This structure is relatively mobile, with different regions of the protein showing different mobilities. The histidine residues, suggested to be involved in metal binding, are all accessible to the solvent, although in slightly different chemical environments.

Regarding the CD spectrum, it is apparent that Fur should be classed as the structural type known as "anti-parallel α-helix bundles" (16). The finding that the protein is largely α-helix is compatible with the NMR data and with the repressor function of the molecule. A "helix-turn-helix" is the most commonly found structural motif in DNA-binding regulatory proteins" (17).

REFERENCES

1. Ernst JF, Bennett RL, Rothfield LI (1978). Constitutive expression of the iron enterochelin and ferrichrome uptake systems in a mutant strain of *Salmonella typhimurium*. J Bacteriol 135:928.

2. Hantke K (1981). Regulation of ferric iron transport in *E. coli*. Mol Gen Genet 182:288.
3. Neilands JB, Konopka K, Schwyn B, Coy M, Francis RT, Paw BH, Bagg A (1987). Comparative biochemistry of microbial iron assimilation. In Winkelmann G, van der Helm D, Neilands JB (eds): "Iron Transport in Microbes, Plants and Animals," Weinheim, Federal Republic of Germany: VCH Publishers, p. 3-30.
4. Hantke K (1987). Selection procedure for deregulated iron transport mutants (*fur*) in *Escherichia coli* K12: *fur* not only affects iron metabolism. Mol Gen Genet 210: 135.
5. Niederhoffer E, Naranjo C, Fee JA (1988). Relationship of the superoxide dismutase genes *sod*A and *sod*B, to the iron uptake (*fur*) regulon in *Escherichia coli* K12 (this volume).
6. Schaffer S, Hantke K, Braun V (1985). Nucleotide sequence of the iron regulatory gene *fur*. Mol Gen Genet 201:204.
7. Wee S, Neilands JB, Bittner ML, Hemming BC, Haymore BL, Seethram R (1988). Expression, isolation and properties of Fur (ferric uptake regulation) protein of *Escherichia coli* K12. Biol Metals 1:in press.
8. Bagg A, Neilands JB (1987). Ferric uptake regulation protein acts as a repressor, employing iron (II) as a cofactor to bind the operator of an iron transport operon in *Escherichia coli*. Biochem 26:5471.
9. de Lorenzo V, Herrero M, Neilands JB (1987). Operator sequences of the aerobactin operon of plasmid ColV-K30 binding the ferric uptake regulation (*fur*) repressor. J Bacteriol 167:350.
10. de Lorenzo V, Herrero M, Giovannini F, Neilands JB (1988). Fur (ferric uptake regulation) protein and CAP (catabolite-activator protein) modulate transcription of *fur* gene in *Escherichia coli*. Eur J Biochem 173:537.
11. Calderwood SB, Mekalanos JJ (1988). Confirmation of the Fur operator site by insertion of a synthetic oligonucleotide into an operon fusion plasmid. J Bacteriol. 170:1015.
12. Wütrich K (1986). "NMR of Proteins and Nucleic Acids", New York: John Wiley & Sons.
13. Dufton MJ, Hider RC (1977). Snake toxin secondary structure predictions. J Mol Biol 115:177.

14. Dalgarno DC, Levine BA, Williams RJP (1983). Structural information from NMR secondary chemical shifts of peptide αC-H protons in proteins. Biosci Reports 3:443.
15. Noggle JH, Schirmer RE (1971) "The Nuclear Overhauser Effect", New York: Academic Press.
16. Richardson JS (1981). The anatomy and taxonomy of protein structure. Adv. Protein Chem 34:168.
17. Ptashne M (1987). "A Genetic Switch", Cambridge, MA and Oxford, UK: Cell Press and Blackwell Scientific Publications.

Metal Ion Homeostasis: Molecular Biology and Chemistry, pages 197–206

NMR INSIGHTS INTO THE DYNAMICS OF METAL INTERACTION WITH METALLOTHIONEIN[1]

James Otvos, Shiow-meei Chen, and Xiucai Liu

Department of Chemistry, University of Wisconsin-Milwaukee, Milwaukee, Wisconsin 53201

ABSTRACT We have investigated the Cd NMR spectral features of four pure isoforms of rabbit liver metallothionein with the aim of determining whether the protein can adopt, via metal-ligand exchange processes, different configurational states in solution. Evidence is presented suggesting the existence of multiple 3-metal cluster stuctures which interconvert very rapidly in MT-2 and slowly in 3 different MT-1 isoforms. The NMR data do not, however, indicate that similar processes are undergone by the Cd_4 cluster. In a separate study, we have obtained direct proof that in all Cd_7MT isoforms, the Cd ions move extremely rapidly between different 3-metal cluster binding sites via inter- rather than intramolecular metal exchange reactions.

INTRODUCTION

Detailed structural information regarding the mode of interaction of mammalian metallothionein (MT) with its bound metals (Zn, Cd) has recently become available from X-ray crystallographic studies of native Cd,ZnMT-2 from rat liver (1) and 2-dimensional ^1H NMR studies of reconstituted Cd_7MT-2 from rabbit liver (2). Despite exhibiting very similar gross structural features (ie., two separate protein domains enfolding the two metal-thiolate clusters),

[1]This work was supported by the National Institutes of Health (ES04095). The NMR facility was supported by NIH Grant RR04095 and NSF Grant DIR-8719808.

the X-ray- and NMR-derived structures are fundamentally different with regard to the identities of the cysteine residues in the protein sequence which serve as ligands to the seven metal ions. Since differences in species origin and protein preparation have been shown unlikely to account for the discrepency (3), we became interested in the possibility that metallothionein might, via intramolecular thiolate ligand exchange processes, be capable of adopting multiple configurational states in solution. If so, amino acid sequence differences among the several MT isoforms or between the same isoform from different species or tissue might modulate the relative energies (and hence concentrations) of these states and/or the rates of interconversion between them, possibly to some functional advantage.

Cd NMR is ideally suited to probe the existence of such configurational states owing to the sensitivity of Cd chemical shifts to very small differences in ligation sphere (4,5). Thus, even if two states differ only slightly, for example by an interchange of two ligands to one metal (ie., bridging $\rightleftharpoons$ terminal), there is a good liklihood that the Cd NMR signal from that metal and perhaps others in the same cluster will be influenced in some way. If ligand exchange is fast on the NMR timescale, only a single averaged resonance will be observed for each metal, whereas a separate signal can be anticipated from each state when ligand exchange is slow. At intermediate rates, broadening of the averaged resonance is expected. By monitoring the number of Cd resonances observed in spectra of several Cd_7-isoMTs and the changes in linewidth they undergo over a range of temperatures, we have been able to infer the existence in solution of several configurational states of the 3-metal cluster domain, some in rapid exchange and some interconverting slowly (if at all). Whether or not any of these corresponds to the structure deduced by X-ray remains to be proven.

Another facile reaction undergone by the metals bound to MT whose existence was discovered by NMR is that of interprotein metal exchange between CdMT and ZnMT leading to the production of mixed-metal clusters (6). Saturation transfer studies of Cd_7MT indicate that in solution there is also a very rapid interchange of Cd among all sites of the 3-metal cluster (7). We are now able to report unequivocal evidence that it is intermolecular rather than intramolecular metal exchange that accounts for this phenomenon, lending support to the idea that the metal cluster structures in MT may be uniquely designed to facilitate metal transfer to metal-requiring proteins *in vivo*.

RESULTS AND DISCUSSION

Temperature Dependence of MT-2.

As shown in Figure 1, the resonances from Cd in the 3-metal cluster of Cd_7MT-2 (numbered 2-4) experience pronounced chemical exchange broadening at low temperature, especially that of resonance #3. This observation offers

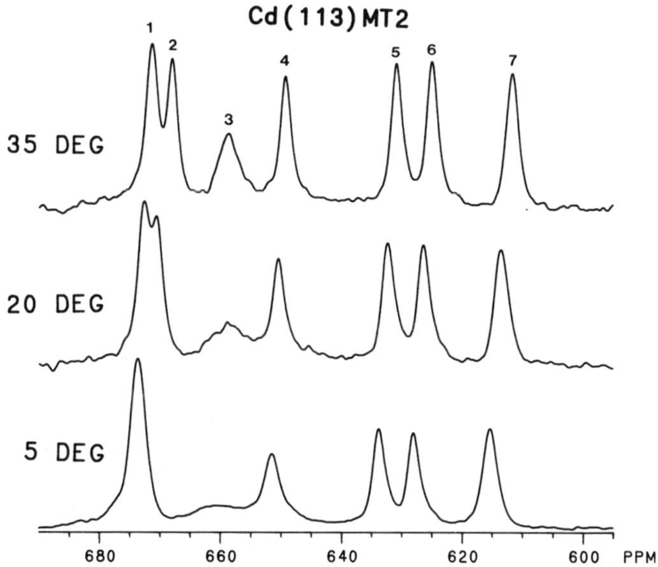

FIGURE 1. ^{113}Cd NMR spectra (110.9 MHz) of $^{113}Cd_7$MT-2 from rabbit liver obtained at the three temperatures indicated.

proof of rapid interconversion in solution of two or more "states" of the 3-metal cluster which differ in terms of the local environments of the bound Cd ions. Although there is currently no direct evidence that these states represent configurational isomers which interconvert via cysteine thiolate ligand exchange, it is a reasonable assumption based on the known lability of metal-thiolate bonds in model inorganic cluster compunds (9). Without the ability to go to low enough temperatures to freeze out the resonances from the equilibrating species, it is impossible to say whether or not the interconverting states are in fact the different 3-metal cluster structures deduced by X-ray and NMR. It is interesting to note, however, that the metal site whose ligand environment is the most different in the two structures is that occupied by the Cd giving rise to the most temperature-dependent resonance (#3).

Cd NMR Properties of MT-1 Isoforms.

For unknown reasons, Cd NMR spectra of Cd_7MT-1 have always been found to be more complex than those of MT-2 (5). Instead of containing one resonance for each of the seven bound metals, MT-1 spectra always contain two sets of 4-metal cluster resonances and variable amplitudes and numbers of 3-metal cluster signals. From the known existence in several mammals of multiple expressed MT-1 genes which differ slightly in amino acid sequence, it was considered probable that such sequence microheterogeneity was the explanation for the Cd NMR spectral complexity. However, another intriguing possibility was that of slowly interconverting configurational states, each with one or more metals coordinated to a different set of cysteine ligands. To test this possibility we employed a ^{1}H-^{113}Cd chemical shift correlation experiment which in essence provides a ^{1}H NMR "fingerprint" of the identities of the cysteine ligands to each metal (10). We were specifically interested in resonances 7A and 7B, which are the most well separated members of the two superimposed sets of 4-metal cluster signals in the MT-1 spectrum (resonances 1, 5, and 6 in Figure 2 are also composed of two superimposed signals which are not resolved under these conditions). As shown in Figure 2, the ^{1}H subspectra of these two Cd resonances are identical, providing direct proof that they are coordinated to the same cysteine residues and thus do not arise from configurational isomers.

To obtain direct evidence for the alternate explanation for MT-1 heterogeneity (ie., multiple isoforms which are not separable by standard ion exchange chromatographic methods), we have developed new reverse phase HPLC protocols to accomplish the preparative-scale isolation of three major isoform components of Cd_7MT-1.

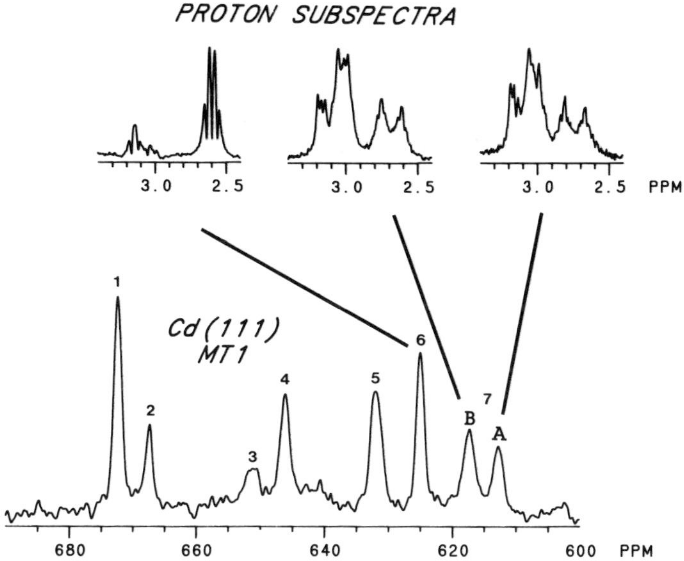

FIGURE 2. Bottom: ^{111}Cd NMR spectrum (106 MHz) of $^{111}Cd_7$MT-1. Top: ^1H subspectra showing the ß-CH_2 proton signals from the cysteine ligands to the Cd ions giving rise to Cd resonances 6, 7A, and 7B. The subspectra were obtained using a one-dimensional variant of a multiple-quantum ^1H-^{111}Cd chemical shift correlation experiment (10) in which selective Cd pulses were applied in turn to Cd resonances 6, 7A, and 7B.

Shown at the top of Figure 3 is the HPLC trace obtained for the $^{111}Cd_7$MT-1 sample whose ^{111}Cd NMR spectrum appears at the bottom. Three major components, named MT-1D, A, and B in order of their elution, can be readily separated and then subjected to NMR analysis. The ^{111}Cd NMR spectra of each of these MT-1 isoforms now clearly indicate that each is homogeneous in terms of containing only a single 4-metal cluster "state" giving rise to a single set of 4 resonances as observed with MT-2. The resonances from MT-1A and D have identical chemical shifts, indicating that their 4-metal cluster environments are very similar or identical. Those from MT-1B all have slightly different shifts, especially #7. With this knowledge, it is now possible to rationalize

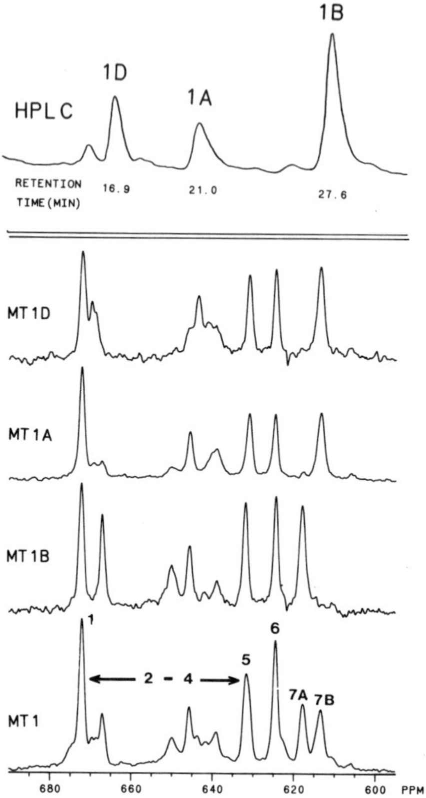

FIGURE 3. ^{111}Cd NMR spectra (106 MHz) of $^{111}Cd_7MT$-1 (bottom) and its three isoform subcomponents separated by reverse phase HPLC. At the top is the HPLC profile obtained using a semi-preparative C4 column (250mm X 10mm) eluted with a 0-20% isopropanol gradient in 50 mM Tris-HCl, pH 7.0.

both the complexity of previous MT-1 specta and their variability, since the spectra would be anticipated to differ if the relative concentrations of the three constituent MT-1 subforms was variable. In fact, we have preliminary evidence that the MT-1D isoform is expressed at higher levels in young rabbits (as were used to obtain the protein whose spectra are in Figure 3) than in adults, where this isoform accounts for only about 5% of the total MT-1. Studies are now underway to document MT-1 isoform profiles as a function of animal development and to determine whether the structural differences among the isoforms confer on them any differences in reactivity in physiologically relevant metal and ligand exchange reactions.

Another significant and unexpected feature of the data in Figure 3 was that all three ostensibly pure MT-1 isoforms exhibit Cd NMR spectra containing more than three resonances from the metals in the 3-metal cluster. In fact, all three isoforms appear to contain two sets of 3-metal cluster signals, possibly arising from two distinct configurational states which interconvert slowly on the NMR timescale. We are in the process of conducting ^{1}H-^{113}Cd correlation experiments similar to those employed in Figure 2 on each MT-1 isoform to confirm this possibility.

Metal Exchange Among Cluster Sites in MT.

We have previously shown using saturation transfer experiments that the Cd ions occupying the 3-metal cluster sites in MT-2 move rapidly among the three sites, but not nearly as rapidly between clusters nor between sites of the 4-metal cluster (7). The spectra in Figure 4 illustrate how these experiments work. The Cd in each of the cluster binding sites is "labeled" by presaturation of its resonance. If exchange of Cd from another binding site into the saturated site occurs on the same time-scale as T_1 relaxation (ca. 0.1-0.7 sec at 106 MHz), a reduction in the intensity of its resonance will be observed. As shown in Figure 4, saturation is transferred to the other two occupants of the 3-metal cluster no matter which of the sites is irradiated (resonances 2-4). A detailed analysis of data such as these indicates that the lifetime of Cd in a particular 3-metal cluster site is only ca. 0.5 sec. at 30° C. Saturation transfer experiments on MT-1 (not shown) give the same result: metal exchange between 3-metal cluster sites is extremely facile.

The mechanism for such a rapid intersite metal transfer reaction might plausibly be either intramolecular (via a series of rapid cysteine ligand exchange reactions) or intermolecular (via bimolecular association between two MT molecules). We have obtained direct evidence for the intermolecular mechanism by conducting saturation transfer studies on solutions containing a mixture of MT-1 and MT-2

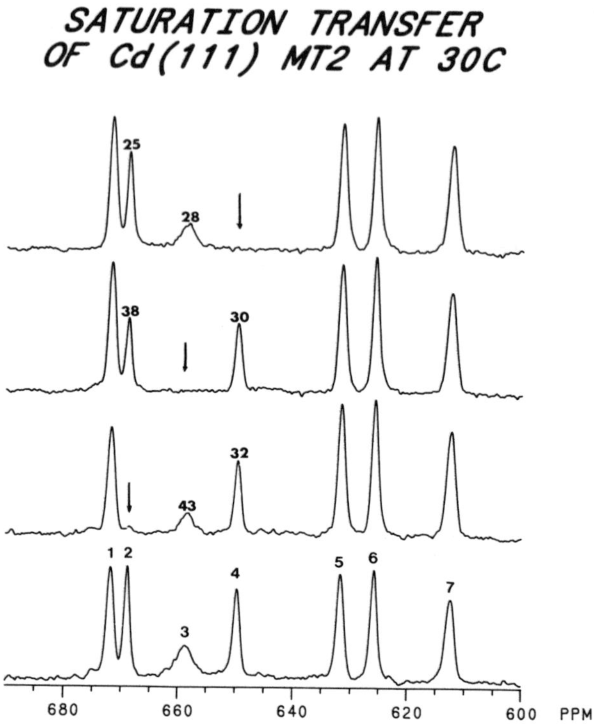

FIGURE 4. ^{111}Cd saturation transfer spectra of CdMT-2 at 106 MHz. The arrows indicate the frequencies at which the saturating pulse was applied and the % reductions in the intensities of the other resonances are given above the peaks.

isoforms. These experiments take advantage of the fact that at least some Cd ions in each isoform have unique resonance positions which do not overlap with those of the other isoform. It is therefore possible with a solution containing a mixture of MT-1 and MT-2 to selectively saturate a 3-metal cluster resonance of one isoform and observe, via intermolecular exchange, transfer of saturation to the 3-metal cluster signals of the other isoform. Figure 5 shows such an experiment in which presaturation was applied selectively to a MT-2 resonance at 658 ppm or a MT-1 resonance at 645 ppm. The results clearly indicate that as a consequence of each

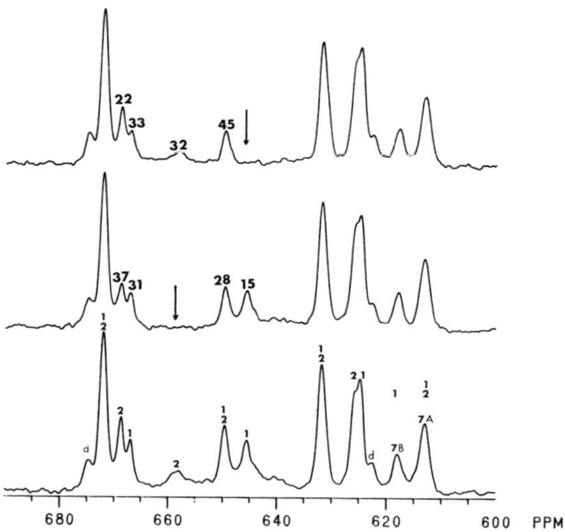

FIGURE 5. ^{111}Cd saturation transfer spectra of a mixture of ^{111}CdMT-1 and ^{111}CdMT-2. On the bottom control spectrum are indicated the isoform origins of the various peaks (1 and 2) as well as two peaks labeled (d) arising from a small amount of dimeric MT in the sample.

preirradiation the 3-metal cluster signals of <u>both</u> isoforms are attenuated by saturation transfer. We believe such remarkably rapid metal transfer between MT molecules is a manifestation of the unique structure of its metal-thiolate clusters, whose primary physiological function appears more and more likely to be intermolecular metal transfer to essential metal-requiring proteins.

REFERENCES

1. Furey WF, Robbins AH, Clancy LL, Winge DR, Wang BC, Stout CD (1986). Crystal Structure of Cd,Zn Metallothionein. Science 231:704.
2. Braun W, Wagner G, Worgotter E, Vasak M, Kagi JHR, Wuthrich K (1986). Polypeptide Fold in the Two Metal Clusters of Metallothionein-2 by Nuclear Magnetic

Resonance in Solution. J Mol Biol 187:125.
3. Vasak M, Worgotter E, Wagner G, Kagi JHR, Wuthrich K (1987). Metal Coordination in Rat Liver Metallothionein-2 Prepared With and Without Reconstitution of the Metal Clusters, and Comparison with Rabbit Liver Metallothionein-2. J Mol Biol 196:711.
4. Armitage IM, Otvos JD (1982). Principles and Applications of ^{113}Cd NMR to Biological Systems. In Berliner LJ, Reuben J (eds): "Biological Magnetic Resonance", New York: Plenum, p. 79.
5. Otvos JD, Armitage IM (1982). Elucidation of Metallothionein Structure by ^{113}Cd NMR. In Bothner-By AA, Glickson JD, Sykes BD (eds) "Biochemical Structure Determination by NMR", New York: Marcel Dekker, p. 65.
6. Nettesheim DG, Engeseth HR, Otvos JD (1985). Products of Metal Exchange Reactions of Metallothionein. Biochemistry 24:6744.
7. Otvos JD, Engeseth HR, Nettesheim DG, Hilt CR (1987). Interprotein Metal Exchange Reactions of Metallothionein. Exp Suppl 52:171.
8. Otvos JD, Engeseth HR, Wehrli S (1985). Preparation and ^{113}Cd NMR Studies of Homogeneous Reconstituted Metallothionein. Biochemistry 24:6735.
9. Hagen KS, Stephan DW, Holm RH (1982). Metal Ion Exchange Reactions in Cage Molecules. Inorg Chem 21:3928.
10. Otvos JD, Engeseth HR, Wehrli S (1985). Multiple Quantum ^{113}Cd-^{1}H Correlation Spectroscopy as a Probe of Metal Coordination Environments in Metalloproteins. J Mag Reson 61:579.

Metal Ion Homeostasis: Molecular Biology and Chemistry, pages 207–225

SOLUTION STRUCTURE DETERMINATION AND PATHWAY OF CLUSTER FORMATION IN METALLOTHIONEIN[1]

Milan Vašák,

Biochemisches Institut der Universität Zürich, Winterthurerstr.190, CH-8057, Zürich, Switzerland

ABSTRACT A comparison of the crystal structure of rat (Zn_2,Cd_5)-MT2 and the NMR solution structure of rabbit $^{113}Cd_7$-MT-2a revealed marked differences in the sequence specific Cd-Cys coordination bonds, even though the handedness of the polypeptide backbone wrapping around both clusters was found to be identical. Since the recently determined solution NMR structure of rat $^{113}Cd_7$-MT-2 showed the same cluster topology as that of rabbit, the differences in primary structure as a cause of the structural anomalies could be ruled out. Heteronuclear $[^{1}H-^{113}Cd]$ NMR studies on biosynthetically ^{113}Cd enriched rat $(Zn_2, ^{113}Cd_5)$-MT-2 established the coexistence of several MT-2 forms in solution with differing relative metal compositions and documented that the major form possessed the same four-metal cluster topology as that found in reconstituted rat $^{113}Cd_7$-MT-2. Moreover, a comparison between the CysS-Cd bond lengths, as revealed by crystal structure and EXAFS studies, showed marked differences. Independently, the ^{113}Cd NMR studies as a function of Cd/apoMT ratios at pH 7.2 and 8.6 established the pH dependent cooperative cluster formation with the Cd_4-cluster being formed first. These results further stressed the dominating role of the four-metal cluster on the overall polypeptide fold. The unequal cluster stability allowed for selective cluster filling with different metal ions.

---

[1]This work was supported by the Swiss National Science Foundation Grant No.3.263-0.85.

INTRODUCTION

Metallothioneins (MTs) represent a class of low molecular weight (6-7 kDa) metal binding proteins widely occurring in nature. MT was first identified and isolated from equine renal cortex almost 30 years ago by Margoshes and Vallee [1] during their search for the protein responsible for the accumulation of cadmium in mammalian kidney. It is still the only protein known which contains this metal. Since then MTs have been isolated from the liver, kidney and other organs of a wide variety of mammalian species and also from several eukaryotic microorganisms, prokaryotes and from plants [2,3]. Despite more than 20 years of intensive investigation, the biological function of this protein is still largely unknown. However, in view of the several metallic constituents of this protein it is believed that MTs play an important role in the metabolism of essential trace metals (Zn,Cu) [4] and in heavy metal detoxification (Cd,Hg)[5]. This conclusion is supported by the increased biosynthesis of this protein in many cells in response to a number of various heavy metal ions, hormones and other agents [6,7].

Because of the wide variation in chemical structure of MTs, a subdivision of these proteins into three classes has been recommended [8]. Class I MTs include all polypeptides whose primary structure is related to that of renal MT. Class II display only very distant evolutionary correspondence to the mammalian forms e.g. MT of yeast, and class III comprises the atypical γ-glutamylcysteinyl units found in plants.

All mammalian MTs (class I) contain a single polypeptide chain of 61 or 62 amino acids of which 20 are cysteines which bind seven divalent metal ions zinc and/or cadmium [9]. Naturally occurring MTs containing monovalent copper, but with higher metal occupation, have also been isolated [2,10]. The common feature of all primary structures of mammalian MTs is the preservation of Cys. Another typical feature is the occurrence of Cys-X-Cys tripeptide sequences, where X stands for any amino acid residue other than Cys. Sequencing studies have established that mammalian MTs display very substantial genetic polymorphism [6,8].

Over the last ten years a number of spectroscopic studies have been conducted on the elucidation of the structure of mammalian MT [9]. Thus, ^{113}Cd NMR [11,12], Co(II) EPR and magnetic circular dichroism [13,14] studies on the respective metalloform of this protein have established the existence of two metal thiolate clusters of three and four metal ions possessing tetrahedral tetrathiolate coordination. Additional support for the two cluster model came from the enzymatic cleavage of this protein and subsequent isolation of the C-terminal cluster domain (cluster A), enfolding four metal ions [15]. Recently, the physicochemical characterization of mammalian MTs was highlighted by the 3D structure determination of two homologous proteins in single crystals by X-ray crystallography [16] and in solution by NMR [17,18]. While both structures agree on the presence of two metal-thiolate clusters confirming a model proposed earlier by Otvos and Armitage [12] they differ markedly in the assignments of the sequence specific cysteine-metal coordination bonds [19].

In this review the strategy applied in the elucidation of the 3D NMR structure of rabbit liver MT-2a in solution as conducted in collaboration with the Wüthrich group (ETH Zürich) is summarized. The recent extension of the NMR work to rat liver MT-2 allows a meaningful comparison of the solution structure with that in single crystal obtained on the same species. Results from our laboratory on studies of the pathway of cluster formation in this protein are also discussed.

SOLUTION STRUCTURE DETERMINATION BY NUCLEAR MAGNETIC RESONANCE

At present NMR is the only technique that allows the determination of the 3D structure of small proteins in solution. 3D structure can be determined by two-dimensional(2D) NMR as was first demonstrated by Wüthrich and coworkers [20]. The strategy developed for the structure determination using 2D ^{1}H NMR can be described as follows: first the non-labile amino acid spin systems are identified based on the J coupling constant that is revealed in a 2D correlation spectrum (COSY) in D_2O. Then the same experiment is carried out in H_2O and the NH and $C^{\alpha}H$ proton pairs within the amino acid residues are assigned. These sets of data afford the through-bond connectivities within the amino acid side chains. In the next step, the NH-αH, NH-NH, and NH-βH proton pairs in adjacent amino acid residues can be assigned due to their dipolar cross-relaxation i.e., through-space connectivities, which are manifested by cross peaks in a 2D cross-relaxation spectrum (NOESY). A combination of the last two steps permits a step by step sequence specific assignments of the backbone protons. In addition, by specific variation of the experimental conditions in the ^{1}H NOESY spectra, a spatial network of proton-proton distances up to ca.5 Å can

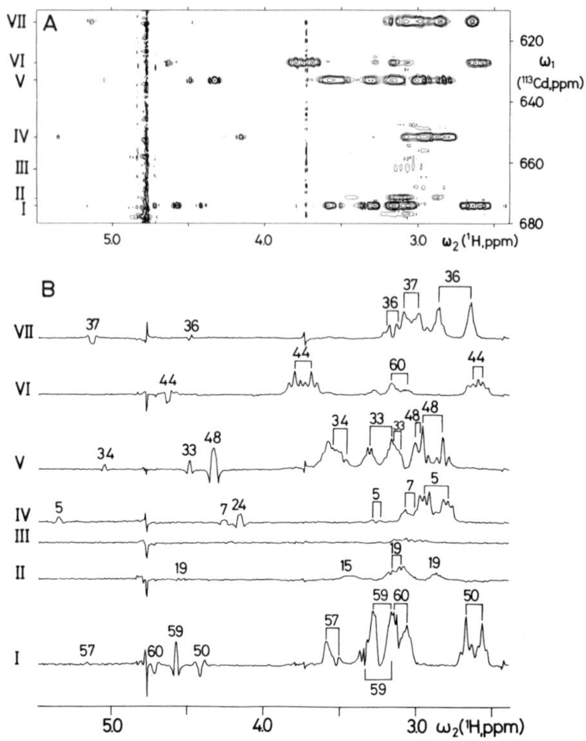

FIGURE 1. Heteronuclear [^{1}H-^{113}Cd] COSY spectrum of rabbit ^{113}Cd$_7$-MT-2a in D$_2$O. (A) contour plot representation. The ^{1}H spectrum is along ω_2, the ^{113}Cd spectrum along ω_1. The position of the seven ^{113}Cd resonances are indicated with roman letters on the left hand site of the spectrum. (B) cross section along ω_2 at the ω_1 positions of the seven ^{113}Cd resonances. The cross-peaks with the cysteine resonances are labelled with the sequence numbers. Labels a and b refer to β-protons with the larger and smaller coupling to the corresponding α-protons, respectively. The size of the heteronuclear ^{1}H-^{113}Cd coupling is indicated with solid brackets. This experiment was designed to emphasize cross-peaks between the ^{113}Cd nuclei and the directly coupled β-protons of the cysteines, whereas relay connectivities to α-protons, which appear between 5.5 and 4.0 ppm, are comparably weak (from Ref.17).

be identified. In the final step the protein structure in

solution is calculated using mainly the through-bond and through-space constraints as input parameters in distance geometry calculations or molecular dynamic calculations [20,21].

Despite the rather low molecular weight of MT (6-7 kDa) the structure determination of rabbit liver MT-2a by NMR was a very difficult task from the start. This was due mainly to the absence of aromatic amino acids i.e.,ring-current shifts, which would give rise to additional signal spread in the spectra, and to the presence of a total of 33 amino acids out of 62 classified as AMX spin-systems i.e.,Cys, Ser, Asp, Asn. The latter made the unambiguous identification of all 20 Cys very difficult. Additional problems arose from the presence of a small amount of isoprotein in the otherwise electrophoretically pure MT sample. Fortunately, the occurrence of Cd in this protein enabled us to take advantage of the existence of the NMR active isotope ^{113}Cd (I=1/2) and NMR inactive ^{112}Cd (I=0). When experimental animals are administrated a cadmium salt, MT containing at most 5 moles of Cd and 2 moles of Zn is isolated. The nonspecific binding of both type of metal ions necessitated a method for protein reconstitution [9,22].

The spin systems of the non-labile protons of the amino acid residues of rabbit Cd_7-MT-2 were elucidated with double-quantum-filtered COSY spectra in D_2O, where all the labile protons had been exchanged against deuterons prior to NMR measurements [23]. A comparison of such spectra of two protein preparations with either ^{112}Cd or ^{113}Cd isotope allowed the identification of the 20 AMX spin systems of the cysteines on the basis of the heteronuclear couplings between ^{113}Cd and cysteine $C^{\beta}H$ protons [24]. Crucial information for the sequential assignments is the identification of the connectivities between the non-labile protons of each amino acid and the corresponding backbone NH's. This information was obtained with COSY and RELAYED-COSY in H_2O [23]. The combination of the above methods enabled the identification of all spin systems present. At first a number of spurious resonances were identified [25], which were subsequently removed by high-performance liquid chromatography (HPLC) sample purification [22]. HPLC showed the electrophoretically homogenous $^{113}Cd_7$-MT-2 form to contain an additional isoform, designated MT-2b (ca. 10%), as well as the major isoform MT-2a.

Sequence-specific resonance assignments were obtained by searching for nuclear Overhauser enhancements between the NH-resonances of residue i+1 and NH, $C^{\alpha}H$ or $C^{\beta}H$ of residue i in the NOESY spectra [25].

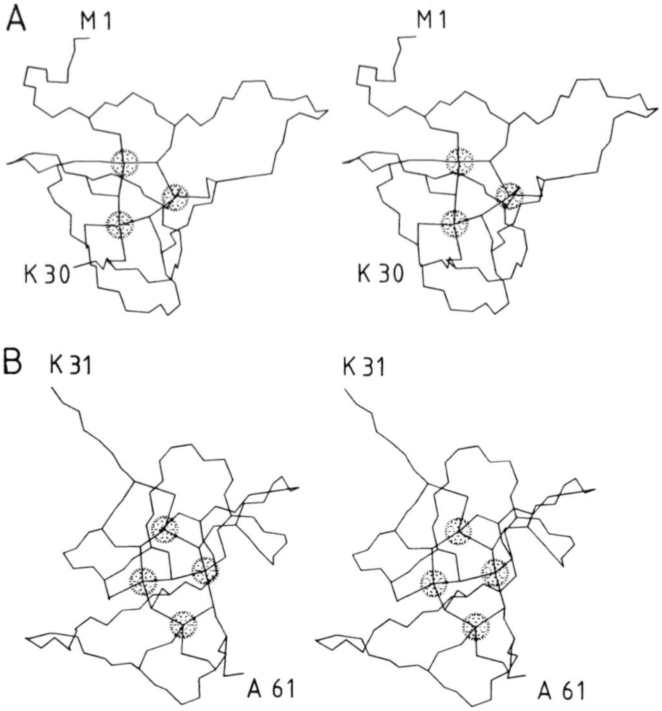

FIGURE 2. Stereo view of the polypeptide backbone and the heavy atom side chain positions in the 3-metal (A) and 4-metal domain (B) structures of rabbit MT-2a each with the lowest residual error function. The Cd(II) ions are represented as a dotted sphere of radius 0.9 Å. For the heavy atom side chain positions in each domain see Fig.3 (from Ref.27).

In view of the presence of two metal-thiolate clusters in this protein the determination of the cysteine-metal bonds was crucial for the solution of the 3D structure. The sequence-specific cysteine-^{113}Cd coordination bonds were obtained by heteronuclear COSY experiments (Fig.1) [17]. The ^{1}H spectrum appears along ω_2-axis, the ^{113}Cd spectrum is along ω_1-axis. This experiment provided the cross-peaks between ^{113}Cd and directly coupled C^{β}H resonances of the cysteine spin systems located between 2.5 and 4.0 ppm. In an independent experiment, employing a relayed coherence transfer step, the cross-peaks between ^{113}Cd resonances and C^{α}H protons of the cysteines were obtained [17]. The identification of the directly coupled C^{β}H and the more remote

$C^{\alpha}H$ protons, on the basis of their chemical shifts, allowed the unambiguous assignment of the cysteine-metal coordinative bonds (Fig.3, top). These connectivities provided 28 constraints as to the polypeptide fold around the metal clusters. Further distance constraints were obtained from NOESY spectra. A combination of both types of data were used as input data into preliminary distance geometry calculations, which yielded the overall polypeptide backbone fold [18]. This structure revealed, short stretches of 3_{10}-helices and a novel secondary structure element designated as half-turns [26]. Recently, the 3D structure was calculated with the program DISMAN [27] using a new more extensive set of NMR data consisting of distance constraints from measurements of NOE's and dihedral angle constraints derived from both coupling constants and NOE's. Since in both data sets no interdomain NOE's have been found the structure calculations were performed separately for the α-domain and the β-domain. A stereo view of the polypeptide backbone and the heavy atom positions of the 9 cysteines in the 3-metal domain structure, and of the 11 cysteines in the 4-metal cluster domain structure, both with lowest residual error function is shown in Fig.2 [27]. The overall chirality of the polypeptide fold was found to be right-handed for the β-domain and left-handed for the α-domain. The three-metal cluster has a boat form, the four-metal cluster a boat-boat form.

CRYSTAL STRUCTURE OF ZN,CD-METALLOTHIONEIN

A detailed account of the crystal structure elucidation is reported in this volume. However, in order to be able to compare both the solution and the crystal structure a brief summary of the latter work is also given.

In contrast to the solution structure which was solved for reconstituted rabbit MT the crystal structure was determined on native rat MT. Isoform MT-2 was isolated from liver of rats which had been subcutaneously administrated of a $CdCl_2$ solution in order to enhance the MT biosynthesis, and subsequently crystallized from a solution of potassium phosphate and sodium formate. Dissolved single crystals contained 5 moles of cadmium and 2 moles of zinc per mole of protein and had the same amino acid composition as the native protein [28]. The crystal structure of (Zn_2,Cd_5)-MT2 was solved at 2.3 Å resolution [16]. Initially, the Cd_4-cluster structure was solved by direct methods. The location of a fifth Cd site was determined relative to the fixed position of the Cd_4-cluster. The five refined Cd sites were used as a basis for iterative single wavelength anomalous scattering (ISAS) calculations. In this structure the N-terminal β-domain of residues 1-29 enfolded a three metal cluster of 1 Cd and 2 Zn coordinated by six terminal and three bridging cysteine thiolates. The C-terminal α-domain of residues 30-61 enfolded the four metal cluster of 4 Cd coordinated by six terminal and five bridging cysteine thiolates (see Fig.4 bottom). All seven metal sites have distorted tetrahedral geometry. Both domains appear to be globular and are linked at residues 30 to 31. The rather limited contact between the α- and β-domain led to the description of the polypeptide fold as two turns wrap around each cluster. In this respect, the spiral of the α-domain is said to be left-handed and the spiral of the β-domain is right-handed [16].

COMPARISON OF THE SOLUTION AND THE CRYSTAL STRUCTURE

Although both the solution structure of rabbit $^{113}Cd_7$-MT-2a and the crystal structure of rat (Zn_2,Cd_5)-MT-2 agree on the organization of both metal-thiolate clusters and reveal the same handedness of the polypeptide chain in two domains, they differ considerably in the sequence specific cysteine-metal coordination bonds. Thus, out of a total of 28 such bonds, only 7 are identical. The following reasons for these differences have been suggested [22]: (1) Differences in the primary structures i.e. there are 10 amino acid substitutions and one insertion in rabbit MT-2a in comparison to rat MT-2; (2) The crystal structure was determined using biosynthetic (Zn_2,Cd_5)-MT-2 while the solution NMR structure was solved on reconstituted $^{113}Cd_7$-MT-2a and (3) Rat MT-2 was crystallized from ca. 4.0 M sodium formate and 0.4 M potassium phosphate at pH 7.5, whereas rabbit MT-2a was studied in 0.02 M Tris.HCl and 0.02 M KCl at pH 7.0.

To clarify whether the differences in the primary structures or different metal populations could be related to different metal cluster topologies in a new round of NMR studies the NMR solution structure of reconstituted rat MT-2 was elucidated. Independently, the effect of the zinc ions on cluster topologies in biosynthetically ^{113}Cd enriched rat $(Zn_2,{}^{113}Cd_5)$-MT2 was investigated [22]. The ^{113}Cd enriched form was prepared in the same manner as that used for crystal growth. In addition, the effects of metal reconstitution on the cluster topology was examined.

The solution structure of rat $^{113}Cd_7$-MT-2 [29,30] yielded

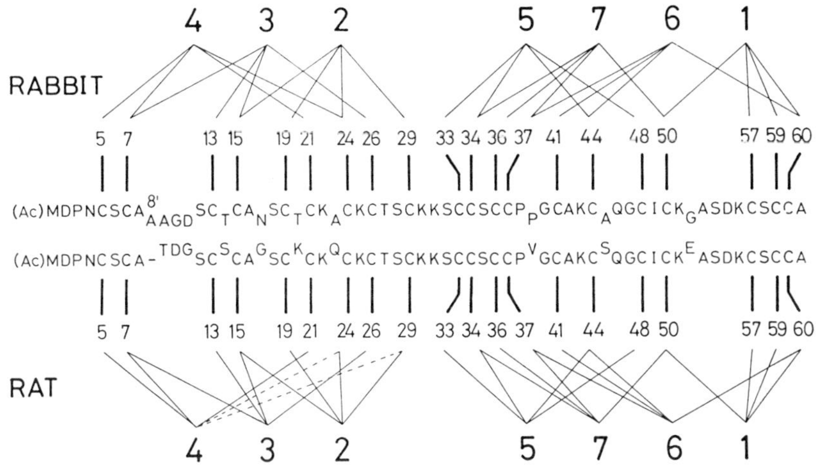

FIGURE 3. Amino acid sequences and cysteine-Cd connectivities in reconstituted rabbit $^{113}Cd_7$-MT-2a and rat $^{113}Cd_7$-MT-2. Amino acid substitutions are indicated by displacement of the one-letter codes above and below the lines. Rabbit MT-2a contains 62 amino acid residues, rat MT-2 61 residues. Since the insertion in rabbit MT-2a is labelled as A8', the numeration of the cysteine residues is identical throughout (from Ref.22).

Cys-Cd connectivities identical with those found in rabbit $^{113}Cd_7$-MT-2 (Fig.3). Also the overall polypeptide fold in both solution structures was found to be similar [30]. Thus the identity of the cluster topologies in both reconstituted rat and rabbit MT forms excludes the differences in primary structure as the cause of the observed differences between the solution and the crystal structures. A comparison of the cluster topologies in rat MT-2 as revealed by NMR and x-ray crystallography is illustrated in Fig.4.

The possible influence of metal reconstitution on the cluster topology was examined using the rabbit MT-2a form [22]. The reconstitution procedure employed in the NMR studies relies on the preparation of metal-free protein at low pH, followed by metal addition and pH adjustment to neutrality, under strictly anaerobic conditions [22]. Independently, in samples of biosynthetic rabbit MT-2a, which usually contain 5 Cd equivs and 2 equivs of Zn, the zinc

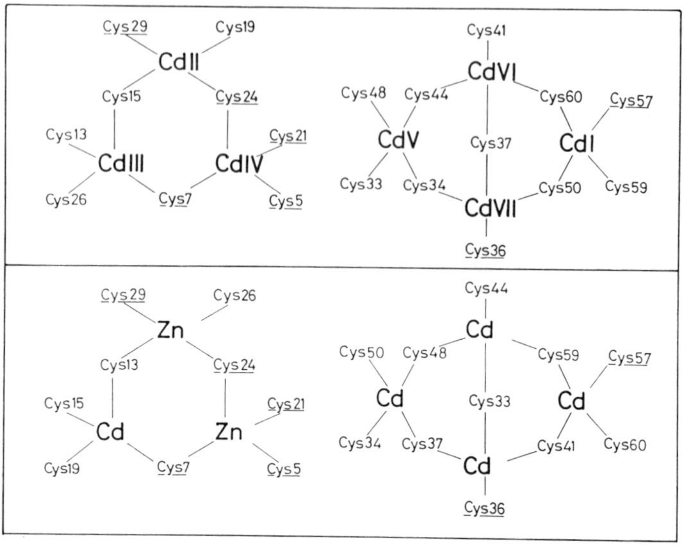

FIGURE 4. Comparison of the cluster topologies in the solution NMR structures of rabbit $^{113}Cd_7$-MT-2a [17,27] and rat $^{113}Cd_7$-MT-2 [22,30] (top) with that in the crystal structure of rat (Zn_2,Cd_5)-MT-2 [16] (bottom). The numbers at the Cd atoms refer to resonance positions in the ^{113}Cd NMR spectra (see Figs. 5a and 5b).

ions were exchanged by excess of cadmium ions at neutral pH. The double-quantum filtered ^1H COSY spectra of samples prepared by either method are indistinguishable, thus indicating that identical cluster architectures prevail [22].

Another crucial point was to investigate the influence of the zinc ions on the cluster topology in mixed rat (Zn_2,Cd_5)-MT-2, a species which was also used in crystallographic studies. The 1D ^{113}Cd NMR spectrum of the biosynthetically enriched $(Zn_2,{}^{113}Cd_5)$-MT-2 form, isolated from liver of rats upon the administration of ^{113}Cd salt, is shown in Fig. 5c. The corresponding ^{113}Cd NMR spectra of Cd reconstituted forms of rat and rabbit MT are illustrated in Fig. 5b and 5c, respectively [22]. Although all samples were homogenous as shown by HPLC, the biosynthetic sample displays at least 15 ^{113}Cd resonances, indicating the co-existence of multiple forms of the protein with different distributions of zinc and cadmium ions in the individual sites of the metal cluster [22]. A similar complexity of the ^{113}Cd NMR pattern has also been reported for biosynthetic rabbit protein [31]. A comparison of Fig. 5b and 5c revealed that the resonances 1,5,6 and 7 of the reconstituted rat MT-2, which correspond to the four-metal cluster, are also present in biosynthetic (Zn_2,$^{113}Cd_5$)-MT-2 with identical chemical shifts and multiplet structures. In contrast, at chemical shifts of the ^{113}Cd signals 3 and 4 in reconstituted rat $^{113}Cd_7$-MT-2 which originate from the three-metal cluster the biosynthetic form contains no resonance lines. This result indicates that there is no detectable protein population in biosynthetic (Zn_2,-$^{113}Cd_5$)-MT-2 in which the three-metal cluster is occupied with the Cd ions and has the same conformation as in the reconstituted protein. A comparison of the heteronuclear [^1H-^{113}Cd] COSY spectra of the reconstituted rabbit and rat MTs with that of biosynthetic rat MT-2 [22] revealed that the ^{113}Cd resonances 1,5,6 and 7 in (Zn_2,$^{113}Cd_5$)-MT-2 (Fig. 5c) correspond to the cluster topology of Cd_4-cluster in reconstituted proteins [22]. That the assigned resonances represent the major protein population in biosynthetic (Zn_2,-$^{113}Cd_5$)-MT-2 has been concluded based on the integration of these ^{113}Cd signals. The ^{113}Cd resonances 1,5,6 and 7 (Fig.5c) accounted for ca. 35% of the total ^{113}Cd resonance intensity detected in the biosynthetic protein as compared to ca. 57% intensity for the corresponding NMR resonances in the reconstituted protein. Additional resonances in Fig. 5c, marked with primed numbers, have been attributed tentatively to the corresponding ^{113}Cd signals based on the similarities in the heteronuclear [^1H-^{113}Cd] COSY patterns. These resonances may originate from different Cd_4-cluster conformations or equally likely from cluster population(s) where both zinc and cadmium ions are concomitantly present [22].

Overall, on the basis of these NMR studies it can be concluded that the metal coordination of the four-cluster in the crystal structure does not correspond to that of the major species in aqueous solution of biosynthetic rat MT-2. Instead, the metal coordination coincides with that found in reconstituted rat and rabbit MTs [22].

So far the structural differences between the crystal structure and the structure in solution, as revealed by NMR spectroscopy, has been discussed. Another highly informative spectroscopic technique which has been successfully applied to various metalloforms of mammalian MTs is X-ray absorption fine structure spectroscopy (EXAFS) [32]. This technique allowed us to probe the local environment (>10 Å) of the protein metal binding sites not only in solid state but also in solution. The metal-ligand coordination bonds can be

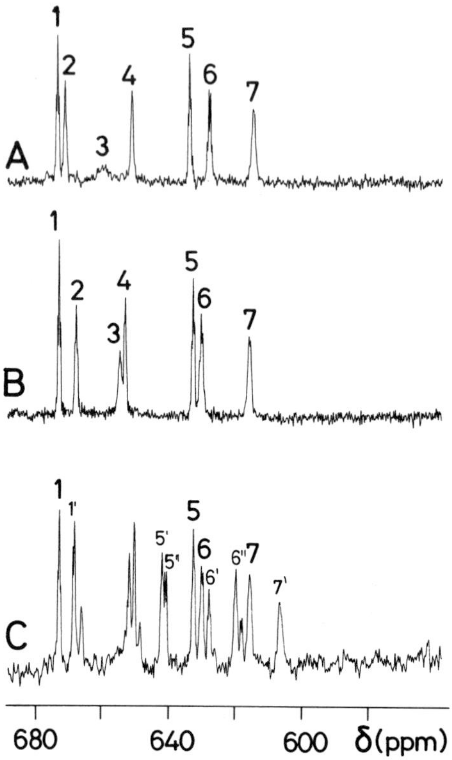

FIGURE 5. (a) ^{1}H decoupled 80 MHz 1D ^{113}Cd NMR spectra of reconstituted rabbit $^{113}Cd_7$-MT-2a, (b) reconstituted rat $^{113}Cd_7$-MT-2 and (c) biosynthetic rat (Zn_2, $^{113}Cd_5$)-MT-2. The numbering of the resonances in (a) and (b) refer to the position in the cluster (see Fig.4 top). The signals 1,5,6 and 7 of (c) have exactly the same chemical shifts as the corresponding signals in (b). The signals marked with primed numbers are discussed in the text (from Ref.22)

determined to an accuracy greater than 0.02 Å [32].

The first data on the crystallographic refinements of rat (Zn_2,Cd_5)-MT-2 have been reported at 2.0 Å resolution [33]. This model differs from the starting model in terms of both the geometry of the metal-thiolate clusters and the positions of amino acid residues, but not in the assignments of the cysteine-metal coordination bonds [33]. However, in the latest model significantly differing metal-sulfur distances were obtained. For example, the average Cd-S distance of the terminal ligands was found to be 3.2 Å while that of the bridging ligands is 2.7 Å. These values are inconsistent with EXAFS data obtained with reconstituted Cd_7-MT and biosynthetic (Zn_2,Cd_5)-MT from both species (rat and rabbit) which showed the average Cd-S distances to be 2.5 Å in solution and in the freeze-dried state. It is interesting to note that the latter value is comparable with distances obtained with inorganic Cd-thiolate clusters [32].

Thus, the solution structure determination of rat MT-2 along with the comparative ^1H COSY studies on various preparations of rabbit MT-2a have excluded the differences in primary structure and the procedure of protein reconstitution as a possible cause for the observed differences in cluster topology in crystal and in solution. Futhermore, the identification of the sequence specific cysteine-Cd coordination bonds provided by heteronuclear [^1H-^{113}Cd] COSY experiments indicated that the biosynthetic $(Zn_2,{}^{113}Cd_5)$-MT-2 contains a Cd_4-cluster of the same topology as found in reconstituted ^{113}Cd forms of both rat and rabbit MTs.

The substantially different ionic strength used in the crystal growth (see above) [28] could play a major role in the preferential crystallization of a rat MT-2 form with a different cluster topology. Note that at least 15 ^{113}Cd signals were observed in the ^{113}Cd NMR spectrum of biosynthetic $(Zn_2,{}^{113}Cd_5)$-MT-2 out of which only the 4 major ^{113}Cd resonances have been unambiguously assigned (see above). The extent to which the crystalline packing forces influence the Cd-S bond lengths and may account for the disagreement between the EXAFS and the crystal structure data is not yet known. Another possible reason for the structural anomalies could be cluster dynamics (see below). Although the mechanism of the intramolecular metal exchange is still unknown a stabilization of a protein structure with different cluster topology in the course of the crystal growth is conceivable. However, since the crystallographic refinement is still in progress [33] a better comparison of the solution and the crystal structure must await a full set of crystallographic data.

DYNAMIC METAL-THIOLATE CLUSTER STRUCTURE IN METALLOTHIONEIN

As previously mentioned the metal-thiolate clusters in this protein are dynamic entities. The evidence for dynamic processes has been provided by ^{113}Cd NMR spectroscopy. Thus, in all ^{113}Cd NMR studies performed so far on the fully metal occupied mammalian $^{113}Cd_7$-MTs the ^{113}Cd signals of the three metal cluster (cluster B) have been found to be of lower intensity (ca.20%) than those of the four-metal cluster (cluster A) resonances [12,34,17,35] (see also Fig. 5a and 5b). It was this effect which prompted detailed ^{113}Cd NMR studies of rabbit liver $^{113}Cd_7$-MT and lead to the suggestion of a conformational flexibility of the cluster structures [34]. Direct evidence for dynamic processes within cluster B has been provided by ^{113}Cd NMR saturation transfer experiments which indicated intramolecular metal exchange [36]. In this connection the temperature behavior of the cluster B resonances between 6 and 44 ^{o}C is relevant [31]. Although at 6 ^{o}C resonance 3 was broadened almost beyond detection its spin-coupling interactions with resonances 2 and 4 remained resolved throughout the temperature range. This effect enabled the estimation of a lower limit of residence lifetimes (>0.01 s) of the ^{113}Cd ions in their respective sites in cluster B. This value excludes a rapid dissociation and reassociation process of ^{113}Cd ions from the cluster B domain [31]. The absence of such NMR detectable processes within cluster A may account for the observed differences in intensity. That comparable processes exist also in cluster A, however with lifetimes in the order of hours, has been suggested based on exchange studies with various Cd isotopes [36].

Recently, a $(Zn_1, {}^{113}Cd_6)$-MT-2 form was prepared in which, as shown by ^{113}Cd NMR, the zinc ions were bound exclusively to the cluster B thus giving rise to protein populations of various mixed $(Zn,Cd)_3$-clusters [37]. Comparison of the overall intensity of the cluster B resonances per Cd in the $(Zn_1, {}^{113}Cd_6)$-MT-2 spectrum with those of the $^{113}Cd_7$-MT-2 form revealed ca.15% intensity gain of the former. This result suggested a stabilization effect of the bound zinc ions upon the previously established intramolecular ^{113}Cd exchange process within the B cluster. It may be noted that attempts to crystallize the metal homogenous Cd_7-MT-2 form have so far been unsuccessful [16].

Structural mobility of the clusters has been deduced also from EXAFS studies of various metalloforms of this protein [38]. While EXAFS spectra revealed back-scattering from thiolate ligands (first shell), substantial movements within the cluster are believed to preclude the observation of the back-scattering from the neighboring metals (second shell) [38]. Second shell back-scattering has been recently observed in Zn_7-MT but only at low temperature (77 K) [32]. Similarly, in perturbed angular correlation of γ-ray (PAC) spectra of the ^{111m}Cd labeled rabbit MT the damping of the major oscillatory component was attributed to inordinately large charge fluctuations in the immediate environment of the ^{111m}Cd nuclei [39].

Although the details of cluster dynamics in MT is as yet unknown substantial structural fluctuations have also been observed in ^{1}H NMR studies of synthetic clusters of Cd(II)-, Zn(II)-, Co(II)- and Fe(II)-thiolates with adamantane type geometry [40]. These results were interpreted in terms of a non-rigid cluster model in which the metal-thiolate bonds are temporarily broken and reformed, giving rise to a number of interchanging cluster substates of comparable stability and structure.

PATHWAY OF CLUSTER FORMATION IN METALLOTHIONEIN

Information on cluster formation in MT has been forthcoming from various chemical and spectroscopic studies. Thus, by monitoring the intensity of the Co(II) EPR signal of Co(II)-MT at pH 8.6 as a function of increasing Co(II) to apoMT ratios it was found that the first four Co(II) equiv are incorporated into separate sites [14]. Above four Co(II) equiv cluster formation sets in as manifested by antiferromagnetic coupling through the bridging thiolates [14]. A similar two-step cluster formation processes under comparable conditions have been observed in the titration of apoMT with Cd(II) ions as monitored by circular dichroism [41] and by differential chemical modification of cysteine residues [42]. In contrast, from analysis of the isolated Cd(II) containing peptides obtained upon the enzymatic digestion of the partially metal-occupied protein, carried out at pH 7.5, it was inferred that each cluster is formed in a cooperative manner [43]. It should be noted that the latter experiments entailed a considerable amount of chemical separation. In order to support their conclusion, the same group [44] performed ion-exchange chromatographic studies at pH 7.0 on apoMT to which one to four mol equiv of Cd(II) ions were added in the absence of the protease. The chromatographic profiles showed an increasing recovery of a charged species containing four metal ions apparently confined to cluster A.

This apparent discrepancy in the pathway of cluster formation has now been reconciled by the ^{113}Cd titration studies of rabbit apoMT-2 in which the formation of metal thiolate clusters was followed by ^{113}Cd NMR measurements both at pH 7.2 and 8.6 [45]. The chemical shifts of the ^{113}Cd resonances

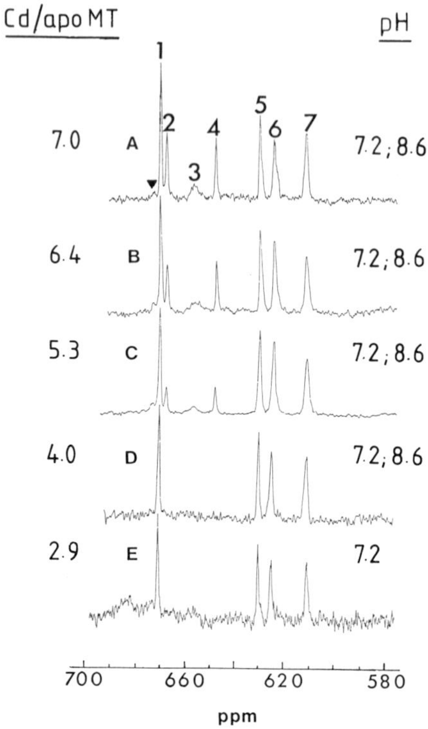

FIGURE 6. ^1H decoupled ^{113}Cd NMR spectra of rabbit ^{113}Cd-MT-2 as a function of ^{113}Cd to apoMT-2 ratio obtained at pH 7.2 and 8.6, except for E. Since the spectra A,B and C were virtually unchanged at both pH values, only spectra at pH 7.2 are shown. The numbering of the resonances in (A) refer to the position in the cluster (from Ref.45).

developing in the course of apoMT-2 titration revealed a clear pH dependence of the cluster formation. Thus, at pH 7.2 and at low metal occupancy (<4), the emergence of the four metal cluster resonances 1,5,6 and 7 (Fig. 6E) was observed. Further addition of $^{113}Cd(II)$ ions generated all the resonances of the three-metal cluster (resonances 2,3 and 4) in succession (Fig. 6B and 6C). In contrast, similar studies at pH 8.6, at low metal occupancy (<4), gave at first rise to a broad NMR signal centered at 688 ppm only [45]. When exactly 4 $^{113}Cd(II)$ equiv were bound to apoMT the ^{113}Cd NMR spectrum changed to the characteristic spectrum of cluster A and further addition of $^{113}Cd(II)$ ions again led to the successive generation of the cluster B resonances (Fig. 6B and 6C). These results demonstrated that the formation of metal-thiolate clusters in MT is sequential and cooperative at physiological pH with the C-terminal cluster (A) being formed first. At elevated pH and below a metal to protein ratio of 4, an apparently entirely different pathway is followed until sufficient Cd(II) is bound to generate the four-metal cluster. Once the four-metal cluster is fully metal saturated, the subsequent addition of ^{113}Cd ions generates cluster B in a cooperative and pH independent manner. The observed pH dependence of the cluster formation in MT has been rationalized in terms of the different degree of deprotonation of the cysteine residues ($pK_a \approx 8.9$), i.e. by the difference in Gibbs' free energy required to bind Cd(II) ions to the thiolate ligands [45]. Moreover, these results stress the determining role of the cluster A domain on the overall protein fold. That the established pH dependence of the cluster formation is not a property of the Cd(II)-MT derivative only but is a general feature of this protein, could be demonstrated by the examination of pH dependence upon cluster formation in Co(II)-MT. Preliminary Co(II) EPR studies performed on the sample of partially metal-occupied $Co(II)_3$-MT at pH 7.2 and 8.6 showed considerable antiferromagnetic coupling i.e., cluster formation, occurring in the former case only (Good & Vasak, unpublished).

In view of the above results it would appear that under appropriate conditions both clusters can be filled independently. Indeed, in the preliminary note [46] the successful preparation of $(Co(II)_3, {}^{113}Cd_4)$-MT has been reported. As shown by ^{113}Cd NMR, electronic absorption and magnetic circular dichroism data, the Co(II) ions selectively occupy cluster B and the Cd(II) ions occupy cluster A.

CONCLUSION

In this review new data on the structure elucidation of mammalian MTs containing divalent metal ions are summarized. Since the published refinement of the crystal structure does not allow for substantial changes in cluster topology it appears that the differences between the solution and the crystal structure will remain unsettled. Hence, a determination of the crystal structure of the reconstituted Cd_7-MT-2 forms or the structure of the biosynthetically ^{113}Cd enriched rat $(Zn_2, {}^{113}Cd_5)$-MT-2, a protein form which subsequent to crystallographic analysis could be studied in solution by NMR, could help greatly to clarify the existing structural differences.

REFERENCES

1. Margoshes M, Vallee BL: J Amer Chem Soc 79:4813,1957.
2. Kägi JHR, Nordberg M (eds): "Metallothionein," Birkhäuser, Basel, 1979.
3. Kägi JHR, Kojima Y (eds): "Proc Int Meet Metallothionein 2nd," Birkhäuser, Basel, N.Y., 1987.
4. Bremner I: In "Proc Int Meet Metallothionein 2nd", Kägi JHR, Kojima Y (eds): Birkhäuser, Basel, N.Y., 1987, p 81.
5. Webb M: In "Proc Int Meet Metallothionein 2nd", Kägi JHR, Kojima Y (eds): Birkhäuser, Basel, N.Y.,1987, p 109.
6. Palmiter RD: In "Proc Int Meet Metallothionein 2nd", Kägi JHR, Kojima Y (eds): Birkhäuser, Basel, N.Y., 1987, p 63.
7. Hamer DH: Ann Rev Biochem 55: 913,1986
8. Kägi JHR, Kojima Y: In "Proc Int Meet Metallothionein 2nd", Kägi JHR, Kojima Y (eds): Birkhäuser, Basel, N.Y., 1987, p 25.
9. Vašák M, Kägi JHR: In "Met Ions Biol Systems", Sigel H (ed): vol.15, Marcell Dekker, N.Y., 1983, p 213.
10. Winge DR: In "Proc Int Meet Metallothionein 2nd", Kägi, JHR, Kojima Y (eds): Birkhäuser, Basel, N.Y.,1987, p 213.
11. Otvos JD, Armitage IM: J Amer Chem Soc 101:7734,1979
12. Otvos JD, Armitage IM: Proc Natl Acad Sci USA 77:7094, 1980.
13. Vašák M: J Am Chem Soc 102:3953, 1980.
14. Vašák M, Kägi JHR: Proc Natl Acad Sci USA 78:6709, 1981.
15. Winge DR, Miklossy KA: J Biol Chem 257:3471, 1982.
16. Furey WF, Robbins AH, Clancy LL, Winge DR, Wang BC, Stout CD: Science (Washington, D.C.) 231:704, 1986.
17. Frey MH, Wagner G, Vašák M, Sørensen OW, Neuhaus D, Wörgötter E, Kägi JHR, Ernst RE, Wüthrich K: J Am Chem Soc 107: 6847,1986.
18. Braun W, Wagner G, Wörgötter E, Vašák M, Kägi JHR, Wüthrich K: J Mol Biol 187:125, 1986.
19. Wagner G, Frey MH, Neuhaus D, Wörgötter E, Vašák M, Kägi JHR, Wüthrich K: In "Pro Int Meet Metallothionein 2nd", Kägi JHR, Kojima Y (eds): Birkhäuser, Basel, N.Y.,1987, p 149.
20. Wüthrich K: "NMR of proteins and nucleic acids", Wiley-Interscience, N.Y., 1986.

21. Braun W: Quart Rev Biophys 19:115,1987.
22. Vašák M, Wörgötter E, Wagner G, Kägi JHR, Wüthrich K: J Mol Biol 196:711,1987.
23. Neuhaus D, Wagner G, Vašák M, Kägi JHR, Wüthrich K: Eur J Biochem 151:257,1985.
24. Neuhaus D, Wagner G, Vašák M, Kägi JHR, Wüthrich K: Eur J Biochem 143:659, 1984.
25. Wagner G, Neuhaus D, Vašák M, Kägi JHR, Wüthrich K: Eur J Biochem 157:275,1986.
26. Wagner G, Neuhaus D, Wörgötter E, Vašák M, Kägi JHR, Wüthrich K: J Mol Biol 187:131,1986.
27. Arseniev A, Schultze P, Wörgötter E, Braun W, Wagner G, Vašák M, Kägi JHR, Wüthrich K: J Mol Biol 1988, in press.
28. Melis KA, Carter DC, Stout CD, Winge DR: J Biol Chem 208: 6255,1983.
29. Wörgötter E, Wagner G, Vašák M, Kägi JHR, Wüthrich K: Eur J Biochem 167:457,1987.
30. Schulze P, Wörgötter E, Braun W, Wagner G, Vašák M, Kägi JHR, Wüthrich K: J Mol Biol 1988, in press.
31. Nettesheim DG, Engeseth HR, Otvos JD: Biochemistry 24:6744,1985.
32. Hasnain S: In "Topics in Current Chem.," Springer-Verlag 1988, in press.
33. Collett SA, Stout CD: Recl Trav Chim Pays-Bas 106:182, 1987.
34. Vašák M, Nicholson JK, Hawkes GE, Sadler PJ: Biochemistry 24:740, 1985.
35. Otvos JD, Engeseth HR, Wehrli S: Biochemistry 24:6735, 1985.
36. Otvos JD, Engeseth HR, Nettesheim DG, Hilt CR: In "Proc Int Meet Metallothionein 2nd", Kägi JHR, Kojima Y (eds): Birkhäuser, Basel, N.Y., 1987, p 171.
37. Vazquez F, Vašák M: 1988, submitted.
38. Vašák M: Environ Health Perspect 65:193,1986.
39. Vašák M, Bauer R: J Amer Chem Soc 104:3236,1982.
40. Hagen K, Stephan DW, Holm RH: Inorg Chem 21:3928,1982.
41. Willner H, Vašák M, Kägi JHR: Biochemistry 26:6287,1987.
42. Bernhard WR, Vašák M, Kägi JHR: Biochemistry 25:1975, 1986.
43. Nielson KB, Winge DR: J Biol Chem 258:13063,1983.
44. Byrd J, Winge DR: Arch Biochem Biophys 250:233,1986.
45. Good M, Hollenstein R, Sadler PJ, Vašák M: Biochemistry 1988,in press.
46. Vašák M, Good M: Recl Trav Chim Pays-Bas 106:184, 1987.

Metal Ion Homeostasis: Molecular Biology and Chemistry, pages 227–235

NEUROSPORA COPPER METALLOTHIONEIN: MOLECULAR STRUCTURE AND REGULATION OF GENE EXPRESSION[1]

Karl Münger,[2] Ursula A. Germann,[3] Mariano Beltramini,[4] Urs Kupper, and Konrad Lerch

Biochemisches Institut der Universität Zürich
Winterthurerstrasse 190
CH-8057 Zürich, Switzerland

ABSTRACT The structure of the copper complex in the low-molecular-weight metallothionein (MT) from the fungus *Neurospora crassa* has been investigated by absorption, circular dichroism, EPR, X-ray absorption edge and EXAFS spectroscopy. The data are most consistent with a three- or four coordinate geometry around the Cu indicating a compact Cu(I)-thiolate cluster $(Cu(I)_6RS_7)^-$. In agreement with the known amino acid sequence of *N. crassa* MT, the gene codes for a polypeptide of 26 amino acids. The promoter region of the gene is devoid of sequences related to the metal regulatory elements of mammalian MT genes. The expression of the *N. crassa* MT gene is strictly copper dependent and is regulated at the transcriptional level. The rapid induction by copper ions of MT mRNA and MT, together with the remarkable stability of Cu-MT intracellularly, suggests that this protein serves an important homeostatic role in the copper metabolism of this fungus.

---

[1]This work was supported by Schweizerischer Nationalfonds grant 3.236-0.85 and by the Kanton of Zürich
[2]Present address: Laboratory of Molecular Biology, NCI, NIH, Bethesda, MD 20892
[3]Present address: Laboratory of Tumor Virus Biology, NCI, NIH, Bethesda, MD 20892
[4]Present address: Department of Biology, University of Padova, Italy

INTRODUCTION

Metallothioneins are low-molecular-weight, cysteine-rich proteins, binding high amounts of metal ions such as Zn, Cd, and Cu (1). Mammalian MTs have a typical chain length of 61 amino acid residues, containing a total of 20 cysteine residues. They bind seven IIB metal ions in the form of two distinct clusters with each ion tetrahedrally coordinated to four cysteinyl residues (2-4). The biosynthesis of these proteins is transcriptionally regulated by heavy metal ions, glucocorticoids and a number of other factors (5).

In contrast to mammalian MTs, fungal MTs contain exlusively copper (6). Cu-MTs have been isolated from yeast (7), *Agaricus bisporus* (8) and *Neurospora crassa* (9). *N. crassa* MT consists of only 25 amino acids and binds 6 Cu ions to seven cysteine residues. Its amino acid sequence shows a striking similarity to those of the N-terminal region of mammalian MTs.

In this paper the spectroscopic properties of *N. crassa* Cu-MT will be presented and compared to those of different Cu(I)-thiolate model complexes. In addition, the regulation of expression of the *N. crassa* MT gene under various growth conditions will be discussed.

SPECTROSCOPIC PROPERTIES

The spectroscopic properties of *N. crassa* Cu-MT arise predominantly from Cu(I)-thiolate complexation. This is clearly borne out of a comparison of the absorption and circular dichroism spectra of freshly isolated Cu-MT and the apoprotein. In agreement with the absence of aromatic residues, the spectra of the apoform are negligible above 250 nm with features in the far UV attributable to amide transitions of the peptide backbone. Binding of Cu(I) intensifies the far UV spectrum, introduces a characteristic shoulder at 250 nm with the broad red-edge extending to 400 nm. At least five Cotton maxima are resolved in the circular dichroism spectrum of Cu-MT in the range of 220-400 nm (10).

N. crassa Cu-MT is characterized by an unusual luminescence upon excitation in the UV (12). The luminescence disappears when the metal is removed from the protein by protonation and is absent in the air-oxidized protein. Hence, the luminescence can be considered as an inherent property of the Cu(I)-thiolate luminophore of the protein.

TABLE 1
EMISSIVE PROPERTIES OF NEUROSPORA CU-MT AND CU(I)-MERCAPTIDE COMPLEXES [a]

Sample	Condition	λ (nm)	Q.Y.	Δλ (nm)
Neurospora Cu-MT	H_2O	565	0.013	62
	D_2O	565	0.026	62
	N_2, H_2O	565	0.033	62
Cu(I)-cysteine	dried	585	-	54
Cu(I)-glutathione	dried	575	-	51
Cu(I)-2-mercapto-ethanesulfonate	77K, Cu/RS 1:1	523	-	86

[a] taken from reference 11.

Table I summarizes some of the emmissive properties of N. crassa Cu-MT. Most characteristic of the luminescence is the largely red-shifted emission with a maximum of 565 nm and the rather broad band (Δλ = 62 nm). The quantum yield of 0.013 undergoes a twofold increase by substitution of H_2O with D_2O and a threefold increase under anaerobic conditions. Table I also contains some emmissive properties of different low molecular weight Cu(I)-thiolate complexes. Their luminescence is expected to be largely quenched at room temperature, due to solvent interactions in aqueous solutions. In agreement, the luminescence is observed only either at low temperature in a glass or in the dried state. The detectable emission of Cu-MT in aqueous solution at 283 K indicates a compact tertiary structure of the protein which would reduce the rate of solvent collisions by an efficient shielding of the Cu(I)-thiolate luminophore.

Using the frequency-domain emission technique, the luminescence lifetimes of Cu-MT have been determined recently (13). Lifetimes of 10.3 and 3.4 μs were found for the protein under anaerobic conditions and in the presence of oxygen, respectively.

Binding of mercurials like $HgCl_2$ or p-chloromercury-benzoate (PCMB) leads to a strong quenching of the luminescence (10). However, as shown in Fig. 1, the two compounds show a markedly different quenching behavior.

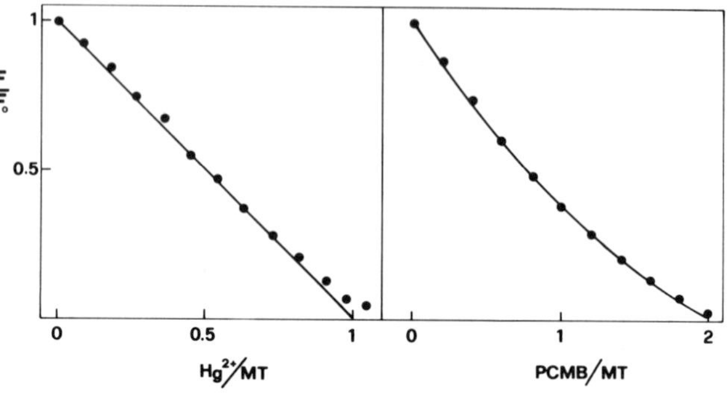

FIGURE 1. Luminescence quenching of _N. crassa_ Cu-MT $HgCl_2$ and p-chloromercurybenzoate (PCMB). Adapted from reference 10.

Addition of 1 mole of $HgCl_2$ per mole protein results in an irreversible, almost complete loss of the luminescence. This quenching process is a linear function of the Hg^{2+}/MT ratio. In contrast, quenching with PCMB is nonlinear, reversible and requires the addition of two molecules per mole MT. These results indicate the presence of two independent binding sites for Hg^{2+} and PCMB. Moreover, they suggest that the 6 Cu(I) ions coordinated to the seven cysteine residues of _N. crassa_ Cu-MT behave as a single metal cluster (10). This hypothesis has been further coroborated by a recent X-ray absorption study (14). Analysis of the EXAFS data by curve fitting showed them to be most consistent with the presence of 3-4 S atoms at 2.20 A and 1-2 Cu atoms at 2.71 A. These results lend strong support for the presence of a very compact Cu_6S_7 cluster in _N. crassa_ Cu-MT

In Fig. 2 are shown the X-ray absorption edge spectra for a two-coordinate linear Cu(I) complex, Cu-MT and a Cu_4S_6 model cluster compound (14). Again, these spectral features clearly argue for a tight, polynuclear Cu(I)-thiolate cluster with a Cu environment of 3-4 S atoms in _N. crassa_ Cu-MT.

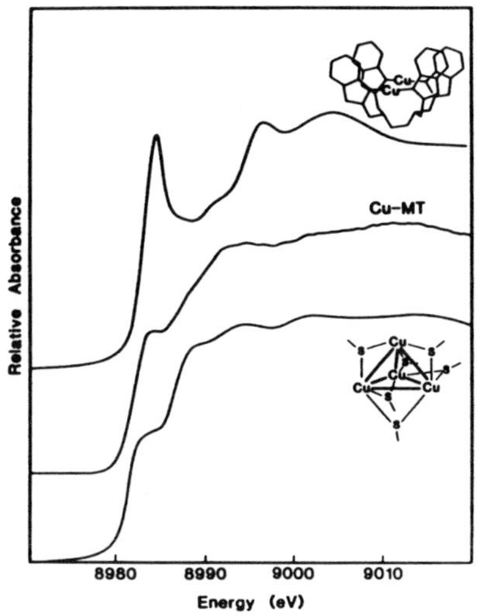

FIGURE 2. Comparison of the X-ray absorption edge spectra for a two-coordinate linear Cu(I) complex (top), *N. crassa* Cu-MT (middle), and a Cu_4S_6 model cluster (bottom). Taken from reference 14.

REGULATION OF GENE EXPRESSION

The structural gene of *N. crassa* Cu-MT has been cloned recently (15). In agreement with the amino acid sequence previously published (9), the gene codes for a polypeptide of 26 amino acid residues (including the N-terminal initiator methionine). The coding region is interrupted by a small intron of 94 bp. Compared to the gene structures of mammalian MTs (5), the intron-exon boundaries are located at different sequence positions. Computer analysis of the promoter region failed to detect sequences similar to the conserved metal regulatory elements of mammalian MTs or the Cu-responsive elements of the yeast Cu-MT gene (5). To investigate the regulation of expression of the *N. crassa* Cu-MT gene, the induction of MT mRNA was followed in cultures inoculated and grown in copper-supplemented media. In a second series of experiments, the induction of MT m-RNA in response to metal shock was investigated (16).

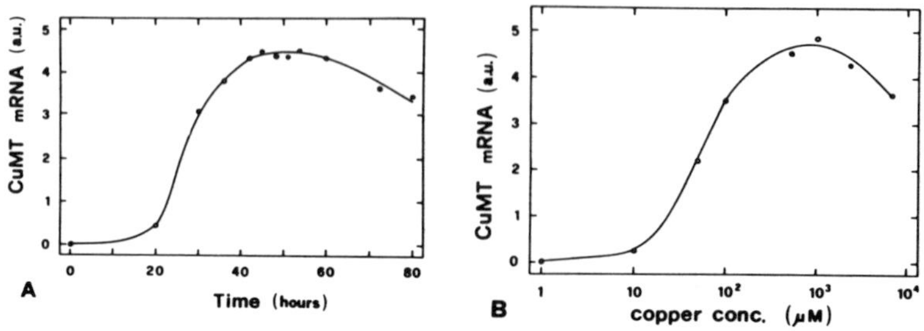

FIGURE 3. Induction of *N. crassa* Cu-MT mRNA. A, the kinetics of mRNA induction are shown for cultures grown in the presence of 0.5 mM $CuSO_4$. B, the dependence of mRNA induction on exogenous copper concentration in the culture medium are presented for cells harvested 55 h after inoculation. Taken from reference 16.

As shown in Fig. 3A, considerable amounts of Cu-MT mRNA can be observed, starting from about 30 h after inoculation. Maximum levels are reached between 45-55 h, with a rather small decline thereafter.

The amounts of Cu-MT mRNA measured 55 h after inoculation as a function of the copper concentration in the culture medium are shown in Fig. 3B. They are strongly dependent on the amounts of copper ions present in the medium up to about 0.25 mM $CuSO_4$ and then level off.

The induction of Cu-MT mRNA in response to metal shock (0.5 mM $CuSO_4$) is shown in Fig. 4. Rapidly increasing Cu-MT m-RNA levels are detectable within minutes after copper administration (Fig. 4A). The peak value is reached after approximately 60 min, with a rapid decay (t 1/2 = 2.5 h) to the initial level (Fig. 4B). The uptake of copper is also a very fast process with 50% of the metal taken up within 60 min after administration (Fig. 4B). About 60 min after induction, Cu-MT begins to appear and approximately 3 h after induction a steady state level of the protein is reached. The Cu-MT concentrations measured over the time period of 17 h remain constant (Fig. 4B). These data are consistent with the expression of the *N. crassa* Cu-MT gene being regulated at the transcriptional level.

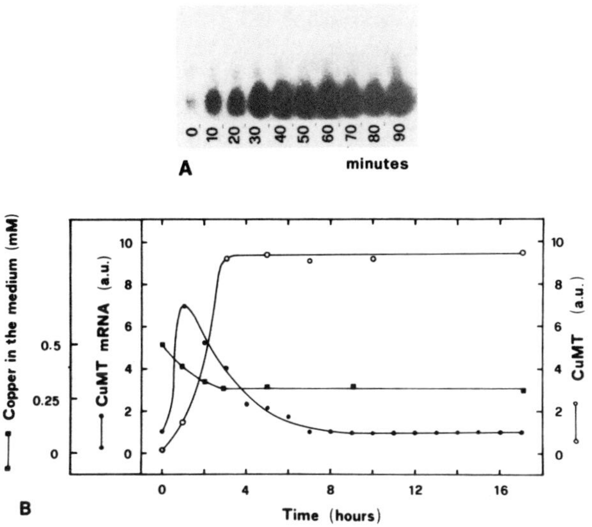

FIGURE 4. Induction of MT mRNA and Cu-MT in N. crassa in response to copper shock (0.5 mM $CuSO_4$). A, kinetics of MT mRNA induction followed by Northern hybridization. B, uptake of copper ions from the medium and induction of MT mRNA and Cu-MT followed over a time period of 17 h. Taken from reference 16.

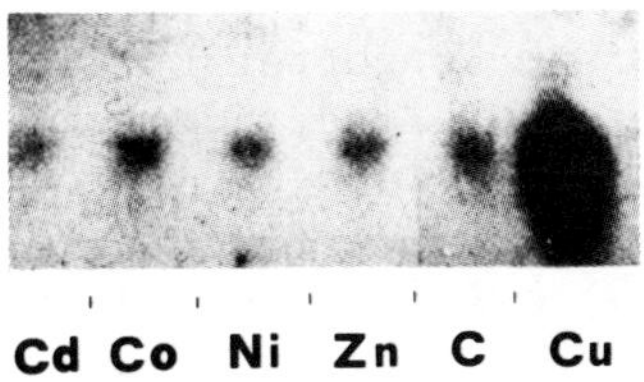

FIGURE 5. Induction of MT mRNA by different metal ions. mRNA levels were estimated by Northern analysis one h after administration of the metal ions. Cd, 0.5 mM $CdSO_4$; Co, 0.5 mM $CoCl_2$; Ni, 0.5mM $NiCl_2$; Zn, 0.5 mM $ZnSO_4$; Cu, 0.5 mM $CuSO_4$; C, control, no metal ions added. Taken from reference 16.

As shown in Fig. 5, the induction of *N. crassa* MT is copper-specific. Additions of Zn^{2+}, Cd^{2+}, Co^{2+}, or Ni^{2+} do not lead to a significant increase of MT mRNA. These results are not surprising since administration of II B metal ions to *N. crassa* does not result in the biosynthesis of the corresponding MTs *in vivo*. In contrast, however, these MT derivatives can be prepared *in vitro* in stable form and are spectroscopically very similar to those of the mammalian MTs (17).

Taken together, the rapid induction of Cu-MT mRNA and Cu-MT by copper ions and the remarkable intracellular stability of Cu-MT strongly suggest that this gene serves an important role in copper homeostasis in *N. crassa*.

ACKNOWLEDGMENTS

We thank Denise Niedermann for excellent technical assistance and Genia de Vallier for typing this manuscript.

REFERENCES

1. Kägi JHR, Kojima Y (1987). "Metallothionein II". Experientia Suppl. Vol. 52. Basel: Birkhäuser Verlag.
2. Otvos JD, Armitage, IM (1980). Structure of the metal clusters in rabbit liver metallothionein. Proc Natl Acad Sci USA 77:7094.
3. Vasak M, Kägi JHR (1981). Metal thiolate clusters in cobalt(II)-metallothionein. Proc Natl Acad Sci USA 78:6709.
4. Braun W, Wagner G, Wörgötter E, Vasak M., Kägi JHR, Wüthrich K (1986). Polypeptide fold in the two metal clusters of metallothionein-2 by nuclear magnetic resonance in solution. J Mol Biol 187:125.
5. Hamer DH (1986). Metallothionein. Annu Rev Biochem 55:913.
6. Lerch K (1981). The Chemistry and Biology of copper metallothionein. Met Ions Biol Syst 13:299.
7. Winge DR, Nielson KB, Gray WR, Hamer DH (1985). Yeast metallothionein, sequence and metal binding properties. J Biol Chem 260:14464.
8. Münger K, Lerch K (1985). Copper metallothionein from the fungus Agaricus bisporus: chemical and spectroscopic properties. Biochemistry 24:6751.

9. Lerch K (1980). Copper metallothionein, a copper-binding protein from Neurospora crassa. Nature (London) 284:368.
10. Beltramini M, Lerch K (1983). Spectroscopic studies on Neurospora copper metallothionein. Biochemistry 22:2043
11. Lerch K, Beltramini M (1983). Neurospora copper metallothionein: molecular structure and biological significance. Chemica Scr 21:109.
12. Beltramini M, Lerch K (1982). Luminescence properties of Neurospora copper metallothionein. FEBS lett 142:219.
13. Beltramini M, Giacometti GM, Salvato B, Giacometti G, Münger K, Lerch K (1988). Luminescence emission from Neurospora copper metallothionein: time resolved studies. Submitted for publication.
14. Smith TA, Lerch K, Hodgson KO (1986). Structural study of the copper sites in metallothionein from Neurospora crassa. Inorg Chem 25:4677.
15. Münger K, Germann UA, Lerch K (1985). Isolation and structural organization of the Neurospora crassa copper metallothionein gene. EMBO J 4:2665.
16. Münger K, Germann UA, Lerch K (1987). The Neurospora crassa metallothionein gene: regulation of expression and chromosome location. J Biol Chem 262:7363.
17. Beltramini M, Lerch K, Vasak M (1984). Metal substitution of Neurospora copper metallothionein. Biochemistry 23:3422.

Metal Ion Homeostasis: Molecular Biology and Chemistry, pages 237–246

EXPRESSION OF A CHINESE HAMSTER METALLOTHIONEIN CODING SEQUENCE AND ITS VARIANTS IN S. CEREVISIAE[1]

Mary Cismowski, Mark Chernaik, In-Koo Rhee[2] and P.C. Huang

Department of Biochemistry, The Johns Hopkins University, Baltimore, MD 21205

ABSTRACT Metallothionein is a low molecular weight cysteine-rich metal binding protein found in all mammalian species studied to date, as well as in the yeast S. cerevisiae. Efficient yeast expression vectors have been developed which express foreign coding sequences in yeast using either a copper inducible or constitutive promoter (1). Wild type and variant Chinese hamster metallothionein cDNA sequences were introduced into this yeast expression system. All constructs studied so far, when introduced into yeast cells, confer increased resistance to the toxic effects of exogenously added metals. Wild type metallothionein, as well as metallothionein variants with single or multiple cysteine to tyrosine amino acid changes, have been purified and characterized from this yeast system.

INTRODUCTION

In mammals, metallothionein (MT) is a highly conserved, 61 amino acid cysteine rich protein which has a high affinity for physiological metal ions such as Cu(I) and Zn(II), as well as toxic metal ions such as Cd(II) and Hg(II) (2). Divalent metal ions bound to MT can be localized to distinct amino-terminal and carboxy-terminal structural domains, designated beta and alpha, respectively (3,4,5). Each domain consists of a cysteine-metal cluster with tetrahedral coordination

---

[1]This work was supported by NIH grant RO1GM32606.

[2]Present address: Department of Agricultural Chemistry, Kyungpook National University, Taegu, Korea.

of divalent metals by cysteine sulfhydryl groups (3,6). The exact role of MT in the metabolism of metal ions, however, is not well understood.

Our lab is studying the in vivo and in vitro properties of the wild type Chinese hamster MT2 isoform and a series of genetically engineered variants of this protein. The variants being studied contain single or multiple cysteine to tyrosine changes (7), insertions of amino acids between the two domains, or a deletion of the alpha domain.

To study the in vivo properties of these proteins, we have utilized a yeast expression system which allows synthesis of wild type and variant MT2 from cloned cDNA sequences in a strain of S. cerevisiae (AB-DE1) which has its endogenous MT locus (CUP1) deleted (1). This host strain is extremely sensitive to exogenous copper and cadmium ions. We have determined the resistance to Cu(I) and Cd(II) conferred to this strain by the expression of wild type and variant Chinese hamster MT2 sequences.

To begin studying the in vitro properties of these proteins, we have used the yeast expression system to purify wild type Chinese hamster MT2 and 3 different cysteine to tyrosine variants. We have also analyzed the antigenic properties and metal binding stoichiometries of the Zn- and Cd-bound forms of these proteins.

METHODS

Construction of Yeast Expression Vectors.

Chinese hamster MT2 cDNA encoding either the wild type protein or MT2 protein containing single or multiple cysteine to tyrosine changes (7) were excised from M13 vectors. An adaptor containing a putative eukaryotic ribosome binding site (8) was ligated 5' to the coding sequences, and the resulting DNA fragments were inserted downstream from the promoter of the yeast CUP1 gene (pYHE vectors) or the yeast TDH gene (pYCHE vectors) in high copy number yeast expression vectors containing the full yeast TRP1 gene and sequences from the 2 micron plasmid (1).

Vectors which express variant MT2 proteins with either amino acids inserted between the two structural domains or lacking the alpha domain were made by the cloning of linkers at an Alul site at codon 32 in the MT2 cDNA sequence, or the deletion of an Alul-Hpal fragment within the coding sequence, respectively.

Yeast vectors were transformed into a $trp1^-$ yeast host lacking endogenous MT sequences (1) by the lithium acetate method (9).

Metal Resistance Assay.

Patches of test cells, growing on fresh agar plates, were streaked onto sectors of minimal media agar plates containing varying levels of $CuSO_4$ or $CdCl_2$ and incubated for 3 days at 30°C. Growth of colonies away from the initial area of application was scored as either +: uninhibited growth, +/-: inhibited growth, or -: no growth.

Purification of Wild Type Hamster MT2 and MT2 Variants.

Transformed yeast cells were grown in minimal media containing either 10 uM $ZnCl_2$ or 10 uM $CdCl_2$ to a final optical density at 600 nm of 1-2. The cell pellet was resuspended in 10 mM TrisHCl, pH 8, 10 mM DTT, 2 mM PMSF, 50 ppm Antifoam A and lysed by vortexing in the presence of 0.5 mm glass beads at 4°C. The lysate was centrifuged at 27000 x g and the supernatant was brought to 0.2 M sucrose. Metallothionein was then purified essentially as described (3), except that reverse-phase C18 HPLC was used as a final purification step. MT was eluted by a linear gradient of 0-60% CH_3CN in 10 mM NaH_2PO_4, pH 7, and elution was followed by absorbance at 214 nm. All purified proteins were treated with iodoacetamide to block sulfhydryl residues (10) prior to analysis on 15% SDS polyacrylamide gels (11).

Determination of Metal Binding Stoichiometries.

Sulfhydryls were quantitated with 2,2'-dithiodipyridine as described (12). Each purified sample was analyzed in duplicate at two different protein concentrations. Samples were then subjected to atomic absorption spectroscopy to determine metal content. Relative metal binding stoichiometries were calculated with the assumption that the mutant proteins lacked cysteine residues as predicted by their DNA sequences.

ELISA Analysis of Purified Metallothioneins.

ELISA analysis was performed as decribed (13) using serial dilutions of purified MT proteins as antigens. Equivalent amounts of protein as judged by sulfhydryl titrations were used, with undiluted samples containing 8 ug protein. Rabbit anti-rat MT antisera (14) was used at a 1:2000 dilution and peroxidase-linked goat anti-rabbit antibody was used at a 1:10000 dilution.

RESULTS

Coding sequences for Chinese hamster MT2 and variants of this protein (Table 1) were cloned into high copy number

TABLE 1

AMINO ACID SEQUENCES OF WILD TYPE AND VARIANT CHINESE HAMSTER METALLOTHIONEIN 2 PROTEINS IN THIS STUDY

Variant	Amino Acid Changes
MT2	Wild type
GA57	Cys50→Tyr50
GA106	Cys13→Tyr13
GA92	Cys13→Tyr13,Cys50→Tyr50
GA77	Cys7→Tyr7,Cys13→Tyr13
GA33	Cys36→Tyr36,Cys44→Tyr44
GA168	Cys29→Tyr29,Cys50→Tyr50,Cys59→Tyr59,Cys60→Tyr60
MT2X61	Met1...Ser32(ProGly)Cys33...Ala61
MT2X62	Met1...Ser32(ProGly)$_2$Cys33...Ala61
MT2X121	Met1...Ser32(ProProGlyAsp)Cys33...Ala61
MT2X122	Met1...Ser32(ProProGlyAsp)$_2$Cys33...Ala61
MT2X123	Met1...Ser32(ProProGlyAsp)$_3$Cys33...Ala61
MT2X124	Met1...Ser32(ProProGlyAsp)$_4$Cys33...Ala61
MT2beta	Met1...Ser32Ala61

yeast expression vectors under the control of a copper-inducible (pYHE- series) or constitutive (pYCHE- series) promoter. The yeast expression vectors were introduced into a yeast strain that lacks endogenous yeast MT sequences, AB-DE1 (1). This yeast strain is sensitive to very low levels of exogenous copper or cadmium. The ability of wild type and variant MT2 to function in metal ion detoxification was determined by testing the resistance of AB-DE1 transformed with our expression vectors to varying levels of copper or cadmium.

Table 2 summarizes the resistance to varying levels of $CuSO_4$ or $CdCl_2$ in minimal media agar plates of AB-DE1 cells

TABLE 2
RESISTANCE CONFERRED TO A CUP1⁻ YEAST STRAIN BY WILD TYPE AND VARIANT CHINESE HAMSTER MT2 EXPRESSION VECTORS

Vector	$CuSO_4$, uM 0	30	100	300	1000	$CdCl_2$, uM 30	100	300	1000
pYHE	+	+/-	-			-			
pYHE-MT2	+	+	+	+	-	-			
pYHE-GA57	+	+	+	+	-	-			
pYHE-GA77	+	+	+	+	-	-			
pYHE-GA33	+	+	+	+	-	-			
pYHE-GA168	+	+	+	+	-	-			
pYHE-MT2beta	+	+	+	+/-	-	-			
pYCHE-MT2	+	+	+	+	-	+	+	+	-
pYCHE-GA57	+	+	+	+	-	+	+	+	-
pYCHE-GA106	+	+	+	+	-	+	+	+	-
pYCHE-GA92	+	+	+	+	-	+	+	+	-
pYCHE-GA77	+	+	+	+	-	+	+	+	-
pYCHE-GA33	+	+	+	+	-	+	+	+	-
pYCHE-GA168	+	+	+	+	-	+	+	+	-
pYCHE-MT2beta	+	+	+	+	-	+	-		
pYCHE-MT2X61	+			+		+	+	+	
pYCHE-MT2X62	+			+		+	+	+	
pYCHE-MT2X121	+			+		+	+	+	
pYCHE-MT2X122	+			+		+/-	+/-	-	
pYCHE-MT2X123	+			+		+/-	+/-	-	
pYCHE-MT2X124	+			+		+/-	-	-	

transformed with our series of expression vectors. If transformed with a vector that lacks MT sequences (pYHE), AB-DE1 failed to grow on plates containing 100 uM $CuSO_4$ or 30 uM $CdCl_2$. Cells transformed with expression vectors for wild type MT2 or variants containing single or multiple cysteine to tyrosine changes (GA57, GA106, GA92, GA77, GA168 and GA33) grew on plates containing up to 300 uM $CuSO_4$ (pYHE- series) and 300 uM $CuSO_4$ or 300 uM $CdCl_2$ (pYCHE- series). All strains failed to grow at 1 mM $CuSO_4$ or $CdCl_2$.

The metal resistance of cells expressing MT2 with insertions of amino acids between the two structural domains differed from cells expressing wild type MT only in that there is a decreased resistance to higher levels of $CdCl_2$. The level of cadmium resistance was lower for cells expressing variant MT with longer insertions (MT2X122, MT2X123 and MT2X124) than for those with shorter insertions (MT2X61, MT2X62 and MT2X121). Cells expressing only the beta domain of MT also showed a decreased resistance to cadmium, failing to grow on plates containing 100 uM $CdCl_2$.

The expression of wild type and variant Chinese hamster MT2 protein in AB-DE1 cells transformed with our yeast expression vectors was confirmed by *in vivo* ^{35}S-cysteine labelling of the appropriate strains, followed by SDS-PAGE of cell extracts (data not shown). This analysis confirmed that our expression vectors directed the synthesis of a small, cysteine rich, copper-inducible (pYHE- series) protein that either comigrated with authentic Chinese hamster MT or had an altered mobility predicted by the different size of the variant protein. The level of MT expressed by our vectors was similar for the wild type and cysteine to tyrosine variants, but was significantly lower for the expansion mutants MT2X122, MT2X123 and MT2X124, and the single beta domain mutant.

Wild type Chinese hamster MT2 and 3 cysteine to tyrosine variants (GA57, GA92 and GA106) were purified from transformed yeast cells expressing protein from the constitutive TDH promoter (Figure 1). The purified proteins comigrated on SDS-polyacrylamide gels with the *in vivo* ^{35}S-cysteine labelled MTs described above. By growing yeast cultures in either $ZnCl_2$ or $CdCl_2$, purified proteins binding predominantly Zn or Cd ions, respectively, were obtained.

Metal binding stoichiometries of the purified proteins were obtained by comparing the sulfhydryl content of the proteins, as determined by reaction with 2,2'-dithiodipyridin to the relative metal content, as determined by atomic absorption spectroscopy. Table 3 shows the results of these

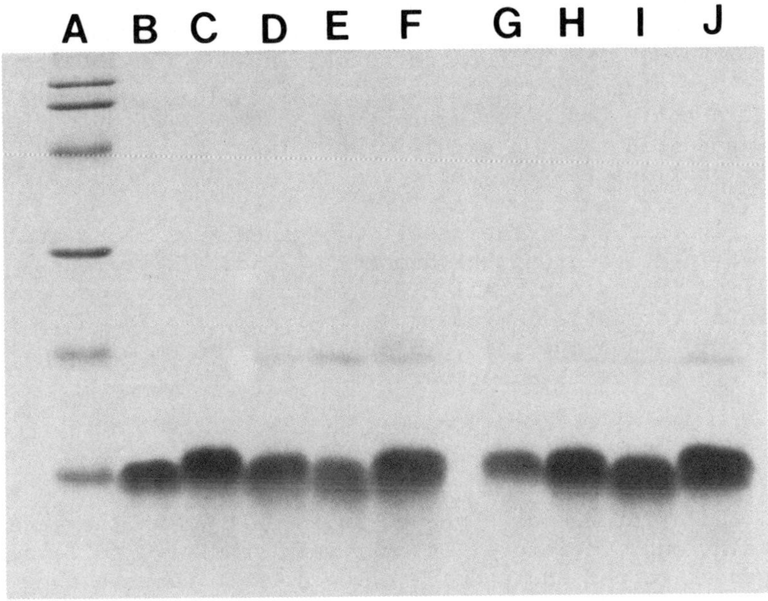

FIGURE 1. SDS-polyacrylamide gel analysis of purified wild type and variant MT2 proteins. Protein was visualized by coomassie staining. A: Molecular weight markers; B: rat Zn-MT1 standard; C: Zn-MT2; D: Zn-GA57; E: Zn-GA92; F: Zn-GA106; G: Cd-MT2; H: Cd-GA57; I: Cd-GA92; J: Cd-GA106.

TABLE 3
METAL CONTENT OF PURIFIED MT2 AND MT2 VARIANTS

Protein	Zn/mol (Cu/mol)	Cd/mol
Rat MT1	6.7 ± 0.2 (0)	---
MT2	6.2 ± 0.2 (1)	6.0 ± 0.4
GA57	6.4 ± 0.3 (0.5)	5.1 ± 0.2
GA106	5.5 ± 0.5 (0.5)	5.7 ± 0.3
GA92	5.0 ± 0.2 (0.5)	6.2 ± 0.4

determinations. MT proteins purified from yeast cells grown in the presence of 10 uM $ZnCl_2$ had significant amounts of copper bound (shown in parentheses) in addition to zinc, while proteins purified from cells grown in 10 uM $CdCl_2$ had no detectable copper or zinc bound. The variant protein GA92 appeared to bind significantly lower zinc than the other proteins tested, while the variant GA57 bound significantly lower cadmium.

Figure 2 shows the results obtained from ELISA analysis of rat Zn-MT1 and Chinese hamster MT2 and MT2 variants purified from yeast. All MT proteins assayed showed strong affinity for antisera raised against purified rat MT. In contrast, a non-specific protein, bovine serum albumin, had no affinity for this antisera.

DISCUSSION

The existence of a yeast strain that lacks the gene for yeast MT and, therefore, is extremely sensitive to levels of exogenous copper and cadmium allowed us to compare the ability of wild type and variant Chinese hamster MT2 proteins to function in metal ion detoxification. As judged by the ability to form colonies on agar plates containing varying levels of $CuSO_4$ or $CdCl_2$ (Table 1), six variants containing single or multiple cysteine to tyrosine changes function as

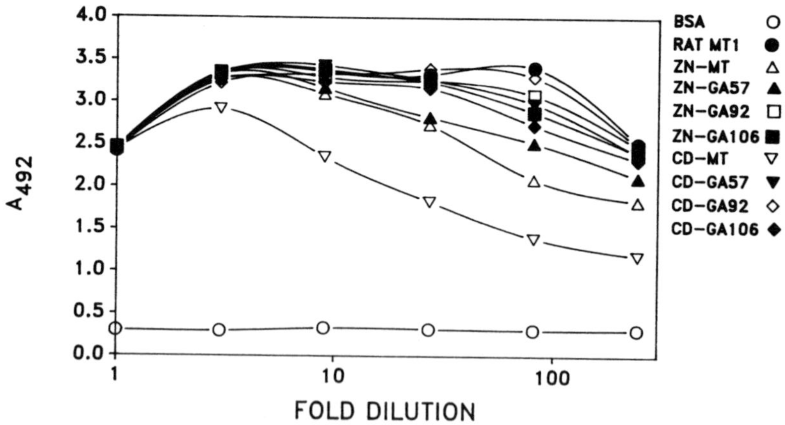

FIGURE 2. ELISA analysis of purified wild type and variant Chinese hamster MT2 proteins.

well as wild type MT2 in metal ion detoxification. Thus, their metal binding properties may not differ greatly from those of wild type MT2.

Variants containing a stretch of amino acids inserted between the two structural domains of MT2 conferred different levels of resistance to cadmium, but not copper. Those variants which had inserts of eight amino acids or more conferred a significantly lower level of resistance. Also, a variant MT2 protein corresponding to the beta domain conferred a significantly lower level of resistance to cadmium, but not copper. The levels of variant MT2 protein expression were reduced for all variants that conferred lower levels of cadmium resistance, indicating that their stability in yeast may be decreased. Therefore we could not detect to what extent, if any, these variants differed from wild type MT2 in their _in vivo_ metal binding properties.

Wild type Chinese hamster MT2 and 3 cysteine to tyrosine variants have been purified from yeast. Several criteria have been used to characterize these proteins as metallothioneins. First _in vivo_ ^{35}S-cysteine labelling was used to establish the MT sequence-dependent accumulation of a cysteine rich protein which comigrated with authentic Chinese hamster MT on SDS-polyacrylamide gels. The purified MT2 and MT2 variants were shown to bind high levels of the metal ions zinc and cadmium (Table 3). Finally, Figure 2 shows that the purified MT proteins obtained could be recognized by antisera raised against purified rat MT.

In vitro binding studies performed on the purified wild type MT2 and cysteine to tyrosine variants showed small but significant differences in their metal ion stoichiometries. The mutant GA92 (cys13,50→tyr13,50) bound less zinc than MT2, but was unaffected in binding to cadmium. In contrast the mutant GA57 (cys50→tyr50) bound less cadmium than MT2, but was unaffected in binding to zinc. Further studies will be conducted to confirm these initial results.

ACKNOWLEDGEMENTS

We thank Mrs. Betty Smith for her excellent technical help, Dr. John Frazier for providing atomic absorption facilities and Dr. David Ecker for kindly providing the yeast strain and parent yeast expression vectors used in this study.

REFERENCES

1. Ecker DJ, Butt TR, Sternberg EJ, Neeper MP, Debouck C, Gorman JA, Crooke ST (1986). Yeast metallothionein function in metal ion detoxification. J Biol Chem 261:16895.
2. Nielson KB, Atkin CL, Winge DR (1985). Distinct metal-binding configurations in metallothionein. J Biol Chem 260:5342.
3. Otvos JD, Armitage IM (1980). Structure of the metal clusters in rabbit liver metallothionein. Proc Natl Acad Sci USA 77:7094.
4. Vasak M, Kagi JHR (1981). Metal thiolate clusters in cobalt(II)-metallothionein. Proc Natl Acad Sci USA 78:6709.
5. Winge DR, Miklossy K-A (1982). Domain nature of metallothionein. J Biol Chem 257:3471.
6. Vasak M, Kagi JHR, Hill HAO (1981). Zinc(II), cadmium(II) and mercury(II) thiolate transitions in metallothionein. Biochemistry 20:2852.
7. Pine R, Cismowski M, Liu SW, Huang PC (1985). Construction and characterization of a library of metallothionein coding sequence mutants. DNA 4:115.
8. Kozak M (1986). Point mutations define a sequence flanking the AUG initiator codon that modulates translation by eukaryotic ribosomes. Cell 44:283.
9. Ito H, Fukuda Y, Murata K, Kimura A (1983). Transformation of intact yeast cells treated with alkali cations. J Bact 153:163.
10. Sone T, Yamaoka K, Minami Y, Tsunoo H (1987). Induction of metallothionein synthesis in Menkes' and normal lymphoblastoid cells is controlled by the level of intracellular copper. J Biol Chem 262:5878.
11. Laemmli UK (1970). Cleavage of structural proteins during the assembly of the head of bacteriophage T4. Nature 227:680.
12. Grassetti DR, Murray JF (1967). Determination of sulfhydryl groups with 2,2'- or 4,4'-dithiodipyridine. Arch Biochem Biophys 119:41.
13. Voller A, Bidwell D, Bartlett A (1976). Microplate enzyme immunoassays for the immunodiagnosis of virus infections. In Rose NR, Friedman H (eds): "Manual of Clinical Immunology," Washington DC: American Society for Microbiology, p 506.
14, Danielson KG, Ohi S, Huang PC (1982). Immunochemical detection of metallothionein in specific epithelial cells of rat organs. Proc Natl Acad Sci USA 79:2301.

Metal Ion Homeostasis: Molecular Biology and Chemistry, pages 247–256

REACTIVATION OF APOHEMOCYANIN BY COPPER METALLOTHIONEIN

Marius Brouwer and Thea Brouwer-Hoexum

Duke University Marine Laboratory
Beaufort, North Carolina 28516

ABSTRACT Hemocyanins are multisubunit proteins, which contain binuclear-Cu(I) centers that can combine with oxygen or carbon monoxide. The mechanism by which Cu(I) is inserted into the active site of apohemocyanin is unknown, but recent studies suggest that the polynuclear metalloprotein, Cu(I)-metallothionein (CuMT), may act *in vivo* as a Cu(I) donor (Engel and Brouwer, 1987, *Biol. Bull.* 173, 239). To further explore the putative function of this metalloprotein, we have used CuMT, isolated from the digestive gland of the American lobster, to study the reactivation of lobster hemocyanin containing reduced amounts of Cu(I). Copper transfer and formation of binuclear-copper sites that can bind CO and oxygen were followed by fluorescence and electronic absorption spectroscopy. These studies showed that formation of CO-binding sites precedes the formation of oxygen-binding sites. Since the binding of oxygen, which bridges between two coppers (Cu-O-O-Cu) is subject to greater stereochemical constraints than the binding of CO, which binds to only one copper (Cu...Cu-CO), we conclude that Cu(I) transfer from MT to apohemocyanin initially results in the formation of distorted binuclear-copper sites which subsequently slowly return to their native stereochemical configuration. CuMT cannot reactivate completely copper-free apohemocyanin, but free Cu(I) can, suggesting that the conformational state of hemocyanin is the determining factor in the Cu(I) transfer process, and that free Cu(I) is not involved in CuMT-mediated copper transfer.

INTRODUCTION

Metallothioneins (MTs) are cysteine-rich metal-binding proteins thought to be involved in interrelated processes associated with metal metabolism, such as metal detoxification, regulation of zinc and copper, and donation of metals to apo-metalloproteins (1-5). Progress in metallothionein research faces two major challenges: (a) establishment of a clear correlation between MT function and a specific physiological process and (b) identification of cellular components that interact with MT.

To examine how natural processes might affect MT turnover and synthesis, we have selected marine crustacea as model systems for study (5,6). This choice is based on the observation that marine crustacea display an enormously active copper metabolism. The levels of hemocyanin, a Cu(I)-containing protein that serves as an oxygen carrier in crustacea, change dramatically during the different stages of molt cycle (6), starvation (see 7 for references), exposure to hypoxic conditions (8), environmental temperature and season (5), and hypo-osmotic conditions (9). Our studies suggest that during the critical stages of the molt cycle of growing blue crabs there is a direct involvement of MT both in copper sequestration and excretion and in copper donation for hemocyanin biosynthesis (5,6). In addition we have shown that CuMT can reactivate apohemocyanin under anaerobic conditions (5,7). Incorporation of copper seems to precede the formation of biologically active oxygen binding sites. It was postulated that the newly-formed binuclear Cu(I) sites cannot combine reversibly with oxygen, but are oxidized instead (5).

Hemocyanins can combine with CO, resulting in the formation of a highly fluorescent Cu(I)-CO complex (10). Whereas oxygen bridges between the coppers in hemocyanin's active site, CO binds to a single copper. It seems therefore likely that the binding of CO is subject to less stringent stereochemical requirements than the binding of oxygen. Moreover, CO is not capable of oxidizing the active-site coppers. These properties have prompted us to study the reconstitution of apohemocyanin with Cu(I)MT in a CO atmosphere.

Our studies show that Cu(I) transfer from MT to apohemocyanin initially results in the formation of binuclear-copper sites which can combine with CO but not with O_2. These sites slowly return to the native stereochemical configuration. Free Cu(I) in solution does not seem to be involved in the CuMT-mediated copper transfer.

METHODS

Hemocyanin, (partially) apohemocyanin and copper-metallothionein of the American lobster Homarus americanus were prepared as described before (5,7). Reconstitution of apohemocyanin (10 μM protein, 3 μM intact binding sites) was carried out by mixing the apoprotein in 50 mM Tris, 10 mM Ca, pH 8.0 with 11 μM Cu(I)MT, in a series of tonometers filled with carbon monoxide. The incubation mixture in one tonometer was excited at 280 nm and the fluorescence emission was recorded from 300-700 nm with a SPEX fluorolog fluorescence spectrophotometer in the ratio mode, as a function of time. Copper insertion into the active site of the apoprotein was estimated from the quenching of the tryptophan fluorescence, which accompanies copper incorporation. CO binding sites were calculated from the fluorescence emission intensity at 580 nm. At different time intervals a tonometer was opened and flushed with oxygen. The concentration of functionally active oxygen binding sites was determined from the intensity of the copper-oxygen charge transfer band at 340 nm (5,7).

RESULTS

Removal of copper from the active site of lobster hemocyanin results in an increase of the intrinsic tryptophan fluorescence of the protein (5). Incubation of apohemocyanin with Cu(I)MT in the presence of CO leads to a decrease of the tryptophan fluorescence emission at 334 nm as a function of time (Figures 1 and 2). Concomitant with this decrease at 334 nm is a strong increase of the fluorescence emission at 580 nm, which is due to the formation of the Cu(I)-CO complex in the reconstituted active site of the hemocyanin molecule (Figures 1 and 3). The fractional saturation of hemocyanin with oxygen and CO, as a function of incubation time with Cu(I)MT, is shown in Figure 4. It is evident that the formation of active sites that can reversibly combine with CO precedes the formation of viable oxygen binding sites.

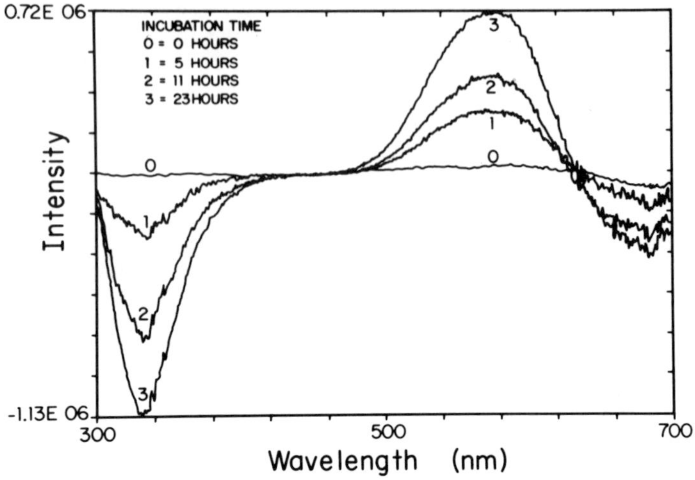

Figure 1. Difference fluorescence emission spectra as a function of incubation time of apohemocyanin with Cu(I)-MT in a CO atmosphere. Excitation was at 280 nm. Conditions are described in Methods.

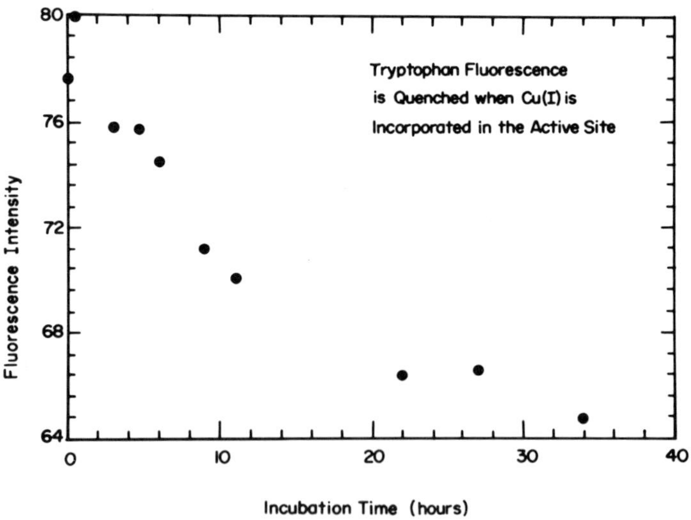

Figure 2. Tryptophan fluorescence emission of apohemocyanin at 334 nm as a function of exposure to Cu(I)-MT. Excitation was at 280 nm. Conditions are described in Methods.

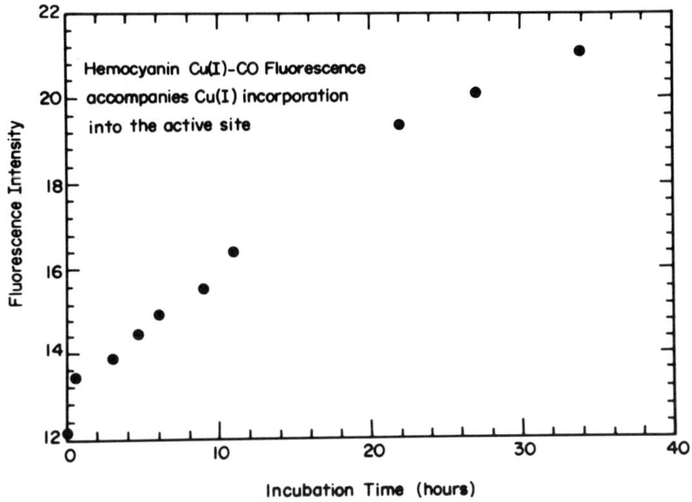

Figure 3. Increase in hemocyanin Cu(I)-CO fluorescence emission at 580 nm as a function of incubation time with Cu(I)-MT in a CO atmosphere. Excitation was at 280 nm. Conditions are described in Methods.

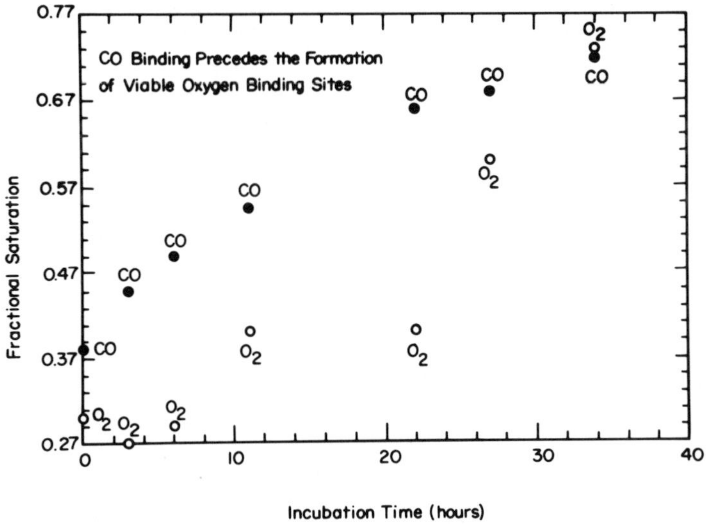

Figure 4. Fractional saturation of hemocyanin with O_2 and CO as a function of exposure to Cu(I)-MT. CO binding sites were calculated from the fluorescence emission at 580 nm. O_2 binding sites were calculated from the absorbance of the Cu-O-O-Cu complex at 340 nm.

DISCUSSION

The possible functions of metallothionein have been studied extensively (1-5). Evidence presented by several investigators is consistent with the hypothesis that MTs supply zinc, and possibly copper, for metabolic processes, in growing, injured or regenerating tissues. Zinc-MT can reactivate apoenzymes that require zinc for their biological activity. It was postulated that the mechanism of metal transfer involved a direct interaction between metallothionein and apoenzyme (11,12).

The mechanism by which copper is incorporated into metalloproteins has not been elucidated (3), but free Cu(II) ions seem not to be involved in this process (13,14). The hypothesis that MT plays a role in copper insertion is therefore tenable. Reports in the literature on this putative function of copper-MT seem conflicting. Cu(I)MT, isolated from rat liver, cannot reactivate superoxide dismutase (15), and Cu(I)MT from _Neurospora_ cannot reactivate tyrosinase (16), but both apoenzymes can be slowly reconstituted with Cu(II) in solution. Hemocyanins, on the other hand, can only be reactivated with Cu(I) (17) and are therefore potential substrates for Cu(I)MT (5,7). Previous studies in our laboratory have shown that hemocyanin from the American lobster _Homarus americanus_, from which part of the Cu(I) was removed by treatment with cyanide, can be reactivated by Cu(I)MT isolated from lobster hepatopancreas, in the absence of oxygen (5,7). It was the purpose of the studies reported here to investigate the Cu(I)MT-mediated reactivation process in more detail. Quenching of tryptophan fluorescence, which accompanies copper insertion, and measurement of the formation of CO and oxygen binding sites were used as probes to dissect the reactivation process into (a) copper transfer, (b) formation of binuclear copper sites that can combine with a non-bridging ligand and (c) formation of native binuclear copper sites.

A number of important conclusions can be drawn from the data presented in Figures 1 through 4.

I. The quenching of the tryptophan fluorescence emission at 334 nm is paralleled by the formation of the hemocyanin-Cu(I)-CO complex which fluoresces at 580 nm (Figure 1). The observed fluorescence quenching is thus due to incorporation of Cu(I) into hemocyanin's active site. The time course of the fluorescence quenching is the same as observed when hemocyanin is incubated with Cu(I)MT in vacuo (5).

II. Mononuclear copper sites prepared by cyanide treatment do not show Cu(I)-CO fluorescence (Brouwer unpublished results, 10). The fluorescence emission at 580 nm represents therefore binuclear copper sites that combine with CO. The formation of these sites precedes the formation of viable oxygen binding sites (Figure 4). The formation of binuclear-copper sites, which combine reversibly with oxygen, takes place after prolonged incubation with Cu(I)MT as reported before (5).

III. The experiment described here validates our original postulate (5) that Cu(I) transfer from MT to apohemocyanin is followed by a rearrangement of the reconstituted, distorted, binuclear-copper site into the native stereochemical configuration. This slow conformational change is the rate limiting step in the formation of active oxygen binding sites. These observations may explain why reconstitution experiments with hemocyanin from the crab Carcinus maenas and Cu(I)MT from Neurospora were unsuccessful in the presence of oxygen (16).

An important question that still remains to be answered is whether Cu(I) transfer requires direct interaction between Cu(I)MT and hemocyanin or whether free Cu(I), released by MT, is involved in the reactivation process. A number of observations strongly support the "direct- interaction" pathway. (a) The postulated mechanism for zinc transfer from ZnMT to apoenzymes requires direct interaction between the two proteins (11,12). (b) Preliminary experiments using ultrafiltration to study the interaction of Cu(I)MT with hemocyanin have shown that Cu(I)MT is retained in the presence of hemocyanin. An apparent dissociation constant for the CuMT-hemocyanin complex of 50 μM was calculated (Brouwer, unpublished results). (c) Cu(I)MT is purified in the presence of beta-mercaptoethanol (7). The presence of this reagent, which has a very high affinity for Cu(I), does not seem to interfere with the reconstitution process, suggesting a direct metal transfer mechanism. (d) As reported earlier, completely copper-free lobster hemocyanin cannot be reactivated by Cu(I)MT (7). Recently we have demonstrated that Cu(I) in solution is effective in the reconstitution of fully apohemocyanin, indicating that copper transfer from Cu(I)MT to apohemocyanin does not involve free Cu(I). Taken together, these observations strongly suggest that Cu(I) transfer from MT to apohemocyanin requires a direct interaction between the two proteins.

It is clear from the data presented here and earlier (5,7) that the conformational state of hemocyanin is the determining factor in the interaction of MT with hemocyanin and in the MT-mediated Cu(I) transfer process. The rearrangement of the reconstituted binuclear-copper sites is the rate limiting step in the formation of viable oxygen binding sites. It is also evident that the copper transfer from MT to apohemocyanin _in vitro_ is rather slow. We will attempt to study copper incorporation into hemocyanin _in vivo_ by means of pulse labeling experiments using hepatopancreas cell suspensions obtained from crabs in which hemocyanin biosynthesis has been induced. (The presence of MT and hemocyanin in these cells has been demonstrated in our laboratory.) To further understand the CuMT-hemocyanin interaction it is relevant to mention that a hemocyanin molecule is folded into three distinct domains, with the central domain containing the binuclear copper site (18). It may well be that CuMT interacts _in vivo_ with the central domain, before the C-terminal domain, that partly enfolds the central domain in the native structure, is folded into its proper three dimensional configuration. To study this hypothesis _in vitro_ we will attempt to prepare the central domain by proteolytically removing the flanking N- and C-terminal domains, and study its interaction with CuMT. Such an approach is feasible, since it has recently been demonstrated that a binuclear-copper fragment of _Limulus_ hemocyanin can be prepared by limited proteolytic digestion of the intact subunit (Dr. Richard Topham, University of Richmond, personal communication).

The data presented here and in previous reports (5,6,7) have shown that the great advantage of the crustacean system in the study of CuMT function lies in the fact that it allows us to follow CuMT and its changes in relation to changes in a single copper-protein, hemocyanin. This has enabled us to make progress toward understanding of MT's natural function in the homeostatic control of metals at the molecular, cellular and organismal levels.

REFERENCES

1. Kagi JHR, Kojima Y (1987). "Metallothionein II." Basel: Birkhauser Verlag.
2. Kagi JHR, Nordberg M (1979). "Metallothionein." Basel: Birkhauser Verlag.
3. Cousins RJ (1985). Absorption, transport, and hepatic metabolism of copper and zinc: special reference to metallothionein and ceruloplasmin. Physiol Rev 65:238.
4. Hamer DH (1986). Metallothionein. Ann Rev Biochem 55:913.
5. Engel DW, Brouwer M (1987). Metal regulation and molting in the blue crab, *Callinectes sapidus*: metallothionein function in metal metabolism. Biol Bull 173:237.
6. Engel DW (1987). Metal regulation and molting in the blue crab *Callinectes sapidus*: copper, zinc and metallothionein. Biol Bull 172:69.
7. Brouwer, M, Whaling P, Engel DW (1986). Copper-metallothioneins in the American lobster, *Homarus americanus*: Potential role as Cu(I) donors to apohemocyanin. Environ Health Perspect 65:93.
8. Senkbeil EG, Wriston JC (1981). Hemocyanin synthesis in the American lobster, *Homarus americanus*. Comp Biochem Physiol 68B:163.
9. Boone WR, Schoffeniels E (1979). Hemocyanin synthesis during hypo-osmotic stress in the shore crab *Carcinas maenas*. Comp Biochem Physiol 63B:207.
10. Finazzi-Agro A, Zolla L, Flamigni L, Kuiper HA, Brunori M (1982). Spectroscopy of (carbon monooxy) hemocyanins. Phosphorescence of the binuclear carbonylated copper centers. Biochemistry 18:4095.
11. Udom AO, Brady FO (1980). Reactivation *in vitro* of zinc-requiring apo-enzymes by rat liver zinc-thionein. Biochem J 187:329.
12. Li TY, Kraker AJ, Shaw CF, Petering DH (1980). Ligand substitution reactions of metallothioneins with EDTA and apocarbonic anhydrase. Proc Natl Acad Sci (U.S.) 77:6334.
13. Holtzman NA, Gaumnitz BM (1970). Studies on the rate of release and turnover of ceruloplasmin and apoceruloplasmin in rat plasma. J Biol Chem 245:2354.

14. Keyhani E, Keyhani J (1975). Cytochrome c oxidase biosynthesis and assembly in *Candida utilis* yeast cells: function of copper in the assembly of active cytochrome c oxidase. Arch Biochem Biophys 167:596.
15. Geller BL, Winge DR (1982). Metal binding sites of rat liver Cu-thionein. Arch Biochem Biophys 213:109.
16. Beltramini M, Lerch K (1982). Copper transfer between *Neurospora* copper metallothionein and type 3 copper apoproteins. FEBS Letters 142:219.
17. Konings WN, Van Driel R, Van Bruggen EFJ, Gruber M (1969). Structure and properties of hemocyanins. V. Binding of oxygen and copper in *Helix pomatia* hemocyanin. Biochim Biophys Acta 194:55.
18. Gaykema WPJ, Hol WGJ, Vereyken JM, Soeter NM, Bak HJ, Beintema JJ (1984). 3.2 Å structure of the copper-containing, oxygen-carrying protein *Panulirus interruptus* haemocyanin. Nature 309:23.

Metal Ion Homeostasis: Molecular Biology and Chemistry, pages 257–267

ZINC METALLOTHIONEIN IN MAMMALIAN BRAINS

M. Ebadi, V.K. Paliwal, T. Takahashi, and P.L. Iversen

Department of Pharmacology
University of Nebraska College of Medicine
42nd Street and Dewey Avenue
Omaha, Nebraska 68105

ABSTRACT Metallothionein-like proteins have been identified in the rat brain, monkey brain, bovine retina, pineal gland and hippocampus, and in the neuroblastoma IMR 32. The rat brain metallothionein is induced following intracerebroventricular administration of zinc and copper, but not cadmium (icv, 0.20 μ mol zinc/μl/hr, for 24-48 hr). Furthermore, the zinc-induced brain metallothionein incorporates a large quantity of ^{35}S cysteine, and is accompanied by enhanced synthesis of poly A^+ RNA. The chromatofocusing of metallothionein-like protein isolated by gel permeation chromatography on Sephadex G-75 exhibits three zinc-binding peaks, which focus at pH 6.8, 6.2, and 5.3, respectively. It is expected that the protein peak focusing at pH 5.3 is a metallothionein-like protein. Purification of the zinc-induced metallothionein by ion exchange chromatography on DEAE-Sephadex A-25 columns, produces two isoforms, eluting at 75 and 137 mM of Tris-acetate buffer, pH 7.5, respectively. The comparative HPLC profiles of the zinc-induced hepatic and brain metallothioneins isoforms I and II are similar. Furthermore, the amino acid compositions of rat brain metallothionein isoforms, which contain 17 and 18 cysteine residues, are identical to hepatic metallothionein isoforms in being devoid of any histidine, arginine, leucine, tyrosine, or phenylalanine. Since the rat brain and the bovine hippocampal metallothioneins do cross react with antibodies formed against both sheep and rat metallothioneins, it is anticipated that the immunologically dominant regions of the NH_2-terminal domain (residues 1-29) of neural metallothioneins are conserved. The neural metallothioneins, in addition to maintaining the homeostasis of zinc and copper in the brain, may also be involved in events associated with synaptic functions.

INTRODUCTION

With the exception of potassium, calcium, and magnesium, zinc is the most abundant metal in the brain and its concentration has been shown to be altered in an extensive number of disorders of the central nervous system including alcoholism, Alzheimer-type dementia, Down's syndrome, epilepsy, Friedreich's ataxia, Guillain-Barre syndrome, hepatic encephalopathy, multiple sclerosis, Pick's disease, retinitis pigmentosa, retinal dystrophy, schizophrenia, and Wernicke-Korsakoff syndrome (see Ebadi and Pfeiffer, 1984; Ebadi and Hama, 1986; and Ebadi and Swanson, 1987 for reviews and references). The status of metallothionein in these conditions, diseases, disorders, or syndromes are not known.

EVENTS LEADING TO THE DISCOVERY OF METALLOTHIONEIN IN THE RAT BRAIN

The discovery of a low-molecular-weight zinc-binding protein in the rat brain (Itoh et al., 1983) and its designation as metallothionein-like protein (Ebadi, 1984) was a fortuitous observation which took place while attempting to comprehend the mechanism of convulsive seizures associated with Pick's disease, a rare neurological disorder, which allegedly results in higher than normal accumulation of zinc in the hippocampus, causing inhibition of γ-aminobutyric acid (GABA) transmission. In a study, Itoh and Ebadi (1982) showed that the icv administration of zinc sulfate (0.3 μmol/10 μl) caused seizures which were prevented by GABA but not by other putative transmitters. During the course of our investigation, some interesting observations were made. The administration of substantially larger doses of zinc sulfate (up to 100 mg/kg), either intravenously or intraperitoneally, did not cause convulsive seizures in rats. Since the convulsive dose of 0.3 μmol/10 μl of zinc sulfate was considerably lower than the concentrations of zinc in 14 regions of rat brain (Ebadi et al., 1981), we postulated that most of the zinc in the brain was bound and did not exist in free form. Based on these observations and conclusions, we searched for and identified a metallothionein-like protein in the rat brain.

Metallothioneins have been isolated from the rat brain (Ebadi, 1984; Ebadi et al., 1984), monkey brain (Gulati et al., 1987), bovine retina (Takahashi et al., 1988), bovine hippocampus (Paliwal et al., 1988), and bovine pineal gland (Awad and Ebadi, 1985; and Awad et al., 1988).

CHARACTERIZATION OF THE METALLOTHIONEIN IN THE RAT BRAIN

The synthesis of the rat brain metallothionein is stimulated only by intracerebroventricular (icv) administration of zinc and copper, but not cadmium delivered at a rate of 0.20 μmol zinc/μl/hr for 24-48 hrs by a surgically implanted Alzet minipump (Ebadi et al., 1984; Fig. 1). The brain metallothionein is not stimulated following a single (Itoh et al., 1983), or repeated (Ebadi, 1986) intraperitoneal (ip) administration of zinc sulfate (Table 1).

TABLE 1
EFFECTS OF ip AND FOR COMPARISON, icv ADMINISTERED $ZnSO_4$ ON THE SYNTHESIS OF HEPATIC AND BRAIN METALLOTHIONEINS

Treatment	Metallothionein in liver, μg Zn/mg protein	Metallothionein in brain μg Zn/mg protein
Saline	0.2 ± 0.01	1.5 ± 0.12
$ZnSO_4$ (7.5 mg/kg/d/ip)	5.4 ± 0.3*	1.4 ± 0.18
$ZnSO_4$ (5.0 mg/kg/d/10 d/ip)	ND	1.6 ± 0.15
$ZnSO_4$ (0.20 μmol/μL/hr for 48 hrs/icv)	1.2 ± 0.18*	7.4 ± 0.60*

*indicates that the experimental values are significantly different from the control groups at P < 0.01. Data from M. Ebadi, Biol. Trace Element Res. 11:101-116, 1986.

The original observation of Itoh et al. (1983) that a single ip administration of zinc sulfate did not stimulate the synthesis of brain metallothionein, has been confirmed by Heilmaier and Summer (1985). Furthermore, Blalock et al. (1987), by measuring mRNA using dot blot hybridization to a ^{32}P-labeled oligonucleotide DNA probe representing the terminal 5' sequence of the metallothionein gene, reported that the number of metallothionein mRNA in the brain did not increase following supplementing diets with increasing concentrations of zinc.

Although the daily ip administration of zinc did not induce the brain metallothionein, the icv-administered zinc induced the hepatic metallothionein (Table 1). The results of this study add further credence to a hypothesis that the steady-state concentration of zinc in the brain is firmly regulated and is not distributed readily (Ebadi et al., 1984; Ebadi and Pfeiffer, 1984; and Ebadi and Hama, 1986). The excess and unbound zinc most probably flows out from the brain via cerebrospinal fluid and reenters into plasma by transport through the choroid plexus, which lies outside of the blood-brain barrier.

The unique patterns of transport, distribution, and retention of zinc in the brain may be related in part to the morphological integrity of the blood-brain barrier. For example, Arvidson and Tjalve (1986) who studied the distribution of iv administered ^{109}Cd within the peripheral and central nervous system in rats 24 h and 1 wk after its injection, reported that the uptake of cadmium in the peripheral sensory and autonomic ganglia (e.g., trigeminal and superior cervical ganglia) was high, whereas, the uptake of cadmium in the brain and spinal cord was low.

CHROMATOFOCUSING OF RAT BRAIN METALLOTHIONEIN

Chromatofocusing studies have shown that icv-administered zinc stimulates the synthesis of three separate zinc-binding proteins, which focus at pHs of 6.8, 6.2, and 5.3, respectively (data not shown). It is expected that the protein focusing between pHs 5.3 and 4.8 is a metallothionein-like protein, whereas the characteristics of the other two zinc-binding proteins are not known and remain to be determined (Ebadi, 1986a).

Purification of the zinc-stimulated metallothionein-like protein by ion exchange chromatography on DEAE-Sephadex 25, using a linear gradient elution technique, produced two isoforms (data not shown) eluting, respectively, at 75 and 137 mM of Tris-acetate buffer, pH 7.5 (Ebadi, 1986).

METALLOTHIONEIN IN DEVELOPING BRAIN

Studies involving high performance liquid chromatography (HPLC) have shown that the adult rat brain contains three zinc-binding proteins. The intracerebroventricular (icv) administration of zinc stimulates the synthesis of only one of these proteins, producing metallothionein-like protein isoforms I and II, with retention times of 17.32 and 18.64 min, respectively. The synthesis of the other two zinc-binding proteins, with retention times of approximately 16.25 and 16.9 min, are not stimulated by icv administration of zinc. All three

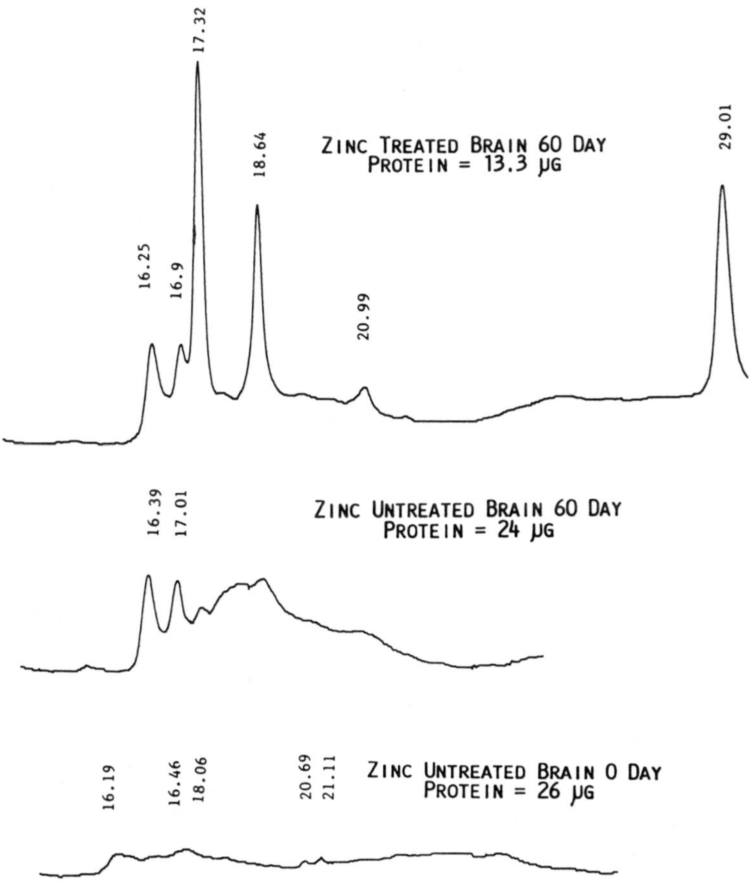

FIGURE 1. Comparative HPLC profiles of metallothionein in developing and adult brains.

zinc-binding proteins are absent in the brains of newborn rats (Fig. 1, Ebadi, 1986b).

METALLOTHIONEINS IN BOVINE RETINA AND HIPPOCAMPUS

The bovine retina contains a low-molecular-weight metallothionein-like protein which exhibits an elution volume (Ve/Vo) of 1.9 on gel permeation chromatography, and produces only one isoform, which on a reverse phase HPLC exhibits a retention time of 16.22 min and is similar to that produced by zinc-induced hepatic metallothionein II (Takahashi et al., 1988).

Similar to the zinc-induced hepatic metallothionein (hep-MT), the hippocampal metallothionein-like protein exhibits an elution volume (Ve/Vo) of 2.0 on gel filtration chromatography, and produces two isoforms, which on a reverse phase HPLC show retention times of 16.72 min (hep-MTI produces 16.53 min) and of 17.94 min (hep-MTII produces 18.45 min), respectively. The hippocampal metallothionein isoform II contains a cysteine to zinc ratio of 2.8 to 1.0, and, as shown by studies involving UV spectral analysis, apparently lacks aromatic amino acids, but possesses metallomercaptide bonds (Paliwal et al., 1988).

AMINO ACID COMPOSITIONS OF BRAIN METALLOTHIONEIN ISOFORMS

The zinc-stimulated metallothionein incorporates ^{35}S cysteine 24-fold higher than the native, unstimulated protein. In addition, when the partially purified protein is first incubated with ^{65}Zn and then subjected to ion exchange chromatography on DEAE Sephadex A-25 columns, the protein peak that binds ^{65}Zn and the one that incorporates ^{35}S cysteine depict identical elution profiles (Ebadi et al., 1984).

The amino acid compositions of rat brain metallothionein isoforms I and II, which resemble those of the hepatic metallothionein, are shown in Table 2 (M. Ebadi, D. Babin, and S. Swanson, 1987).

TABLE 2
THE AMINO ACID COMPOSITION OF RAT BRAIN METALLOTHIONEIN ISOFORMS I AND II

Amino acids	Isoform I Residues	Isoform I % Total	Isoform II Residues	Isoform II % Total
Aspartic acid	5	8.4	4	6.5
Threonine	3	5.0	3	4.9
Serine	8	13.3	9	14.8
Glutamic acid	3	5.0	4	6.6
Proline	2	3.3	2	3.3
Glycine	7	11.7	5	8.2
Alanine	4	6.7	5	8.2
Valine	2	3.3	1	1.6
Cysteine	17	28.3	18	29.5
Methionine	1	1.7	1	1.6
Isoleucine	–	–	–	–
Lysine	8	13.3	9	14.8
Total residues	60		61	

Since the rat brain and the bovine hippocampal metallothioneins do cross react with antibodies formed against both sheep and rat metallothioneins, it is anticipated that the immunologically dominant regions of the NH_2-terminal domain (residues 1-29) of neural metallothioneins are conserved.

The amino acid sequence for rat liver metallothionein 2 (as reviewed by Kagi and Kojima, 1987) was compared with the amino acid composition for rat brain metallothionein (Fig. 2). The sequence was generated by determining the sequence variability for metallothionein 2 from mouse, rat, hamster, calf, monkey, and human and assigning amino acids to sequence positions based on probability. A space was inserted to separate the rat exon 1, 2, and 3 products. This method of comparison indicates an overall identity, represented by a colon, of 89% (54/61 amino acids), and 67% (6/9 amino acids), 91% (20/22 amino acids), and 93% (28/30 amino acids) identity for exons 1, 2, and 3, respectively. A single period indicates amino acids which are highly conserved between the two sequences, a glutamic acid for glutamine in exons 2 and 3 and a threonine for serine in exon 2. There are fewer cysteines in the rat brain protein than the rat liver, and conversely there is a glycine and an aspartic acid in the brain protein that are not in the liver protein.

```
1          10         20         30          40          50          60
Rat Liver Metallothionein 2
MDPNCSCAT DGSCSCAGSCKCKQ CKCTSCKK SCCSCCPVGCAKCSQ GCICKEASDKCSCCA
:::. : :: :::::::::::::.::::.::: :::::::::::::::.::.:::::::::::::
MDPd S AT DGSCSCAGSCKCKeCKCTtCKK SCCSCCPVGCAKCSeGCkCKEASDKCSCCA
Rat Brain Protein (Putative sequence)
```

FIGURE 2. The proposed sequence of the rat brain metallothionein isoform II.

THE PROPOSED FUNCTION OF BRAIN METALLOTHIONEIN

The function(s) of brain metallothionein need to be established. Although the metallothionein-like proteins isolated from the brain share certain properties with those isolated and characterized in the peripheral organs such as liver and kidney, there is no reason to believe that these proteins have identical functions. Since the synthesis of brain metallothionein is stimulated following icv administration of zinc (Ebadi et al., 1984) and is depressed in zinc deficient state (Ebadi and Wallwork, 1985), one may assume that the levels of unbound zinc regulate the synthesis of brain metallothionein. Furthermore, the basal levels of metallothionein in rat brain, bovine pineal gland, retina, and hippocampus are quite high, an indication that zinc in sufficient amounts may be released to stimulate the synthesis of its binding protein. Unlike the hepatic metallothionein, the synthesis of brain metallothionein increases postnatally (Fig. 1) suggesting that the regulation of the synthesis of the two proteins may not be identical. Since cadmium is extremely neurotoxic in nature, it does not accumulate in neuron in sufficient amounts to stimulate the synthesis of metallothionein. Therefore, it is doubtful that the brain metallothionein plays a major role in cadmium detoxification, a role which is carried out admirably by the liver and the kidneys. Metallothionein by donating zinc may stimulate the activity of pyridoxal phosphate requiring enzymes such as glutamic acid decarboxylase (50% stimulation in hippocampus = 1.7×10^{-7} M), and by the virtue of its inducibility, it may bind the excess zinc and prevent from inhibition an extensive number of sulfhydryl containing enzymes such as glutamic acid decarboxylase (50% inhibition in hippocampus = 6.5×10^{-4} M) and receptors such as glutamate and dopamine (Fig. 3, see also Ebadi and Hama, 1986; and Awad et al., 1988).

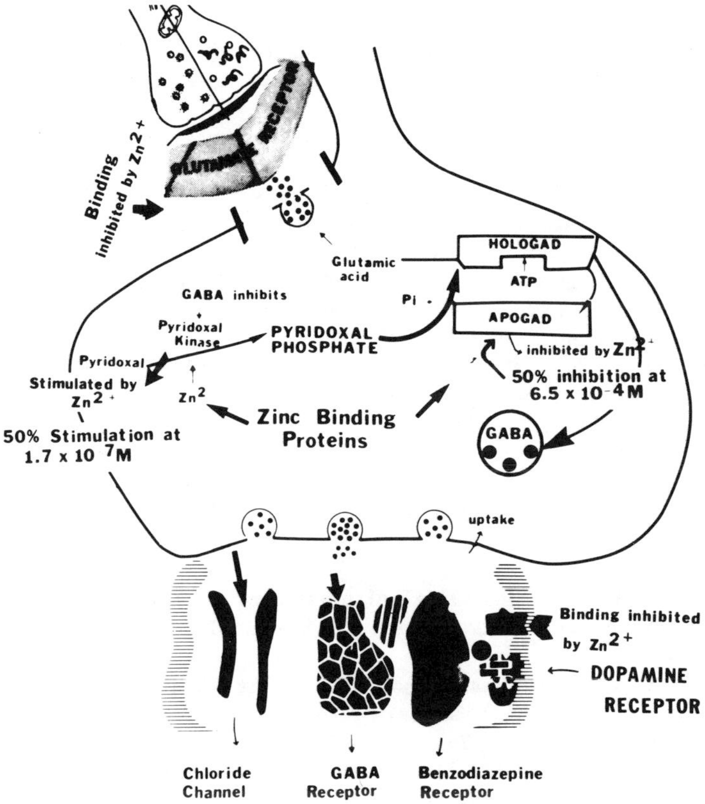

FIGURE 3. A model depicting the proposed roles of zinc and zinc-binding protein in events associated with GABAergic transmission in the brain. Data from M. Ebadi, regulation and function of pyridoxal phosphate in CNS in ''Selected Topics from Neurochemistry,'' N.N. Osborne (ed.), Pergamon Press, London, pp. 341-376, 1985.

ACKNOWLEDGMENT

The authors gratefully acknowledge the editorial and secretarial assistance of Mrs. Vickie Neal Stuetelberg.

REFERENCES

1. Arvidson B, Tjalve H (1986). Distribution of ^{109}Cd in the nervous system of rats after intravenous injection. Acta Neuropathol (Berl) 69:111–116.
2. Awad A, Ebadi M (1985). The characterization of metallothionein in bovine pineal gland. Fed Proc 44:3053.
3. Awad A, Govitrapong P, Hama Y, Hegazy M, Ebadi M (in press). Presence of metallothionein-like protein in the bovine pineal gland. J Neural Trans.
4. Blalock L, Dunn MA, Cousins RJ (1987). Metallothionein gene expression in rats: tissue specific regulation by dietary copper and zinc. J Nutr 118:222–228.
5. Ebadi M (1984). The presence of metallothionein-like protein (MTL) in rat brain. Fed Proc 43:3317.
6. Ebadi M (1985). Regulation and function of pyridoxal phosphate. In Osborne NN (ed): ''Selected Topics from Neurochemistry,'' London: Pergamon Press, pp 341–376.
7. Ebadi M (1986a). Biochemical characterization of a metallothionein-like protein in rat brain. Biol Trace Elem Res 11:117–128.
8. Ebadi M (1986b). Biochemical alteration of a metallothionein-like protein in developing rat brain. Biol Trace Elem Res 11:117–128.
9. Ebadi M, Babin D, Swanson S (1987). The structure and antigenicity of metallothionein-like protein in rat brain. Soc Neurosci Abs 17:1417.
10. Ebadi M, Hama Y (1986). Zinc-binding proteins in the brain. Adv Exp Med Biol 203:557–570.
11. Ebadi M, Itoh M, Bifano J, Wendt K, Earle A (1981). The role of Zn^{++} in pyridoxal phosphate mediated regulation of glutamic acid decarboxylase in brain. Int J Biochem 13:1107–1112.
12. Ebadi M, Pfeiffer R (1984). Zinc in neurologial disorders and in experimentally-induced epileptic seizures. Neurol Neurobiol 11B:307–324.
13. Ebadi M, Swanson S (1987). Characterization of metallothionein-like protein in rat brain. Experientia 52(Supplement):289–291.
14. Ebadi M, Wallwork JC (1985). Zinc-binding proteins (ligands) in brains of severely zinc-deficient rats. Biol Trace Elem Res 7:129–139.

15. Ebadi M, White RJ, Swanson S (1984). The presence and functions of zinc-binding proteins in developing and mature brains. Neurol Neurobiol 11A:39-57.
16. Gulati S, Paliwal VK, Sharma M, Gill KD, Nath R (1987). Isolation and characterization of a metallothionein-like protein from monkey brain. Toxicol 45:53-64.
17. Heilmaier HE, Summer KH (1985). Metallothionein content and zinc status in various tissues of rats treated with iodoacetic acid and zinc. Arch Toxicol 56:247-251.
18. Itoh M, Ebadi M (1982). The selective inhibition of hippocampal glutamic acid decarboxylase in zinc-treated epileptic seizures. Neurochem Res 7:1287-1298.
19. Itoh M, Ebadi M, Swanson S (1983). The presence of zinc-binding proteins in brain. J Neurochem 41:823-829.
20. Kagi JHR, Kojima Y (1987). Chemistry and biochemistry of metallothionein. Experientia 52(Supplement):25-61.
21. Paliwal VK, Takahashi T, Ebadi M (1988). The biochemical characterization of a metallothionein-like protein from bovine hippocampus. J Cell Biochem 12D(Supplement)353.
22. Takahashi T, Paliwal VK, Ebadi M (1988). The presence of a metallothionein-like protein in bovine retina. J Cell Biochem 12D(Supplement):353.

This study was supported by a grant from USPHS ES-03949.

Metal Ion Homeostasis: Molecular Biology and Chemistry, pages 269–280

AN EXAFS STUDY OF THE COPPER ACCUMULATED BY MAMMALIAN CELLS[1]

Jonathan H. Freedman[*+], J. Peisach[#+] and L. Powers[!]

[*]Institute for Structural and Functional Studies, University City Science Center, Philadelphia, PA, [+]Departments of Molecular Pharmacology and [#]Physiology and Biophysics, Albert Einstein College of Medicine, Bronx, NY and [!]AT&T Bell Laboratories, Murray Hill, NJ

ABSTRACT X-ray absorption spectroscopy has been applied to the _in vivo_ examination of a hepatoma cell line that accumulates copper and is resistant to metal toxicity. The _in vivo_ structure of the metal binding site of the accumulated copper, complexed by metallothionein in these cells, was compared to purified copper-metallothionein. Analysis of the K-absorption edge suggest that the accumulated copper is univalent and is complexed in a distorted tetrahedral geometry, as in the purified protein. Analysis of the extended X-ray absorption fine structure spectrum for cellular copper indicates that the absorbing metal is coordinated by four sulfur atoms at a distance of 2.28 ± 0.02 Å. In addition, a copper atom is located in the outer coordination shell 3.69 ± 0.05 Å from the absorbing metal. These copper and sulfur atoms may be arranged in a cluster similar to that observed in other metallothioneins.

---

[1]This investigation was supported in part by U.S. Public Health Service Grants AM-17702, GM-40168 and ES-05406. Synchrotron time was provided by the Stanford Synchrotron Radiation Laboratory, supported by National Science Foundation Grant DMR 77-27489, in cooperation with the Stanford Linear Accelerator Center and the U.S. Department of Energy.

INTRODUCTION

The mechanism of cellular resistance to copper toxicity was previously investigated in a series of copper-resistant rat hepatoma cell lines which were developed by subculturing wild type cells in media supplemented with increasing concentrations of copper (1). Resistant cells accumulate copper at concentrations proportional to their level of resistance. In addition, the majority of the accumulated metal is complexed by a cytoplasmic, cysteine-rich protein identical to rat liver copper-metallothionein (Cu-MT). Further, the cellular level of this protein is directly proportional to the amount of accumulated copper (1,2).

It has been shown by a variety of spectroscopic techniques that the chelated metals in purified Cu-MT and Cd,Zn-MTs are coordinated exclusively by cysteinyl sulfur atoms in a tetrahedral or distorted-tetrahedral geometry (3,4,5,6,7). Several of the sulfur atoms form thiolate bridges between adjacent metal atoms, producing highly organized metal-thiolate clusters (3,4,8,9). It has been argued that these clusters may become altered during protein purification, as they are readily oxidized due to their high cysteine and metal content (11). For this reason we have applied X-ray absorption spectroscopy (XAS) (12,13) to whole cells in order to determine the *in vivo* structural environment of accumulated copper bound to the Cu-MT produced by copper-resistant hepatoma cells.

EXPERIMENTAL PROCEDURES

Sample Preparation

Rat liver hepatoma cells were grown as monolayers at 37 °C under a 5-7% CO_2 atmosphere in RPMI-1640 medium in 10 mM HEPES buffer, pH 7.4, supplemented with 10% fetal bovine serum, 2 mM glutamine, non-essential amino acids, penicillin G (10 units/ml), and streptomycin (10 μg/ml). The growth medium for the resistant cell line, HAC_{800}, contained 800 μM Cu(II), added as a Cu(II)-histidine complex (1). Samples for XAS were prepared from wild type cells, designated HAC, and resistant cells grown on 8 x 100 mm dishes, after removing the growth media and washing twice with phosphate-buffered saline, pH 7.4 (1). Cells were removed from dishes with a rubber policeman and centrifuged at 3,000 x g for 10 min. The cell pellet was then washed twice with phosphate-buffered saline containing 1.0 mM EDTA, and the final pellet was transferred to a lucite XAS sample holder and immediately frozen in liquid nitrogen. Approximately 3 x 10^7 cells were packed into a 0.1 ml XAS sample holder, making the copper concentration in the HAC_{800} cell sample ~3 mM, based on the copper content previously determined by atomic absorption spectroscopy (0.9 μg copper/10^6 cells) (1).

Edge and EXAFS Measurements

X-ray absorption measurements were carried out at the Stanford Synchrotron Radiation Laboratory during dedicated operation of the SPEAR storage ring (40-80 mA at 3.0 GeV). Edge and extended X-ray absorption fine structure (EXAFS) measurements were made at beam line I-5, which is an unfocused beam providing ~1 x 10^{10} photon/sec, with 1-eV resolution at 9 keV. The X-ray signal was measured by monitoring the fluorescence emitted by the sample with the fast-plastic scintillation detector array previously described (14). The incident intensity of the X-ray beam was determined from the scattering off a mylar window with a scintillation detector placed 90° from the X-ray beam, between the sample and the defining slits (15).

X-ray absorption spectra of frozen cell samples were obtained at -100 °C, to minimize radiation damage, with the cryostat system previously described (16,17,18). In order to improve signal-to-noise, approximately 20 scans of each sample were collected and averaged. Comparison of the absorption edge spectrum obtained at the end of the experiment with that obtained at the beginning indicated that radiation damage had not occurred at the metal binding site.

Data Analysis

Absorption edge data were analyzed after removing a linear background and standardizing the energy scale of the 4p peak of cupric chloride (8996 eV). To determine the oxidation state and geometry of the accumulated metal, electronic transition energies of corrected sample data were compared to those of copper complexes of known structure including Cu(I)-diethyldithiocarbamate, Cu(II)-diethyldithiocarbamate, and purified dog liver Cu-MT (5).

The EXAFS of the copper accumulated in resistant cells was obtained by removing the contribution of copper normally present in cells by subtracting the XAS of HAC cells from that of HAC_{800} cells. Since the copper accumulated by HAC_{800} cells is complexed primarily by MT (2), the resulting difference spectrum contains the EXAFS of copper bound to this protein.

The EXAFS portion of the sample absorption spectrum was isolated from the background and/or isolated atom contributions by a cupric spline fit of the data, then multiplied by the cube of the photoelectron wave vector, k, to remove the $1/k^3$ dependence of the EXAFS, and finally Fourier transformed (14,19). The contributions of the first and outer coordination shells in the Fourier transform were backtransformed using a Fourier filter and resolved into their phase and amplitude functions. These functions were compared by the two-atom type data analysis procedure previously described (20) to those for a copper-nitrogen model, Cu(II)-tetraphenylporphyrin, with an average copper-nitrogen distance of 1.984 Å (21) and a copper-sulfur model, Cu(II)-diethyldithiocarbamate, with an average copper-sulfur distance of 2.316 Å (22). To determine copper-copper distances in the higher shell data Cu(II)-N,N,N,N-tetramethylethylenediamine with an absorber-scattered distance of 3.000 Å (23) was used.

RESULTS AND DISCUSSION

The K-absorption edges of copper for HAC_{800} and HAC cells are shown in Figure 1. The intensity of the absorption in the 8995 eV region, which is directly proportional to the metal concentration, was 10-fold greater for resistant cells than for HAC cells. This ratio is in good agreement with the relative copper concentrations previously measured by atomic absorption spectroscopy (1). Energy transitions in the absorption edge of resistant cells at 8984, 8988 and 8993 eV, are identical to those for Cu(I)-diethyldithiocarbamate and purified Cu-MT (Fig. 1), where the complexed copper has been shown to be univalent and coordinated in a distorted tetrahedral geometry (5). Since the absorption edge for HAC_{800} cells can be superimposed on that of purified Cu-MT (5), the accumulated copper in resistant cells is

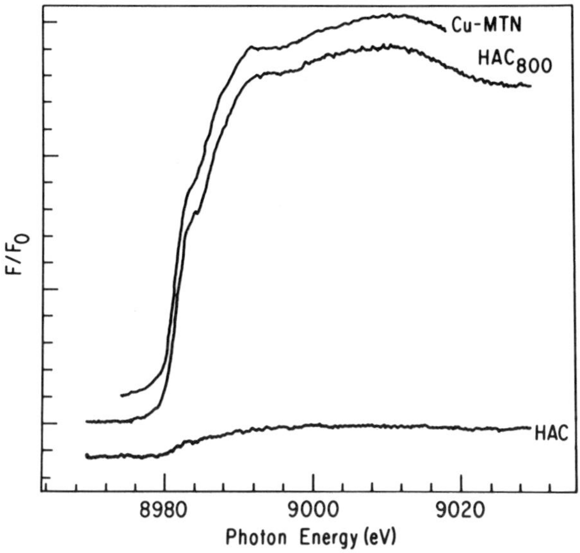

Figure 1. X-ray K-absorption edges for HAC_{800} cells grown in medium containing 800 μM copper and HAC cells. A linear background has been subtracted from the data setting the absorption below the edge equal to zero. Data are expressed as the ratio of the fluorescence intensity (F) to the intensity of the incident radiation (F_0). For comparison, the absorption edge for purified Cu-MT is presented (from ref 5). All three traces have been arbitrarily displaced vertically for clarity.

also univalent and in a site of comparable symmetry.

The background subtracted k^3-multiplied EXAFS spectra of accumulated copper in HAC_{800} cells and Cu-MT are shown in Figure 2. The signal to noise ratio of the EXAFS data from cells is comparable to that previously observed for a variety of other purified proteins (16,24,25).

In the Fourier transform of the EXAFS data (Fig. 3) the predominant peaks near $R + \alpha(k) = 1.8$ and 3.6 Å, in the cell data, represent backscattering contributions from the first and outer coordination shell atoms, respectively. This transform is similar to that from Cu-MT with peaks near $R + \alpha(k) = 1.8$ and 3.5 Å (5), suggesting that the structural environment of the copper accumulated by resistant cells and in the purified protein are comparable.

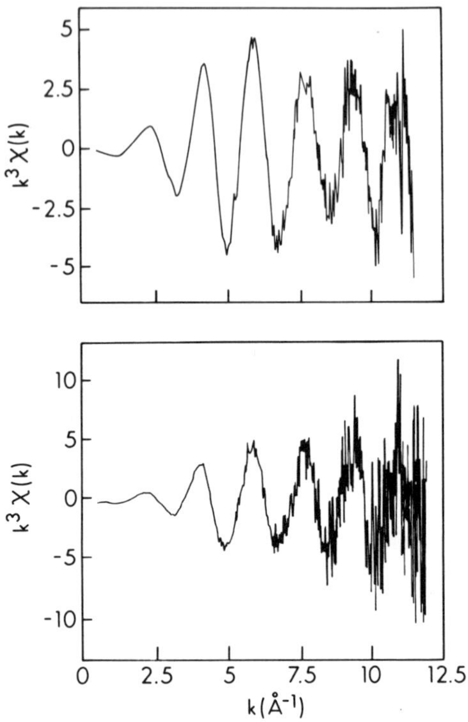

Figure 2. Background subtracted k^3-multiplied EXAFS of the copper accumulated by HAC_{800} cells (upper trace) and purified Cu-MT (lower trace) (from ref 5). k is the wave vector calculated from the photon energy (13). The copper XAS for wild type cells has been subtracted for the HAC_{800} data. The spectral difference is attributed to copper accumulated by HAC_{800} cells. The data have been truncated to k = 12.0 1/Å to remove the contributions of the zinc K-absorption edge (9660 eV).

The Fourier transform for the copper accumulated by resistant cells has a clearly resolved peak at 3.5 Å that is not well resolved in the data from purified Cu-MT. The differences between these transforms suggests that the structure of the metal binding site in the purified protein may have become distorted during its isolation.

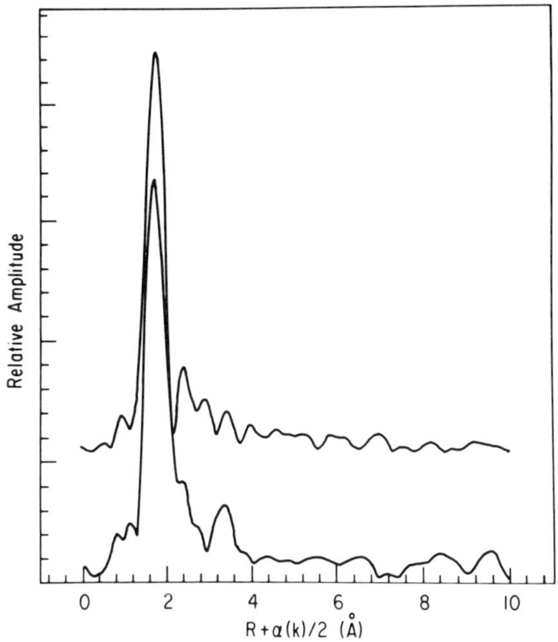

Figure 3. Magnitude of the Fourier transform of the k^3-multiplied data of the copper accumulated by HAC_{800} cells presented in Fig 2 (lower trace) and Cu-MT (upper trace) (from ref 5). R is the radial distance between the absorbing atom and the backscattering atoms, and $\alpha(k)$ the energy dependent absorber-scatterer phase shift (30).

The Fourier filter backtransformed data for copper accumulated by HAC_{800} cells were analyzed by comparing these data to those for copper-nitrogen, copper-sulfur, and copper-copper model compounds to determine the numbers and types of atoms coordinated to the absorbing copper atom, and their distances (Table 1), The results of this analysis suggest that the absorbing copper atom is coordinated by four sulfur atoms in the first coordination shell at an average distance of 2.28 ± 0.02 Å, similar to the 2.27 ± 0.02 Å copper-sulfur distance observed for purified Cu-MT and in Cu_6Zn_3-MT, 2.25 ± 0.01 Å (5,26,27). Introduction of nitrogen contributions into the model causes a significant decrease in the quality of the fit, i.e. increase in the value of χ^2, indicating an

Table 1

FIT PARAMETERS FOR FIRST SHELL FILTERED DATA FOR HAC_{800} CELLS

Model	r_N (Å)	r_S (Å)	$\Delta\sigma_N^2$ ($x10^3$)	$\Delta\sigma_S^2$ ($x10^2$)	$\Delta E_{o,N}$ (eV)	$\Delta E_{o,S}$ (eV)	χ^2
0N4S	-	**2.28**	-	**1.82**	-	**4.79**	**1.5**
1N3S	2.10	2.31	5.44	1.25	-5.59	-7.58	2.5
2N2S	2.13	2.31	1.02	1.91	-6.84	-7.49	3.1
3N1S	2.13	2.33	1.25	4.41	-7.15	-8.49	4.3
4N0S	2.16	-	1.40	-	-9.37	-	10.2

Parameters were calculated with the number of nitrogen (N) and sulfur (S) atoms held constant by the fitting procedure previously described (5). The terms r_i, $\Delta\sigma_i^2$, and $\Delta E_{o,i}$ are the absorber-scatterer distance, change in Debye-Waller factor, and change in the threshold energy, relative to the model complex, respectively, χ^2 is the sum of residuals squared, which represent the goodness of fit. The parameters shown in **bold face** are the best fits of the data, by the criteria previously described (5). The errors for r_i and $\Delta\sigma_i^2$ are ± 0.02 Å and ± 1.5 x 10^{-3} $Å^2$, respectively.

absence of nitrogen or oxygen atoms directly coordinated to the absorbing metal atom. In addition, the cell data are best fit with a single copper-sulfur distance rather than to multiple distances, as reported for purified yeast Cu-MT, 2.22 and 2.36 Å (28), and in this way is similar to the findings for Cu-MT from dog and pig (5,26,27).

The resolved outer shell peak in the HAC_{800} Fourier transform corresponds to a copper atom at 3.69 ± 0.05 Å, although it may contain contributions from lower atomic number atoms (e.g., nitrogen, oxygen or sulfur). This copper-copper distance is within the range of distances observed by XAS of Cu-MT and corresponds to an outer shell backscattering copper atom found in the purified protein (5).

Spectroscopic studies of the metal centers in purified mammalian Cu-MT and Cd,Zn-MT have shown that the metal ions share bridging sulfur atoms forming metal-thiolate clusters (3,4,5,6,9,10). Assuming a similar structural arrangement for the copper bound in the metallothionein from HAC_{800} cells, a Cu-S-Cu bond angle of 108 ± 4° can be calculated from absorber-scatterer distances for Cu-S and Cu-Cu of 2.28 ± 0.02 and 3.69 ± 0.05 Å respectively. A similar Cu-Cu distance for metals sharing a bridging sulfur atom, and metal-sulfur- metal bond angle are observed in Cd_5,Zn_2-MT (3,4), suggesting that the backscattering copper atom near 3.69 Å and the absorbing metal are bonded to the same sulfur and may be in a clustered arrangement. In addition, the computed bond angle of 108° suggests sp^3 hybridization of the sulfur atom, similar to that previously observed for Cu(I) bound to thiol containing ligands (29).

REFERENCES

1. Freedman JH, Weiner RJ, Peisach J (1986). Resistance to copper toxicity of cultured hepatoma cells. J Biol Chem 261:11840-11848.
2. Freedman JH, Peisach J Manuscript in preparation.
3. Furey WF, Robbins AH, Clancy LL, Winge DR, Wang BC, Stout CD (1986). Crystal structure of Cd,Zn metallothionein. Science 231:704-710.
4. Furey WF, Robbins AH, Clancy LL, Winge DR, Wang BC, Stout CD (1987). Crystal structure of Cd,Zn metallothionein. Experientia, Suppl 52:139-148.

5. Freedman JH, Powers L, Peisach J (1986). Structure of the copper cluster in canine hepatic metallothionein using X-ray absorption spectroscopy. Biochemistry 25:2342-2349.
6. Otvos JD, Armitage IM (1982). Elucidation of metallothionein structure by ^{113}Cd NMR. In Sykes BD, Glickson J, Bothner-By AA, (eds.): "Biochemical Structure Determination by NMR," New York: Marcel-Dekker, pp. 65-96.
7. Wagner G, Frey MH, Neuhaus D, Wörgötter E, Braun W, Vasak M, Käji JHR, Wuthrich K (1987). Spatial structure of rabbit liver metallothionein-2 in solution by NMR. Experientia Suppl 52:149-157.
8. Otvos JD, Armitage IM (1980). Structure of the metal clusters in rabbit liver metallothionein. Proc Natl Acad Sci, USA 77:7094-7098.
9. Otvos JD, Engeseth HR, Wehrli S (1985). Preparation and ^{113}Cd NMR studies of homogeneous reconstituted metallothionein: Reaffirmation of the two-cluster arrangement of metals. Biochemistry 24:2427-2431.
10. Vasak M, Overnell J, Good, M (1987). Spectroscopic and chemical approaches to the study of metal-thiolate clusters in metallothionein (MT). Experientia, Suppl. 52:179-190.
11. Minkel DT, Poulsen K, Wielgus S, Shaw III CF, Petering DH (1980). On the sensitivity of metallothionein to oxidation during isolation. Biochem J 191:475-485.
12. Eisenberger P, Kincaid BM (1978). EXAFS: New horizons in structure dererminations. Science 200:1441-1447.
13. Powers L (1982). X-ray absorption spectroscopy application to biological molecules. Biochim Bioiphys Acta 683:1-38.
14. Powers L, Blumberg WE, Chance B, Barlow CH, Leigh JS, Smith J, Yonetani T, Vik S, Peisach, J (1979). The nature of the copper atoms of cytochrome c oxidase as studied by optical and x-ray absorption edge spectroscopy. Biochim Biophys Acta 546:520-538.
15. Garcia-Iniguez L, Powers L, Chance B, Sellin S, Mannervik B, Mildyan, AS (1984). X-ray absorption studies of the Zn^{2+} site of glyoxalase I. Biochemistry 23:685-689.
16. Peisach J, Powers L, Blumberg WE, Chance B (1982). Stellacyanin. Studies of the metal-binding site using X-ray absorption spectroscopy. Biophys J 38:277-285.

17. Chance B, Angiolillo P, Yang EK, Powers L (1980). Identification and assay of synchrotron radiation-indiced alterations on metalloenzymes and proteins. FEBS Lett 112:178-182.
18. Powers L, Chance B, Ching Y, Angiolillo P (1981). Structural features and the reaction mechanism of cytochrome oxidase. Biophys J 34:465-498.
19. Eisenberger P, Lengeler B (1980). Extended X-ray absorption fine-structure determinations of coordination numbers: Limitations. Phys Rev Part B 22:3551-3562.
20. Brown JM, Powers L, Kincaid BM, Larrabee JA, Spiro TJ (1980). Structural studies of the hemocyanin active site. 1. Extended X-ray absorption fine structure (EXAFS) analysis. J Am Soc 102:4210-4216.
21. Fleischer EB, Miller CK, Webb LE (1964). Crystal and molecular structure of some metal tetraphenyl-porphyrins. J Am Chem Soc 86:2342-2347.
22. O'Connor BH, Maslen EN (1966). Crystal structure of Cu(II) succinate dihydrate. Acta Crystallogr 21:828-830.
23. Mitchell TD, Bernard WH, Wasson JR (1970). Crystal structure of di-μ-hydroxobis(N,N,N',N'-tetramethyl-ethylenediamine)dicopper(II) bromide. Acta Crystallogr Sect B 26:2069-2101.
24. Woolery GL, Powers L, Peisach J, Spiro TG (1984). X-ray absorption study of Rhus laccase: Evidence for a copper-copper interaction, which disappears on type 2 copper removal. Biochemistry 23:3428-3434.
25. Woolery GL, Powers L, Winkler M, Solomon EI, Spiro TG (1984). EXAFS studies of binuclear copper site of oxy-, deoxy-, metaquo-, metfluoro-, and metazido-hemocyanin from arthropods and molluscs. J Am Chem Soc 106:86-92.
26. Abrahams IL, Bremner I, Diakun GP, Garner CD, Hasnain SS, Ross I, Vasak M (1986). Structural study of the copper and zinc sites in metallothioneins by using extended X-ray absorption fine structure. Biochem J 236:585-589.
27. Hasnain SS, Diakun GP, Abrahams I, Ross I, Garner CD, Bremner I, Vasak M (1987). EXAFS studies of metallothionein. Experientia Suppl 52:227-236.
28. Bordas J, Koch MHJ, Hartmann J-H, Weser U (1982). Tetrahedral copper-sulphur coordination in yeast Cu-thionein. An EXAFS study. FEBS Lett 140:19-21.

29. Vortisch V, Kroneck P, Hemmerich P (1976). Model studies on the coordination of copper in enzyme. IV. Structure and stability of cuprous complexes with sulfur-containing ligands. J Am Chem Soc 98:2821-2826.

30. Lee PA, Citrin PH, Eisenberger P, Kincaid BM (1981). Extended X-ray absorption fine structure-its strengths and limitations as a structural tool. Rev Mod Phys 53:769-806.

III. CHEMISTRY AND BIOLOGY OF METAL BINDING PEPTIDES

Metal Ion Homeostasis: Molecular Biology and Chemistry, pages 283–300

PHYTOCHELATINS IN PLANTS

Erwin Grill

Institut für Pharmazeutische Biologie,
Universität München, West Germany

ABSTRACT Phytochelatins and homo-phytochelatins are a class of heavy metal-binding peptides which are ubiquitous in plants from algae to orchids. The cellular function of phytochelatins appears to be detoxification and homeostasis of heavy metal ions. These peptides are composed of repetitive γ-glutamyl-cysteine units with a carboxyl-terminal glycine in phytochelatin or β-alanine in homo-phytochelatin. The homologous peptides, ranging from 5 to 23 amino acids, are induced by ions such as zinc, cadmium, copper and lead and complex these ions via thiolate coordination. On account of the γ-glutamyl linkages in their primary structure the peptides are not primary gene products, but are formed by a ribosome-independent reaction involving glutathione or homo-glutathione.

Metal-tolerant and non-tolerant plants accumulate heavy metals as the result of a nonspecific uptake system which does not exclude toxic ions (1). In fact, metal analysis of tolerant plants led Jowett in 1958 to postulate a specific mechanism by which heavy metals are rendered nontoxic (2). Several strategies of metal tolerance were suggested including compartmentation of ions in vacuoles by complexation to organic acids (3), ion complexation at cell walls (4), reduction in ion uptake (5), metal-resistant enzymes (6) and chelation by specific metal-binding proteins (7,8).

The absorption of heavy metal ions by roots is thought to occur as an active metabolic process by a carrier system for which the ions of zinc, copper, nickel and cadmium may compete (9,10). It seems unlikely that the specific demands for essential heavy metal ions are regulated at the level of uptake. Thus, plants should possess a system capable of regulating the availability of these micronutrients within the cell. Such a homeostatic and also a detoxifying function could be fulfilled by heavy metal-binding molecules like the phytochelatins (PCs). PCs are a class of metal-inducible peptides which complex metal ions by thiolate coordination (11). A variety of heavy metals such as zinc, copper, cadmium, lead and mercury are able to induce the synthesis of these peptides. The primary structure of phytochelatins was elucidated as (γ-Glu-Cys)n-Gly with index n ranging from 2 to 11. Phytochelatins are ubiquitous in plants with the exception of soybean and a few other plants of the order _Fabales_ which synthesize homo-phytochelatins (h-PCs) upon heavy metal exposure (12, 13). H-PCs are also poly(γ-glutamyl-cysteinyl) peptides but the C-terminal glycine of phytochelatin is replaced by β-alanine.

ISOLATION AND CHARACTERIZATION

In 1977 Casterline and Barnett reported the existence of a low molecular weight, cadmium-binding protein in soybean which was present in roots and leaves of cadmium-treated plants (14). Subsequent investigations of _Agrostis_ (7, 15), beans (16), tomato (17), tobacco (18), maize (19), _Silene_ (8), red cabbage (20) and spinach (21) revealed the general occurrence of a low molecular weight protein, or peptides complexing copper or cadmium ions in higher plants. W. Rauser (15) and G. Wagner (20) demonstrated the inducible nature of the metal-binding component characterized by a high cysteine and glutamine/glutamic acid content.

Our lab attempted the isolation and characterization of the metal-binding material by employing cell suspension cultures of higher plants. In general, these cultures are dedifferentiated and, therefore, more active in their primary metabolism than intact plants. Exposure of _Rauvolfia serpentina_ cell suspension cultures to 200 μM Cd^{2+} resulted in an almost complete (>90%) uptake of the metal within 5 days (22). The cadmium uptake was accompanied by a 6 to 7 fold increase of the total sulfhydryl content in the cells during the first 2 days of exposure. Crude extracts of those metal-treated cells revealed, after gel filtration, a fraction of about 3.6 kDa with which more than 90% of the extracted cadmimum was associated and which was subsequently identified as PC (11). The cadmium fraction contained a very high amount of sulfhydryl groups as demonstrated by reaction with Ellman's reagent and was absent in control cells to which cadmium was administered during extraction. A similar elution profile was obtained with extracts from soybean roots exposed to 20 μM Cd^{2+} for 4 days (Fig. 1A).

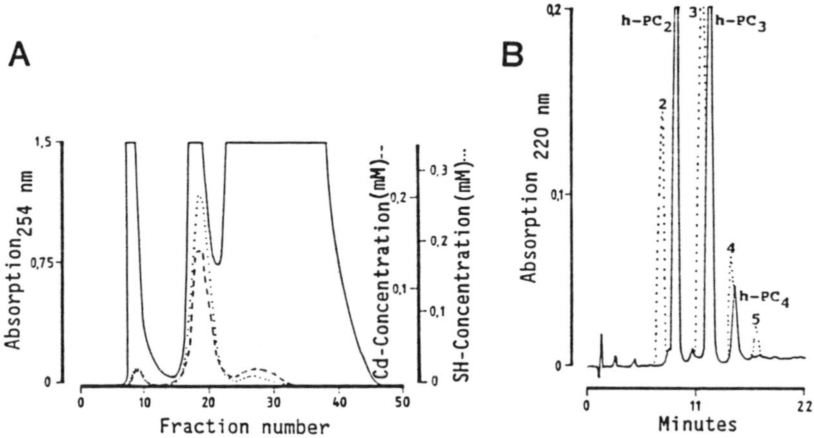

Figure 1. Homo-phytochelatin isolation from roots of soybean roots exposed to 20 μM $Cd(NO_3)_2$ for 4 days. Elution profile (A) of the crude extract separated by a Sephadex G-50 column and (B) of the purified complex under metal-dissociating conditions employing reversed HPLC. The dotted line represents the elution pattern obtained with phytochelatin and the individual peptides are signified by their index number n of the formula $(\gamma\text{-Glu-Cys})_n\text{Gly}$. Conditions were as described (11).

The 3-4 kDa complex containing about 60 % of the detectable cadmium was very hydrophilic since it was not precipitated in 80% saturated ammonium sulfate solution. By combining anion exchange chromatography, ammonium sulfate fractionation, ultrafiltraton and gel filtration, PC and h-PC were isolated with an average yield of approximate 50-60% of the extracted cadmium (Table 1). Typically, from 10 g of dry cell mass about 60 mg cadmium-PC was isolated with a metal content of 14.8 % (w/w) corresponding to about 5 g metal-free PC/ kg dry weight. The UV-absorption and

CD-spectrum (Fig. 2A) were indicative of a cadmium thiolate coordination. Amino acid analysis of PC revealed more than 40 mole percent of cysteine, equal numbers of glutamine/ glutamic acid and only glycine.

TABLE 1
ISOLATION OF PHYTOCHELATIN

purification step	volume (ml)	protein (mg)	cadmium (mg)	yield (% Cd)
crude extract	330	225	18.6	100
DEAE eluate	18	91.2	17.6	95
salt fractionation	21.8	1.7	16.4	88
ultrafiltration	6.1	1.1	14.0	75
gel filtration	52.4	0.02	11.1	60

Cell suspension cultures of Rauvolfia serpentina were exposed to 200 μM $Cd(NO_3)_2$ for 5 days and the cadmium-binding peptides purified as described (11). Phytochelatin does not react in standard protein assays.In this case protein was determined according to Bradford (51). Metal analysis was performed by atomic absorption spectrometry employing a Perkin Elmer 300 (flame mode). The lyophylised fraction yielded 59.1 mg pure cadmium phytochelatin from 280 g cells (fwt; 10.5 g dwt). The plant tissue contained 21.4 mg cadmium.

Gel filtration of the isolate revealed metal complexes of 3.6 kDa and 2.5 kDa (Fig. 2B). Under acidic, metal-dissociating conditions the complex was resolved into several distinct peptides by reversed phase HPLC (11). Such a separation is shown in Fig. 1B for h-PCs isolated from soybean.

The individual peptides of PC and h-PC metal-complexes were isolated by preparative HPLC and the sequence was established by two independent lines of evidence. Edman degradation of the molecules had failed even though the N-terminal amino acid was not blocked, and subsequently was identified as glutamic acid. Incubation of the peptides with the enzyme γ-glutamyltranspeptidase liberated a glutamic acid and demonstrated the presence of an isopeptidic linkage between the γ-carboxyl group of glutamic acid and the second amino acid.

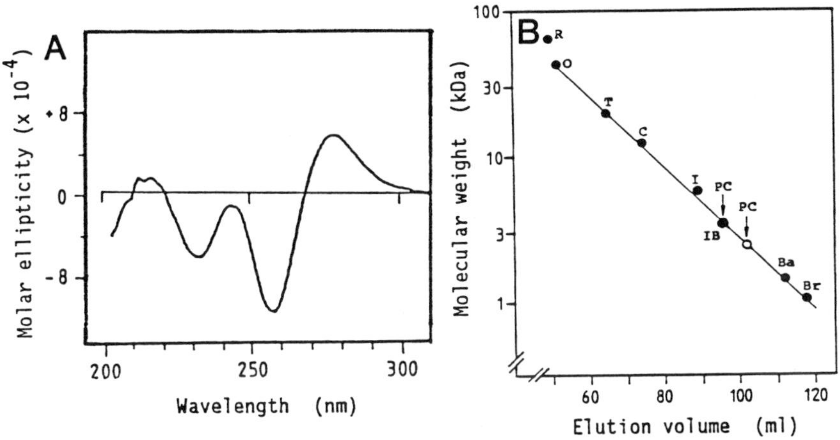

Figure 2. Characterization of cadmium phytochelatin (A) by CD-spectrum and (B) by determination of the molecular weight. (A) The CD spectrum of purified cadmium complex from <u>Rosa canina</u> was recorded in 50 mM sodium phosphate buffer, pH 7.0. Molar ellipticity is given in units of degrees per square centimeter per decimole. (B) Gel filtration of Cd-phytochelatin and molecular weight standards on a Sephadex G-50 column (1.5x90 cm, flow rate 11.6 ml/h) in 0.1 M KCl, 50 mM potassium phosphate buffer, pH 7.5. R, bovine serum albumin 67 kDa; O, ovalbumin 44 kDa; T, trypsin inhibitor 20.1 kDa; C, cytochrome C 12.3 kDa; I, insulin 6 kDa; IB, insulin B chain 3.45 kDa; Ba, bacitracin 1.45 kDa and Br, bradykinin 1.06 kDa. The phytochelatin (PC) was resolved into complexes of 3.6 kDa and 2.5 kDa.

The primary structures of the peptides were elucidated, first, by a combination of γ-glutamyltranspeptidase treatments and Edman degradation steps and , second, by partial hydrolysis and two-dimensional thin layer chromatography of the peptides which had been dinitrophenylated at the N-terminus (Fig 3). The peptides consist of a linear chain of reiterated γ-glutamyl-cysteinyl units with a C-terminal glycine or β-alanine in PCs and h-PCs, respectively, (Fig 4) ranging from 5 to 23 amino acids in length.

PC_2 and PC_3 (the index signifies the number of glutamyl-cysteinyl units in the individual peptides) were also sequenced by Kondo et al. (23), and PC_3 and PC_4 were sequenced elegantly by Steffens et al. (24) employing fast atom bombardment mass spectrometry.

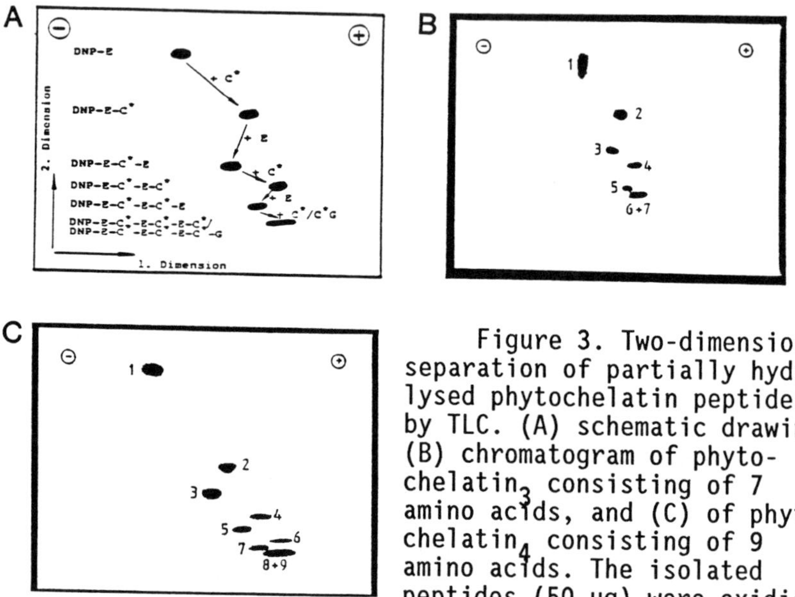

Figure 3. Two-dimensional separation of partially hydrolysed phytochelatin peptides by TLC. (A) schematic drawing, (B) chromatogram of phytochelatin$_3$ consisting of 7 amino acids, and (C) of phytochelatin$_4$ consisting of 9 amino acids. The isolated peptides (50 ug) were oxidised by performic acid and the N-terminus dinitrophenylated. The yellow peptides were partially hydrolysed (11) and the generated fragments separated in the first dimension by electrophoresis (water: formic acid:acetic acid, 854:26:120; 20 V/cm, 1h) on a silica plate, and in the second dimension by chromatography (n-propanol:30% ammonia:water, 50:15:10). The spots were photographed under UV light (360nm) and their amino acid composition determined in relation to the concentration of the dinitrophenyl moiety. DNP, dinitrophenyl; E, glutamic acid; C*, cysteic acid, G, glycine. The numbers correspond to the number of amino acids in the fragments. The uncleaved phytochelatin and the peptide lacking only glycine (B, 6+7; C, 8+9) were not fully separated.

The repeated isopeptidic linkages in these peptides imply that PCs are not formed on ribosomes and, thus, can not be considered as primary gene products.

A Phytochelatin

B Homo-Phytochelatin

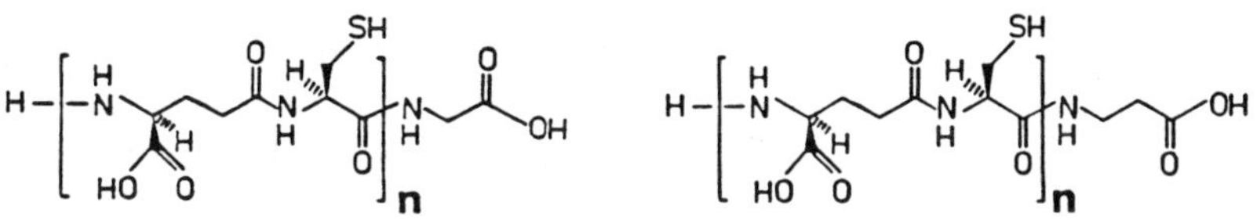

$(\gamma\text{-Glu-Cys})_n\text{Gly}$, n = 2-11

$(\gamma\text{-Glu-Cys})_n\beta\text{-Ala}$, n = 2-5

Figure 4. Primary structure of (A) phytochelatin and (B) homo-phytochelatin.

Another pecularity of PC is the dependence of the apparent molecular weight of the metal complex on the ionic strength. In gel filtration experiments using low ionic strength buffers (eg. 5 mM Tris-Cl pH 7.8) the complex behaved like a globular protein of almost 10 kDa (12). A similar observation was reported by Rauser and Lue-Kim measuring complexes of up to 20 kDa (25). The 2-5 fold increase in the apparent molecular weight is not readily explained by a solely intramolecular conformational change. Rather it probably reflects an intermolecular aggregation of PC-metal clusters.

INDUCING METALS

Despite the detailed biochemical characterization of PC and h-PC we have no idea how the molecules are induced. We can postulate that a kind of receptor senses the heavy metal ion status or, less likely, the sulfhydryl status of the cell. Such a sensor could be a low affinity heavy metal-binding site (in comparison to PC, but with a higher affinity than glutathione) which triggers the synthesis of complexing peptides when occupied by metal ions. In order to gain a first insight into such a mechanism we investigated a wide range of metal ions capable of inducing PC. In cell suspension cultures of Rauvolfia serpentina PC formation was observed after exposure to ions from the group of the periodic system I B, Cu-Ag-Au, II B, Zn-Cd-Hg, IV B, Sn-Pb, V B, As-Sb-Bi and VI B, Se-Te (12). The elements As and Se were applied to the plant culture as the anions AsO_4^{3-} and SeO_3^{2-}. It is not known whether, or how As and Se are complexed to PC but complexes similar to the analogous glutathione complexes are likely. Glutathione (GSH) reduces these anions to form $(GS)_3As$ and $(GS)_2Se$ (26). Interestingly, metal complexes are more stable with dithiol ligands, which provide a ring structure, than with monothiols. Therefore, a PC-As complex should be far more stable than the corresponding complex with GSH.

In addition to cadmium, complexes of copper, zinc and lead were verified by isolating and characterizing the corresponding complexes (Table 2).

TABLE 2
ISOLATION OF PHYTOCHELATIN METAL COMPLEXES

metal-salt	concen-tration uM	crude extract (mg metal)	purified phytochelatin (mg metal)	Yield (% metal)	ratio Me:SH
$CuSO_4$	50	3.7	1.3	35	1:1
$Zn(NO_3)_2$	1000	52.8	9.5	18	1:3[a]
$Pb(NO_3)_2$	1000	11.3	4.3	38	1:2
$Cd(NO_3)_2$	200	18.5	11.1	60	1:2

Cell suspension cultures of Rauvolfia serpentina were exposed for 4-5 days to the heavy metal ions, then harvested and PC isolated (11). The ratio of metal to sulfhydryl groups (Me:SH) is given in even numbers.
[a] The zinc phytochelatin complex is labile to weak acids causing substantial metal loss during purifation and resulting in a low metal:sulfur ratio. Reconstituted zinc phytochelatin revealed a ratio of 1:2.

Metal-PC complexes of cadmium, lead and zinc (PC reconstituted with zinc) revealed a metal to sulfhydryl ratio of 1:2 whereas the copper-PC showed a stoichiometry of 1:1, resembling copper-(I)metallothionein of Neurospora (27) and indicating a monovalent redox state of these ions. PC was not induced by alkaline and alkaline earth ions, which supports the idea that ions with a high affinity to sulfhydryl groups are required to induce a response. In general, more than 80% of the ions of copper, zinc, lead and cadmium in crude extracts of Rauvolfia serpentina cell suspension cultures were bound to PC. In tissue of whole plants 60-80% of the cadmium was associated with PC (12, 13, 20, 25), and in maize roots up to 95% (28).

UBIQUITY OF PHYTOCHELATINS

PCs are induced by a wide variety of heavy metals and their formation is not restricted to cell suspension cultures. In fact, these peptides are found in plants throughout the plant kingdom from unicellular algae to the most developed plants, the orchids.

To screen for PC induction in plants, we used cadmium ions since these metal ions proved to be the strongest inducer of PC biosynthesis. The assay was based on HPLC separation of deproteinized extracts with sulfhydryl-specific detection by employing a post column derivatisation (13, 14). Up to now almost 0.1% of the approximately 300,000 known plant species have been examined, and all of the plants investigated either synthesize PC or h-PC under cadmium stress (11, 12, 29, unpublished results). Among these plants were several species of the green algae, like Euglena and Chlorella, and red and brown algae (36). The cadmium-PC of Chlorella was isolated and unequivocally identified by peptide sequencing. In addition to ferns and mosses, gymnosperms and more than two hundred different angiospems revealed a cadmium-induced synthesis of PC or h-PC. These data imply that the poly-glutamyl-cysteinyl peptides are ubiquitous in the plant kingdom. Thus, the metal-binding proteins reported earlier (7, 8, 16-18) were most likely PC, and the protein isolated from soybean (14) h-PC since these molecules were induced by heavy metals and complexed them predominantly. The discrepancy in the molecular weight can be explained by the tendency of PC to form metal complexes of higher apparent molecular weight in low ionic strength buffers (see above).

Additionally, certain fungi are capable of PC synthesis. Murasugi et al. (30) isolated and characterized a cadmium-binding complex from the fission yeast _Schizosaccharomyces pombe_. The primary structure was subsequently determined to be identical to PC_2 and PC_3 (23). We confirmed these results and showed that the whole set of PC peptides is formed in the yeast and is induced by various metal ions (31). In contrast, in the yeast _Saccharomyces cerevisiae_ the heavy metal-binding metallothionein has been identified (32, 33). In this organism, and in the fungus _Neurospora crassa_, no PC formation was observed after Cd exposure (Fig.5).

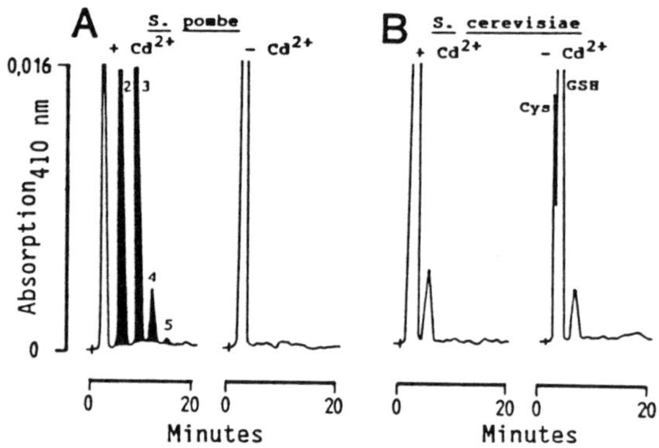

Figure 5. Assay for phytochelatin in cadmium-exposed yeasts, (A) _Schizosaccharomyces pombe_ and (B) _Saccharomyces cerevisiae_ by HPLC. Cells were cultivated in 1 mM $Cd(NO_3)_2$ for 24 hours and extracted. The deproteinized crude extract was separated by HPLC with sulfhydryl-specific detection via post column derivatisation with Ellman reagent (12). Phytochelatin peptides are marked in black and signified by the index number n of PC_n.

In both microorganisms the metallothioneins are only induced by copper ions. Recently D. Winge and coworkers (34) have found that the fungus _Torulopsis glabrata_ produces metallothionein and PC under copper and cadmium exposure, respectively. We have investigated cadmium-treated fungi for PC formation and only a small fraction of the tested organisms synthesized PC. In general, these results suggest that plants and some fungi use PCs as a heavy metal-complexing system whereas animals and most fungi utilize metallothioneins. Both systems are analogous in that they complex heavy metal ions via thiolate coordination, but they differ completely in their biosynthesis since PC are not directly encoded by genes.

BIOSYNTHESIS AND FUNCTION

The repetition of γ-Glu-Cys units in PC implied the involvement of GSH in their formation (11). The first support of this idea was obtained by the isolation of h-PC. This alanine analogue of PC was identified in some cadmium-exposed plants of the *Fabaceae*, such as mungbean and soybean known to have homo-glutathione (γ-Glu-Cys-β-Ala) instead of GSH (35, 36). In plants like *Lathyrus ochrus* which contain both GSH and homo-glutathione the synthesis of PC and h-PC was observed (13).

A second indication for a role of the tripeptide in PC formation was revealed by monitoring the intracellular GSH and PC levels during heavy metal induction. Addition of cadmium to a cell suspension culture of *R. serpentina* resulted in a rapid and linear decrease in the GSH concentration and in a corresponding incorporation of γ-Glu-Cys units into PC during the first 4 hours of exposure (12). At the beginning of the induction period predominantly PC_2 was synthesized, the peptide containing 2 Glu-Cys units, followed after a lag phase by the formation of PC_3 and the longer peptides. In the fission yeast an identical induction characteristic was observed (31) consistent with a subsequent transfer of γ-Glu-Cys units onto GSH and onto the growing PC chain, thus, giving rise to the whole set of peptides.

Upon heavy metal induction PC formation always proceeded at a maximal rate with no detectable lag period (12, 22, 31), even with a resolution of a few minutes of the induction process (Fig. 6). This finding suggests that PC is synthesized by constitutive enzyme(s). P. Goldsbrough and coworkers demonstrated that inhibition of the protein biosynthesis by cycloheximide does not prevent PC formation (37). However, PC synthesis was inhibited by blocking the GSH biosynthesis with buthionine sulfoximine (12, 31, 37, 38). Buthionine sulfoximine specifically inhibits the γ-glutamyl-cysteine synthetase (39), the first enzyme of the GSH biosynthetic pathway. The buthionine sulfoximine-treated plant cells were more metal-sensitive than control cells (Table 3) indicating a role of either GSH, PC or both in metal detoxification. Metal distribution analysis of crude plant extracts have shown that a metal-PC complex is more favored than a metal-GSH complex and, thus, the detoxification is probably due to PC.

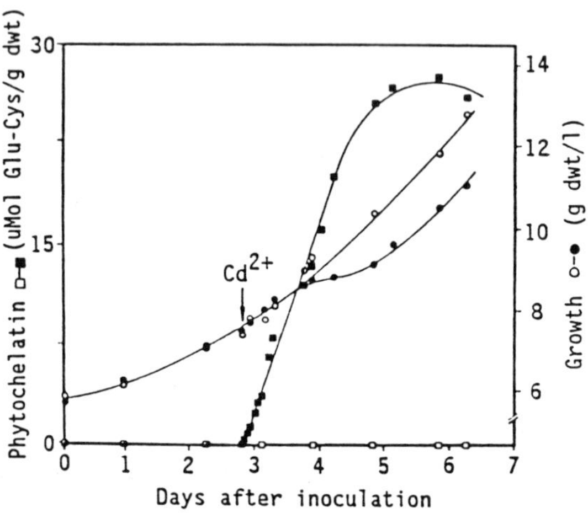

Figure 6. Induction of phytochelatin in cadmium-treated Rosa canina cell suspension cultures. Unexposed cells (open symbols) and exposed cells (closed symbols, 200 μM $Cd(NO_3)_2$) were grown in zinc- and copper-free nutrient solution (50).

This assumption is supported by the finding that growth retardation of plant cell suspension cultures caused by copper ions ceased after most of the toxic ions had been taken up and complexed to the set of metal-binding peptides (12). A similar phenomenon is revealed in Fig.6 employing cadmium ions. In this case the growth arrest does not happen immediately after the addition of cadmium ions, as observed with 50 μM copper ions, probably reflecting a less toxic effect. Cadmium uptake and PC synthesis is strictly correlated (22), thus, as soon as the applied (and taken up) cadmium ions were complexed by the peptides the growth of the cells resumed.

A large body of evidence for a metal detoxifying role

of PC stems from cadmium-resistant cell suspension cultures of tomato (24, 37, 40) and Datura innoxia (41, 42) since these cultures overproduce PC after metal exposure in comparison to non-tolerant cells. Cadmium-resistant tomato cells also revealed an increased tolerance towards copper ions (37).

TABLE 3
GROWTH INHIBITION OF METAL-EXPOSED RAUVOLFIA SERPENTINA CELLS IN THE PRESENCE OF BUTHIONINE SULFOXIMINE

	buthionine sulfoximine (mM)	phytochelatin level in (Glu-Cys) units μmol/g dwt	%	growth g dwt/l	%
without $AgNO_3$	0	0	0	4.4	100
	0.2	0	0	4.9	111
	2	0	0	4.1	93
35 μM $AgNO_3$	0	12.3	100	4.5	103
	0.2	3.9	32	3.5	79
	2	0.6	5	1.4	32

Growing cultures were preincubated with the inhibitor for 6 hours prior administration of silver ions. Cells were harvested 3 days after metal exposure. During this period the control cultures were grown from 9.0 g dwt/l to 13.4 ± 0.4 g dwt/l. The culture medium lacked zinc and copper.

In view of these data from cell suspension cultures a heavy metal detoxifying function has to be assigned to PC, implicating a role of the peptides in metal-tolerant plants. Metal tolerance based on PC should not be restricted to a specific ion, but rather should result in co-tolerance for a number of metals, although a resistance towards lead, zinc and copper was not observed in a cadmium-tolerant tomato cell line (43). In fact, the copper-tolerant monkey flower, Mimulus guttatus, is co-tolerant to zinc, lead and nickel ions (44) and analysis of the copper-exposed plant indicated the presence of copper-binding PC (45). A role of copper-complexing proteins in resistant Agrostis tenuis (46) and Silene cucubalus (8) in metal tolerance was suggested by their increased synthesis rate in tolerant compared to non-tolerant species. The metal-binding proteins were most likely PCs (12).

In zinc-tolerant Silene grown on a zinc-rich mine tailing, a low level of PC induction was observed (47). A few percent of the accumulated zinc could only be complexed to these PCs. The zinc ions were primarily associated with low molecular weight material (<1 kDa) consistent with a complexation of zinc to organic acids or phytate in vacuoles (3, 48). In this tolerant plant, zinc ions may be transiently complexed to PC during cytoplasmic transfer to their final storage site in vacuoles.

Thus, it seems that PC is involved in copper tolerance and could function as a nonspecific metal tolerance mechanism which was assumed to exist in Thlaspie goesingense (49). It is quite clear that PC can not account for the zinc tolerance of Silene, and for the resistance of some plants towards specific metal ions (50). However, metal-inducible peptides can represent a general mechanism of heavy metal ion tolerance in plants which is also used for homeostasis of essential metal ions such as zinc and copper.

An indication for a homeostatic function of PC was observed by the induction of PC by 'physiological' concentrations of micronutrients (12). In cell suspension cultures grown for one or two growth cycles in the absence of zinc and copper, PC could not be detected indicating that PC is not present in the cells or is present at a concentration beyond detection limit of the assay (approx. 20 nMol PC per gram fresh tissue). Transfer of the suspension cells to the complete medium supplemented with zinc (37 μM) and copper (0.1 μM) resulted in an induction of PC. During the onset of cell growth the total PC content of the culture decreased, obviously caused by degradation of PC. Presumably, as the heavy metal ions of the PC-complex are needed in biosynthesis or as zinc ions get stored in the vacuole, a metal transfer occurs from the PC-complex towards acceptors such as apoforms of metalloenzymes. The metal-free PCs can subsequently be degraded. Thus, the heavy metal-binding peptides could serve as a degradable storage form of essential metal ions such as zinc, copper and nickel.

The data accumulated so far support the involvement of PC in the detoxification and homoestasis of heavy metals in plants. Without doubt several aspects of heavy metal metabolism in plants are poorly understood, but in the last few years substantial progress has been made. It seems likely that the biosynthesis of phytochelatin peptides will soon be clarified and their homeostatic function established.

ACKNOWLEDGMENTS

This work was supported by a grant from the Bundesministerium fur Forschung und Technologie, Bonn, West Germany, to the Zentrum fur Molekulare Biologie, München. I would like to thank Dr. Tom Johnson for his help in preparing this manuscript.

REFERENCES

1. Ernst WHO (1974). Schwermetallvegetation der Erde. G. Fischer Verlag, Stuttgart.
2. Jowett D (1958). Populations of Agrostis spp. tolerant to heavy metals. Nature 182:816.
3. Mathys W (1977). The role of malate, oxalate, and mustard oil glucosides in the evolution of zinc resistance in herbage plants. Physiol Plant 40:130.
4. Turner RG (1970). The subcellular distribution of zinc and copper within the roots of metal-tolerant clones of Agrostis tenuis Sibth. New Phytol 69:725.
5. Baker AJM (1978). Ecophysiological aspects of zinc tolerance in Silene maritima With. New Phytol 80:635.
6. Wainwright SJ, Woolhouse HW (1975). Physiological mechanism of heavy metal tolerance in plants. In Chadwick MJ, Goodman GT (eds): "The ecology of resource degradation and renewal," Oxford: Blackwell, p 231.
7. Rauser WE, Curvetto NR (1980). Metallothionein occurs in roots of Agrostis tolerant to excess copper. Nature 287:563.
8. Lolkema PC, Donker MH, Schouten AJ, Ernst WHO (1984). The possibly role of metallothioneins in copper tolerance of Silene cucubalus. Planta 162:174.
9. Bowen JE (1981). Kinetic of uptake of boron, zinc, copper and manganese in barley and sugarcane. J Plant Nut 3:215.

10. Cataldo DA, Garland TR, Wildung RE (1983). Cadmium uptake kinetics in intact soybean plants. Plant Physiol 73:844.
11. Grill E, Winnacker E-L, Zenk MH (1985). Phytochelatins: The principal heavy metal-complexing peptides of higher plants. Science 230:574.
12. Grill E, Winnacker E-L, Zenk MH (1987). Phytochelatins, a class of heavy metal-binding peptides from plants, are functionally analogous to metallothioneins. Proc Natl Acad Sci USA 84:439.
13. Grill E, Gekeler W, Winnacker E-L, Zenk MH (1986). Homo-phytochelatins are heavy metal-binding peptides of homo-glutathione containing _Fabales_. FEBS Lett 205:47.
14. Casterline JL, Barnett NM (1977). Isolation and characterization of cadmium-binding components in soybean plants. Plant Physiol 59:S 124.
15. Rauser WE, Hartmann H-J, Weser U (1983). Cadmium-thiolate protein from the grass _Agrostis gigantea_. FEBS Lett 164:102.
16. Weigel HJ, Jager HJ (1980). Subcellular distribution and chemical form of cadmium in bean plants. Plant Physiol 65:480.
17. Bartolf M, Brennan E, Price CA (1980). Partial characterization of a cadmium-binding protein from the roots of cadmium-treated tomato. Plant Physiol 66:438.
18. Wagner GJ, Trotter MM (1982). Inducible cadmium-binding complexes of cabbage and tobacco. Plant Physiol 69:804.
19. Rauser WE, Glover J (1984). Cadmium-binding protein in the roots of maize. Can J Bot 62:1645.
20. Wagner GJ (1984). Characterization of a cadmium-binding complex of cabbage leaves. Plant Physiol 76:797.
21. Tukendorf A, Lyszcz S, Baszynski T (1984). Copper-binding proteins in spinach tolerant to excess copper. J Plant Physiol 115:351.
22. Grill E. Winnacker E-L, Zenk MH (1985). Induction of heavy metal-sequestering phytochelatin by cadmium in cell cultures of _Rauvolfia_ _serpentina_. Naturwissenschaften 72:432.
23. Kondo N, Isobe M, Imai K, Goto T (1984). Cadystin A and B, major unit peptides comprising cadmium binding peptides induced in a fission yeast - separation, revision of structure and synthesis. Tetrahedron Lett 25:3869.
24. Steffens JC, Hunt DF, Williams BG (1986). Accumulation of non-protein metal-binding polypeptides (γ-glutamyl-cysteinyl)n-glycine in selected cadmium-

resistant tomato cells. J Biol Chem 261:13879.
25. Lue-Kim H, Rauser WE (1986). Partial characterization of cadmium-binding protein from roots of tomato. Plant Physiol 81:896.
26. Jocelyn PC (1972). Biochemistry of the SH-group. New York: Academic Press.
27. Lerch K (1980). Copper metallothionein, a copper-binding protein from Neurospora crassa. Nature 284:368.
28. Bernhard WR, Kagi JHR (1987). Purification and partial characterization of atypical cadmium-binding polypeptides from Zea mays. In Kagi JHR, Kojima Y. (eds): "Metallothionein II," Basel: Birkhauser Verlag, p 309.
29. Gekeler W, Grill E, Winnacker E-L, Zenk MH (1988). in press.
30. Murasugi A, Wada C, Hayashi Y. Cadmium-binding peptide induced in fission yeast, Schizosaccharomyces pombe. J Biochem 91:222.
31. Grill E, Winnacker E-L, Zenk MH (1986). Synthesis of seven different homologous phytochelatins in metal-exposed Schizosaccharomyces pombe cells. FEBS Lett 197:115.
32. Prinz R, Weser U (1975). A natural occurring Cu-thionein in Saccharomyces cerevisiae. Z Physiol Chem 356:767.
33. Butt TR, Sternberg EJ, Gorman JA, Clark P, Hamer D, Rosenberg M, Crooke ST (1984). Copper metallothionein of yeast, structure of the gene, and regulation of expression. Proc Natl Acad. Sci 81:3332.
34. Winge DR, Reese RN, Mehra RK, Tarbet EB, Hughes A (1988) this volume.
35. Price CA (1957). A new thiol in legumes. Nature 180:148.
36. Carnegie PR (1963). Isolation of a homologue of glutathione and other acidic peptides from seedlings of Phaseolus aureus. Biochem J 89:459.
37. Scheller HV, Huang B, Hatch E, Goldsbrough PB (1987). Phytochelatin synthesis and glutathione levels in response to heavy metals in tomato cells. Plant Phys 85:1031.
38. Reese RN, Wagner GJ (1987). Effects of buthionine sulfoximine on Cd-binding peptide levels in suspension-cultured tobacco cells treated with Cd, Zn, or Cu. Plant Physiol 84, 574.
39. Griffith OW, Meister A (1979). Potent and specific inhibition of glutathione biosynthesis by buthionine sulfoximine. J Biol Chem 254:7578.

40. Bennetzen JL, Adams TL (1984). Selection and characterization of cadmium-resistant suspension cultures of the wild tomato Lycopersicon esculentum. Plant Cell Rep 3:241.
41. Jackson PJ, Roth EJ, McClure PR, Naranjo CM (1984). Selection, isolation and characterization of cadmium-resistant Datura innoxia suspension cultures. Plant Physiol 75:914.
42. Jackson PJ, Unkefer CJ, Doolen Ja, Watt K, Robinson NJ (1987). Poly(γ-glutamylcysteinyl)glycine: Its role in cadmium resistance in plant cells. Proc Natl Acad Sci USA 84:6619.
43. Huang B, Hatch E, Goldsbrough (1987). Selection and characterization of cadmium tolerant cells in tomato. Plant Sci 52:211.
44. Allan WR, Sheppard PM (1971). Copper tolerance in some Californian populations of the monkey flower, Mimulus guttatus. Proc Roy Soc London, B 177:177.
45. Thurman DA, Salt D, Sewall A, Thomas M, Rodrigues de Miranda J, Tomsett AB (1988). Copper-inducible proteins and genes in Mimulus guttatus. J Cell Biochem S 12D, abs. Z118.
46. Rauser WE (1984). Copper-binding protein and copper tolerance in Agrostis gigantea. Plant Sci 33:239.
47. Grill E, Winnacker E-L, Zenk MH (1988). in press.
48. Van Steveninck RFM, Van Steveninck ME, Fernando DR, Horst WJ, Marschner H (1987). Deposition of zinc phytate in globular bodies in roots of Deschampsia caespitosa ecotypes; a detoxification mechanism? J Plant Phsiol 131:247.
49. Reeves RD, Baker AJM (1984). Studies on metal uptake by plants from serpentine and non-serpentine populations of Thlaspie goesingense. New Phytol 98:191.
50, Baker AJM (1987). Metal tolerance. New Phytol 106:93.
51. Bradford MM (1976). A rapid and sensitive method for the quantification of microgram quantities of proteins utilizing the principle of protein-dye binding. Anal Biochem 72:248.

Metal Ion Homeostasis: Molecular Biology and Chemistry, pages 301–311

STRUCTURAL ASPECTS OF METAL-γ-GLUTAMYL PEPTIDES

D.R. Winge, R.N. Reese, R.K. Mehra, E.B. Tarbet, A.K. Hughes and C.T. Dameron

Departments of Medicine and Biochemistry
University of Utah Medical Center
Salt Lake City, Utah 84132

INTRODUCTION

Cellular resistance to heavy metal ion cytotoxicity occurs by multiple mechanisms. Certain prokaryotes either attenuate metal ion uptake or facilitate an efflux pathway to reduce the intracellular free metal ion concentration (1-3). Eukaryotic organisms regulate the intracellular free metal ion concentration by sequestering metal ions bound to macromolecules. Two widely distributed metal-binding molecules that function in metal ion resistance are cysteine-rich polypeptides in the metallothionein family and short cysteine-containing γ-glutamyl peptides (4-8). The γ-Glu peptides supplant metallothionein as the major metal ion detoxification mechanism in Plantae and certain species within Fungi and Protista (7-10). Peptides within Plantae and Fungi have the same general structure of $(\gamma\text{-Glu-Cys})_n$-Gly (11,12). The linkage between Glu and Cys is an isopeptide bond involving the γ-carboxyl group of Glu. The peptides have an obvious structural relationship to the tripeptide glutathione (γ-Glu-Cys-Gly). Heterogeneity is observed in peptide length with peptides of n_2 through n_5 being most common (13,14). Synthesis of the peptides is facilitated by metal ions in an undefined pathway. Metal ions are subsequently bound within a metal-thiolate cluster composed of an oligomer of peptides (8,14), but detailed information on the cluster is scant. In this report structural features of the metal-γ-Glu peptide complex are presented.

PROPERTIES OF METAL γ-GLU PEPTIDE COMPLEXES

S. pombe and various plant cells have been used by

numerous investigators as sources of the γ-Glu peptides (7-11,15,16). The γ-Glu peptides have been referred to as phytochelatin cadystin, or metallothionein III (7,8,17), but these names are too restrictive or vague. The term chelatin was originally erroneously used to designate a Cu-binding protein later shown to be metallothionein (18). Many reports still linger concerning a chelatin polypeptide in various species. Due to the lack of consensus we have chosen to designate the peptides as simply γ-Glu peptides.

We have studied metal-γ-Glu peptide complexes from two fungi, Schizosaccharomyces pombe and Candida glabrata, and a plant source, Daucus carota (carrot cells). We demonstrated accumulation of γ-Glu peptides in each species grown in the presence of Cd salts (14,19). Cu salts induce formation of γ-Glu peptides in S.pombe and D.carota, but not C.glabrata. Two polypeptides with properties resembling metallothionein (MT) are synthesized in C.glabrata cultures incubated with Cu salts (19). This is the first organism conclusively shown to express both γ-Glu peptides and MT, and the expression of each molecule is metal specific.

Metal-γ-Glu peptide complexes are isolated as oligomers of multiple peptide units (8,14). The peptide stoichiometry per complex is uncertain as the molecular weight(M_r) of the complex has not been rigorously determined. On gel filtration the observed Stokes radius is dependent on the ionic strength of the elution buffer (9,14). This is presumably the result of electrostatic repulsion between glutamyl carboxylates. At high salt concentrations the M_r of the complex is nearly 3,100 suggestive of a trimeric or tetrameric complex (9,14). The oligomer of γ-Glu peptides does not consist of peptides with unique chain lengths (14,20). Peptides from S.pombe and carrot cells are predominantly n_2, n_3 and n_4 with smaller quantities of n_5, in agreement with two other investigations on S.pombe (8,13). In addition to length variants of the structure $(\gamma\text{-Glu-Cys})_n$-Gly, we documented minor desGly forms of each peptide (20). The mol ratios of the various peptides vary within a range of 1:30 and the ratios are constant across the elution profile of gel filtration. This implies that the peptide constituents of each oligomer vary without affecting the Stokes radius of the complex. In C.glabrata peptides of n_2 and n_2-desGly are the major forms.

Multiple metal ions bind within the γ-Glu peptide

cluster. The exact stoichiometry is unclear due to the mentioned ambiguity in the M_r of the oligomer, but 2.3 Cu ions are bound per peptide in _S.pombe_. Absorption spectroscopy of the metal-complex reveals electronic transitions in the ultraviolet that are typical of metal-thiolate charge-transfer transitions. Excitation of the Cu-γ-Glu peptide complex with ultraviolet radiation results in luminescence with a corrected emission maximal at 619 nm (14). Luminescence seen with Cu complexes is indicative of d_{10} diamagnetic Cu(I) coordination to thiol groups that interact only slightly with the solvent (21,22). Similar electronic transitions and luminescence are apparent in Cu-metallothioneins from _Neurospora crassa_, _Saccharomyces cerevisiae_ and mammalian liver (23-25). One interpretation of the luminescence is that the Cu-thiolate cluster is shielded from solvent interaction by an outer cage of γ-Glu peptides. This situation would be analogous to the known structure of mammalian metallothionein in which the interior core of the molecule is the metal-sulfur polynuclear cluster enfolded by a monolayer of polypeptide (26). It is unclear whether the Cu(I) stoichiometry varies in oligomers heterogeneous in peptide length, although it is conceivable that peptide length is not as critical as the number of cysteinyl residues within a complex. Each metal-thiolate cluster may have a similar metal-cysteine stoichiometry and this stoichiometry may be attained with different permutations of peptides varying in 'n'. Known Cu-metallothioneins have Cu:Cys ratios of 6:9, 6:7 and 8:12, whereas polynuclear clusters formed with simple organic thiolates having Cu:S ratios of 4:6, 5:7, and 8:12 have been characterzied (27-32). The Cu-γ-Glu peptide complex may exhibit one of these preferred stoichiometries. If the driving force for cluster formation is a given Cu:S stoichiometry, cluster heterogeneity would be limited to a varying number of peptide chains providing the critical number of cysteinyl residues.

The luminescence emission is independent of ionic strength, unlike the Stokes radius of the complex. If the Stokes radius dependency on ionic strength relates to electrostatic effects on peptide chain conformation, the conformational perturbation does not affect solvent quenching of the luminophore. The structural basis for this is unclear. An alternative explanation of the ionic strength effect on Stokes radius is that the aggregation state of the complex changes. However, it is unlikely that ionic strength would alter the number of metal-thiol coordinate bonds. The electrostatic hypothesis is supported by the known shift in Stokes radius of polyglutamic acid with changes in the ionic strength (14).

LABILE SULFUR IN THE CADMIUM-γ-GLU PEPTIDE COMPLEX

Metal clusters synthesized in cultures containing cadmium salts but not copper salts contain labile sulfur (14,33). This has been reported in γ-Glu peptide complexes from plants and fungi (11,33). Incorporation of sulfide within Cd clusters occurs by virtue of a cadmium-dependent stimulation of sulfide generation within the cells. Cadmium salts stimulate the production of sulfide ion several fold in logarithmically growing cultures of *S.pombe* and *C.glabrata* (19,34). A portion of the sulfide is incorporated within the Cd-γ-Glu peptide complexes and an additional quantity forms an extracellular and cell associated CdS colloid to be described later. Cluster formation with Cd(II) and γ-Glu peptides is independent of sulfide in that the initial Cd-γ-Glu peptide complex formed after addition of cadmium salts to the culture medium is devoid of sulfide. Sulfide-containing complexes predominate at later times (35). Sulfide-containing peptide complexes do not consist of a unique sulfide content; rather a spectrum of molecules with sulfide levels ranging from 0.1 to 1.5 mol per mol peptide are recovered (19,34). There is, therefore, not one unique Cd-γ-Glu peptide complex. The sulfide content is dependent on the time of cell harvesting and on growth conditions. Clearly distinct processes are involved in γ-Glu peptide synthesis and metal-stimulated sulfide production. The low sulfide form of the complex can be readily converted to the high sulfide form *in vitro* by the addition of sodium sulfide to either the purified peptide complex or to a crude cell extract containing the γ-Glu peptide complex.

Incorporation of sulfide ions markedly alters three properties of the Cd-γ-Glu peptide complex. First, sulfide ions enhance the stability and metal content of the Cd-γ-Glu peptide cluster. Cd(II) levels vary between 1.7 and 3.1 g atoms per mol peptide in low and high sulfide native isolate forms respectively (34). In the absence of bound sulfide, Cd(II) binds maximally at 1 g atoms per mol peptide. Cd contents as high as 5.5 mol eq. are observed with *in vitro* sulfide-treated Cd-γ-Glu peptide complexes. The range of the molar sulfide:cadmium ratio is from $\leq$ 0.06 in the lowest sulfide forms to 0.8 in the high sulfide forms. Secondly, high sulfide forms of the complex bind Cd(II) more tenaciously as shown by the greater concentration of hydrogen ions necessary for dissociation of the high-sulfide complexes. Thirdly, sufide binding increases the Stokes radius of the γ-Glu peptide complex. Apparent M_r of low and high sulfide complexes were determined to be 3100 and 9000 respectively. The effect of sulfide on the Stokes radius is responsible for the gel filtration resolution of the spectrum of Cd-γ-Glu peptide complexes varying in sulfide.

Sulfide-containing Cd-γ-Glu peptide complexes exhibit an electronic transition in the ultraviolet (34). The wavelength maximum of the transition varies depending on the sulfide content with high sulfide levels producing a transition maximum at 318 nm. The transition is clearly related to the presence of bound sulfide ions as acidification of a high sulfide complex abolishes the 318 nm transition and subsequent re-neutralization yields an absorption spectrum resembling the low sulfide form. Optical properties of Cd-sulfide complexes are well known (36,37). Pure synthetic CdS colloids exhibit optical properties that vary with the size of the granule (36,37). Sulfide-containing Cd-γ-Glu peptide complexes may consist of a CdS granular core coated with γ-Glu peptide units. Spectral transitions observed in the Cd-γ-Glu peptide complexes may reflect changes in the CdS core size. Variation in the Stokes radius of the complex as a function of the sulfide content is also consistent with this model.

The enhanced stability of the high sulfide forms of the Cd-γ-Glu peptide complex may increase the effectiveness of the complex in maintaining a low free intracellular concentration of Cd(II) ions. Cellular tolerance to Cd(II) appears to be accomplished by sequestration of metal ions in a stably bound state complexed to the γ-Glu peptides. Sulfide incorporation within the complex reduces Cd(II) dissociation from the metal-cysteinyl thiolate cluster and may limit Cd-induced cytotoxicity.

Sulfide acts not only to enhance the effectiveness of the γ-Glu peptide system of Cd(II) detoxification but also in a secondary role to increase metal tolerance through precipitation of Cd(II) ions. _S.pombe_ and _C.glabrata_ grown in the presence of cadmium salts contain an anionic CdS colloid associated both with the cell and in the extracellular growth medium. The location of the CdS colloid within the cell is unclear. Ultracentrifugation at 100,000 x g of the growth medium collected from logarithmic cultures resulted in the sedimentation of a CdS complex containing a molar sulfide:Cd(II) ratio of 0.7 ± 0.2. The amount of Cd(II) complexed as the CdS colloid varies with growth conditions. Accumulation of the CdS colloid is dependent on the initial concentration of Cd(II) and the composition of the growth medium. Cells grown in minimal salt medium contain the low sulfide form of the Cd-γ-Glu peptide complex and the CdS colloid only associated with the cell. Greater amounts of CdS colloid are apparent in cultures grown in rich YTD growth medium, and these cultures produced the high sulfide forms of the Cd-γ-Glu peptide complex. Conditions that favor CdS colloid formation yield the high sulfide form of the γ-Glu peptide complex. Up to 50% of the Cd(II) in the YTD growth medium can be complexed in this CdS colloid in stationary cultures of _S.pombe_ and _C.glabrata_, but the nutrients responsible for modulating the sulfide response are unknown.

METAL RECONSTITUTION OF γ-GLU PEPTIDES

Metal reconstitution studies were carried out on metal-depleted γ-Glu peptides to understand the significance of the peptide length heterogeneity in the oligomeric metal-thiolate cluster (20). Peptides of n_2, n_3, n_4 and n_3-desGly prepared by reverse phase HPLC were titrated with Cu(I). Cu-γ-Glu peptide complex formation was demonstrated with each of the γ-Glu peptides as determined by the characteristic 260 nm shoulder in the ultraviolet absorption spectrum and the luminescence emission. Metal binding stoichiometries within complexes formed by unique 'n' peptides differ. Peptides of n_3 and n_4 bind maximally 1.5 mol eq. Cu(I) ions whereas n_2 shows maximal binding near 0.5 mol eq. The criterion of maximal binding stoichiometry was the mol equivalency that exhibited both the highest absorption in the 260 nm shoulder and greatest luminescence emission. Metal addition beyond the maximal stoichiometry abolishes the 260 nm shoulder and reduces the luminescence emission presumably by disruption of the Cu-thiolate complex. The maximal binding stoichiometry with n_3 and n_4 peptides is markedly less than the maximal Cu(I) stoichiometry of a heterogeneous mixture of peptides. The stoichiometry observed in native isolates of the cluster is reproduced only if a heterogeneous mixture of peptides is used. Cluster formation also occurs with desGly peptides so the carboxyl-terminal Gly is not critical for cluster formation.

Cd-γ-Glu peptide complexes can be formed by <u>in vitro</u> reconstitution of apo-peptides. The maximal extent of Cd(II) binding to a peptide mixture is nearly 1 mol eq. in the absence of added sulfide but greater than 2 mol eq. in samples with > 0.2 mol eq. sulfide added. The ultraviolet transition at 318 nm is apparent in reconstituted samples only in the presence of added sulfide. Addition of sodium sulfide to Zn(II) reconstituted γ-Glu peptides reveals sulfide-dependent electronic transitions near 255 nm. The different spectral transitions related to sulfide in the Zn(II)- and Cd(II)-peptide complexes imply direct metal-sulfide binding.

SYNTHETIC PEPTIDES

The use of γ-glutamyl-cysteine dipeptide repeats in certain organisms for metal detoxification may have evolved for two reasons. If glutathione is a precursor of γ-Glu peptide synthesis, the abundant concentration of glutathione in most cells (1-5 mM) may provide ample quantities of the substrate. The γ-glutamyl linkage may be explained therefore by the realtionship to the putative substrate, glutathione. Alternatively, the isopeptide repeats may have evolved by virtue of the greater inherent flexibility of peptide chains containing γ-isopeptide bonds rather than the usual α-peptide bonds. There are 4 single bonds with free rotation within the γ-Glu-Cys repeat, whereas only 2 bonds occur in a typical α-peptide linkage as shown below:

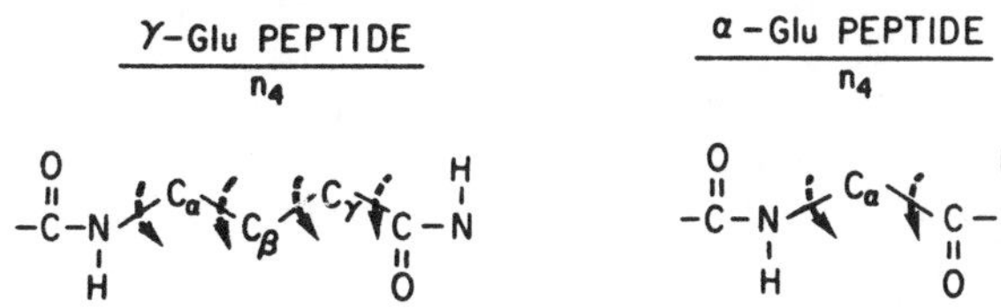

The multiple single bonds allow for additional free rotation. Spontaneous folding in polypeptides usually occurs by virtue of the constrained flexibility in α peptide bonds. It is conceivable that the high flexibility in γ-Glu peptides allows for the metal-thiolate polynuclear cluster to form. To explore this question we synthesized peptides of $(Glu\text{-}Cys)_n$-Gly with either α or γ linkages in the dipeptide repeat. The two peptides purified through reverse phase HPLC were used in metal ion reconstitution studies. Clusters form with both peptides as judged by absorption spectroscopy and luminescence. Maximal binding stoichiometries are 1.5 and 2 Cu(I) ions and 1 and 2 Cd(II) ions for the γ-Glu peptide and α-Glu peptide respectively. The luminescence emission yield (relative luminescence/Cu concentration) is 1.4 greater for the α-Glu peptide compared to the γ-Glu peptide. Each metal-peptide complex chromatographs as an oligomer with K_d of 0.6 and 0.45 for the Cu(I) complex with γ-Glu and α-Glu peptides respectively. Addition of sodium sulfide to the two Cd-peptide complexes results in sulfide-dependent electronic transitions with each peptide. At maximal sulfide levels, the transition in the α-Glu peptide is blue shifted from the usual 318 nm maxima to nearly 295 nm. Incorporation of sulfide ions in each peptide complex enhances the Stokes radius of the complexes as judged by the 0.2 change in the distribution function K_d. Clearly, the greater flexibility in the γ-Glu peptides is not critical for metal-thiolate cluster formation. We also demonstrated metal-thiolate cluster formation in an oligomer of peptides of structure Lys-Cys-Thr-Cys-Cys-Ala. Cluster formation appears to be merely a consequence of having multiple cysteinyl residues in a given peptide.

CONCLUSIONS

The metal-γ-Glu peptide complex appears to participate in metal detoxification in numerous species. The γ-Glu peptides spontaneously associate with metal ions to generate the metal-peptide oligomeric complex. The peptides do not appear to provide any significant constraints in folding other than providing the cysteinyl thiolates involved in metal ligation. The metal-peptide complex appears to consist of multiple metal ions bound in a metal-thiolate polynuclear core cluster enfolded with multiple γ-Glu peptides. Sulfide ions associate with the Cd-peptide cluster in a complicated manner altering both the Stokes radius and the Cd(II) content of the cluster. It is conceivable that the Cd-γ-Glu peptide core is a CdS matrix with γ-Glu peptide units coating the complex. The spectral transitions may be related to the size of the CdS core. Clearly, the γ-Glu peptide system represents a heterogeneous mixture of peptides and sulfide contents.

REFERENCES

1. Silver, S., Budd, K., Leahy, K.M., Shaw, W.V., Hammond, D., Novick, R.P., Willsky, G.R., Malamy, M.H. and Rosenberg, H. (1981). Inducible Plasmid-Determining Resistance to Arsenate, Arsenite, and Antimony (III) in Escherichia coli and Staphylococcus aureus. Journal of Bacteriology 146:983.
2. Silver, S. and Keach, D. (1982). Energy-dependent Arsenate Efflux: The Mechanism of Plasmid-mediated Resistance. Proc. Natl. Acad. Sci. USA 79:6114.
3. Tynecka, Z., Gos, Z., and Zajac, J. (1981). Energy-Dependent Efflux of Cadmium Coded by a Plasmid Resistance Determinant in Staphylococcus aureus. Journal of Bacteriology 147:313.
4. Hamer, D.H. (1986). Metallothionein. Ann. Rev. Biochem. 55:913.
5. Kagi, J.H.R. and Kojima, Y. (1987). Chemistry and Biochemistry of Metallothionein. Experientia Supplement 52:25.
6. Bremner, I. (1987) Nutritional and Physiological Significance of Metallothionein. Experientia Supplement 52:81.
7. Grill, E., Winnacker, E.-L. and Zenk, M.H. (1985). Phytochelatins: The Principal Heavy-Metal Complexing Peptides of Higher Plants. Science 230:674.
8. Hayashi, Y., Nakagawa, C.W. and Murasugi, A. (1986), Unique Properties of Cd-Binding Peptides Induced in Fission Yeast, Schizosaccharomyces pombe. Environ. Health Perspec. 65:13.
9. Grill, E., Winnacker, E.-L. and Zenk, M.H. (1987). Phytochelatins, a Class of Heavy Metal-binding Peptides from Plants are Functionally Analogous to Metallothioneins. Proc. Natl. Acad. Sci. USA 84:439.
10. Weber, D.N., Shaw, C.F.III, and Petering, D.H. (1987). Euglena gracilis Cadmium-binding Protein-II Contains Sulfide Ions. J. Biol. Chem. 262:6962.
11. Steffens, J.C., Hunt, D.F. and Williams, B.G. (1986). Accumulation of Non-Protein Metal-binding Polypeptides (γ-Glutamyl-Cysteinyl) n-Glycine in Selected Cadmium-resistant Tomato Cells. J. Biol.

Chem. 261:13879.

12. Kondo, N., Isobe, M., Imai, K. and Goto, T. (1985). Synthesis of Metallothionein-like Peptides Cadystin A and B Occurring in a Fission Yeast, and their Isomers. Agric. Bio. Chem. 49:71.
13. Grill, E., Winnacker, E.-L., and Zenk, M.H. (1986). Synthesis of Seven Different Homologous Phytochelatins in Metal-exposed _Schizosaccaromyces pombe_ Cells. FEBS Lett 197:115.
14. Reese, R.N., Mehra, R.K. Tarbet, E.B. and Winge, D.R. (1988). Studies on the γ-Glutamyl Cu-Binding Peptide from _Schizosaccharomyces pombe_. J. Biol. Chem. 263:4186.
15. Reese, R.N. and Wagner, G.J. (1987). Properties of Tobacco (_Nicotiana Tabacum_) Cadmium-binding Peptide(s). Biochem. J. 241:641.
16. Bernhard, W.R. and Kagi, J.H.R. (1987). Purification and Characterization of Atypical Cadmium-binding Polypeptides from _Zea Mays_. Experientia Supplement 52:309.
17. Fowler, B.A., Hildebrand, C.E., Kojima, Y. and Webb, M. (1987). Recommendations on Nomenclature of Metallothionein. Experientia Supplement 52:19
18. Winge, D.R., Geller, B.L. and Garvey, J. (1981). Isolation of Cu-thionein from Rat Liver. Arch. Biochem. Biophys. 208:160.
19. Mehra, R.K., Tarbet, E.B., Gray, W.R. and Winge, D.R. (1988). Metal-Specific Synthesis of Two Metallothioneins and γ-Glutamyl Peptides in _Candida glabrata_. Submitted.
20. Mehra, R.K. and Winge, D.R. (1988). Cu(I) Binding to the _Schizosaccharomyces pombe_ γ-Glutamyl Peptides Varying in Chain Length. Arch. Biochem. Biophys. in press.
21. Anglin, J.H., Batten, W.H., Raz, A.I. and Sayre, R.M. (1971). Fluorescence of Cu, Au and Ag Mercaptides. Photochem. and Photobiol. 13:279.
22. Lytle, R.E., (1970). Solution Luminescence of Metal Complexes. Applied Spectroscopy 24:319.
23. Lerch, K. and Beltramini, M. (1983). Neurospora Copper Metallothionein: Molecular Structure and Biological Significance. Chemica Scripta 21:109.
24. Byrd J., Berger, R.M., McMillin, D.R., Wright, C.F, Hamer, D.H. and Winge, D.R. (1988). Characterization of the Cu-Thiolate Cluster in Yeast Metallothionein and Two Truncated Mutants.

J. Biol. Chem. 263: in press.

25. Mehra, R., and Bremner, I. (1984). Special Differences in the Occurrence of Copper-Metallothionein in the Particulate Fractions of the liver of Copper-loaded Animals. Biochem J. 219:539-546.

26. Furey, W.F., Robbins, A.H., Clancy, L.L. Winge, D.R. Wang, B.C. and Stout, C.D. (1986). Crystal Structure of Cd,Zn Metallothionein. Science 231:704.

27. Smith, T.A., Lerch, K. and Hodgson, K.O. (1986). Structural Studies of the Cu Sites in Metallothionein from *Neurospora crassa*. Inorganic Chem. 25:4677.

28. George, G.N., Winge, D., Stout, C.D. and Cramer, S.P. (1986). X-ray Absorption Studies of the Copper-Beta Domain of Rat Liver Metallothionein. J. Inorganic Biochem. 27:213.

29. George, G.N., Byrd, J. and Winge, D.R. (1988). X-ray Absorption of Yeast Cu Metallothionein (1988). J. Biol. Chem. 263: in press.

30. Dance, I.G. and Calabrese, J.C. (1976). The Crystal and Molecular Structure of the Hexa (μ-Benzenethiolato) Tetracuprate (I) Dianion. Inorganica Chimica Acta 19:L41.

31. Dance, I.G. (1978). The Hepta (μ-Benzenethiolato) Pentametallate (I) Dianions of Copper and Silver: Formation and Crystal Structures. Aust. J. Chem. 31:2195.

32. McCandlish, L.E. Bissell, E.C., Coucouvanis, C., Fackler, J.P. and Knox, K. (1968). A New Metal Cluster System Containing a Cube of Metal Atoms. J. Am. Chem. Soc. 90:7357.

33. Murasugi, A., Wada, C. and Hayashi, Y. (1983). Occurrence of Acid-Labile Sulfide in Cadmium-binding Peptide from Fission Yeast. J. Biochem. 93:661.

34. Reese, R.N. and Winge, D.R. (1988). Sulfide Stabilization of the Cadmium-γ-Glutamyl Peptide Complex of *Schizosaccharomyces pombe*. submitted.

35. Murasugi, A, Wada Nakagawa, C. and Hayashi, Y. (1984). Formation of Cadmium-Binding Peptide Allomorphs in Fission Yeast. J. Biochem. 96:1375.

Metal Ion Homeostasis: Molecular Biology and Chemistry, pages 313–324

INORGANIC STUDIES OF THE CADMIUM-BINDING PEPTIDES FROM *EUGLENA GRACILIS*[1]

C. Frank Shaw III, David H. Petering, Daniel N. Weber, and David J. Gingrich

The Department of Chemistry and
The Aquatic Biomedical Research Center
The University of Wisconsin-Milwaukee,
PO Box 413, Milwaukee, WI, 53201

INTRODUCTION

Euglena gracilis, a unicellular flagellate, has been used as a model for studying the cadmium-zinc interactions which occur in virtually all higher organisms (1). *Euglena gracilis* can be grown in well defined media with precisely controlled levels of nutrients and toxicants. As reported recently, these cells form two cadmium binding peptides (Cd-BP-I and -II) in response to toxic levels of Cd^{+2} (>2 µg/ml), and have a large pool of very low molecular weight zinc species (1,2).

The formation of the *E. gracilis* Cd-BPs in response to cadmium resembles in some respects the formation of cadmium metallothionein (MT) in Cd-treated mammals. Both are cysteine rich metal-binding proteins or peptides. Our previous research (1-3) demonstrates that Cd-BP-I and -II have the following properties (contrasted to those of MT, where appropriate)

a. lower M_r values than MTs,
b. more negatively charged than MTs,
c. reactive with 5,5'-dithiobis-(2-nitrobenzoic acid),
d. not cross-reactive with mammalian MT antibodies,
e. relatively low Zn^{+2} content, even with high Zn^{+2} concentrations in the media, and
f. high content of sulfide ions, which are not present in metallothionein.

The finding that Cd-BP-II contains large amounts of sulfide ion (3) suggested that it might be a metallochelatin

---

[1]This research was supported by the UWM Marine and Freshwater Biomedical Core Center, funded by NIEHS Grant ES04184.

related to the sulfide containing γ-glutamyl-cysteinyl peptides of *Schizosaccharomyces pombe* (4) and similar plant peptides (5,6). These peptides, known variously as phytochelatins, cadystins, metallochelatins and most descriptively as poly(γ-glutamylcysteinyl)glycine, are enzymatically synthesized in response to metal challenges and have a simple repeating sequence (4-6):

$$(\gamma\text{-glu-cys})_n\text{-gly}$$

In this report we will present evidence establishing the similarity of Cd-BP-II to various metallochelatins. We will also describe the use of various reagents (DTNB, EDTA, PAR and H^+) as chemical probes for comparing the relative reactivities of Cd-BP-I, -II and mammalian MT's.

RESULTS

Cd-BP-I and -II can be resolved using HPLC ion exchange resins. As shown in Fig 1, the elution of BP-II is dependent on the choice of buffer (50 mM Tris-HCl or 20 mM phosphate) under otherwise identical conditions of pH (7.9) and salt gradient (0-800 mM KCl). BP-I elutes reproducibly at 19-20 mS, while BP-II elutes in a broad band between 50 and 70 mS using tris buffer and in a sharper band with pronounced shoulders centered at 48 mS using phosphate buffer. This variability may result from the unusual composition and structure of the sulfide containing BP-II.

To ascertain whether BP-II contains a single large peptide chain or multiple smaller fragments, a sample was treated with EDTA (to remove the cadmium ions) and iodoacetamide (to alkylate the thiols, thereby preventing inter-peptide disulfide bond formation). As shown in Figure 2, the peptide components of the Cd-BP-II complex elute with an M_r value smaller than that of the intact complex, and larger than the Cd-EDTA complex which is also formed. In contrast to the behavior observed here, the single 60-unit peptide chain of apo-thionein elutes with a larger M_r than the intact metallothionein complex.(7)

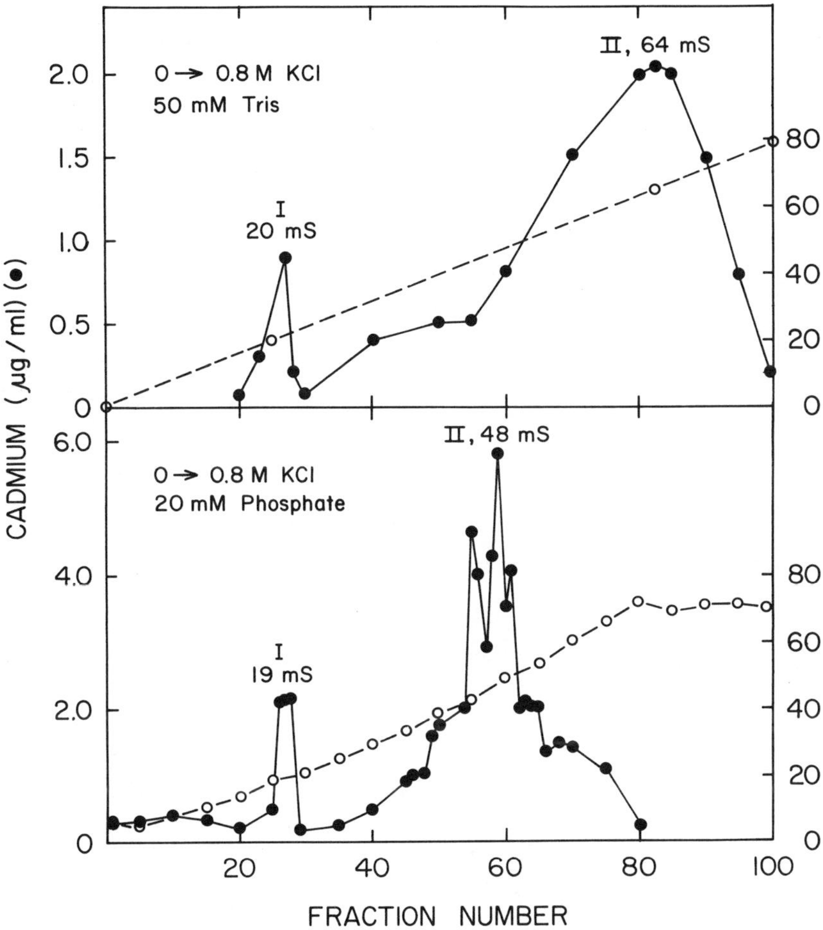

Figure 1. Buffer Effects on HPLC Ion-Exchange Chromatography of Cd-BP-II. Cd-BP (~10 μg Cd/mL) was eluted in (**TOP**) 50 mM Tris-HCl, pH 7.9 and (**BOTTOM**) 20 mM Phosphate, pH 7.9. Both gradients were 0-800 mM KCl on a semi-preparative DEAE column.

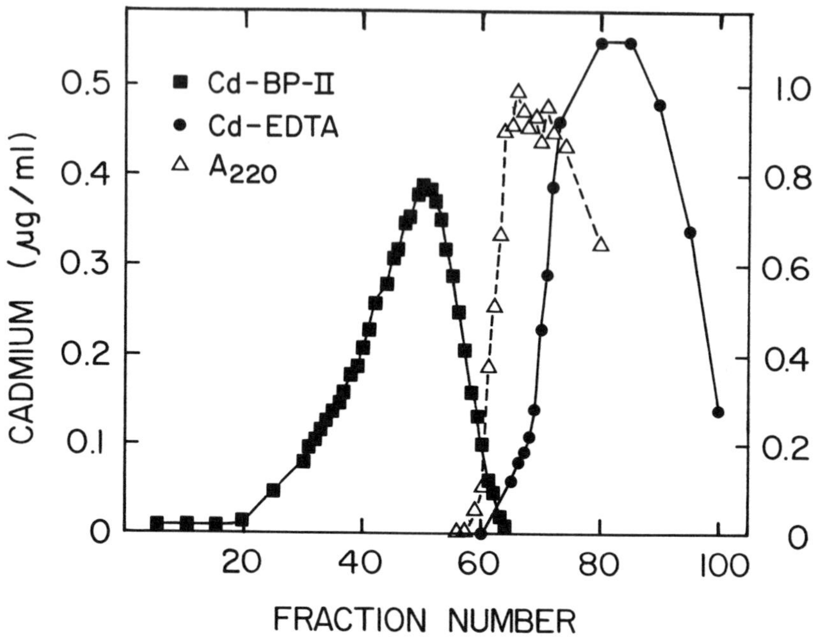

Figure 2. Isolation of the Cd-BP-II Peptide Component. Cd-BP-II (10 mL; ~5 µg Cd/mL) was incubated with excess iodoacetamide (1.0 mg) and EDTA (4.35 mg) for 24 h, then eluted over Sephadex G-75 in 20 mM Tris-HCl, pH 8.6 in the presence of 2 mM $HSCH_2CH_2OH$. The elution of an untreated sample over the same column is shown for comparison.

To explore the effect of the sulfide ions on the Cd-binding capacity of BP-II, a sample was divided into 2 equivalent aliquots and treated as follows. Each was acidified to pH 2 to dissociate the peptide and to convert the sulfide to H_2S. One was immediately neutralized with hydroxide, reestablishing the original pH 8.6. The second was exhaustively evacuated to remove the H_2S, then returned to pH 8.6. Each was chromatographed over Sephadex G-25 as shown in Fig. 3. The sample reconstituted with the sulfide ion intact eluted with a M_r similar to that of the original sample, although the structure may be dramatically altered. The "desulfided" sample, in contrast, contained mainly a low molecular weight Cd species; less than 20% of the original Cd^{+2} was bound in a peptide complex with a M_r similar to the intact complex. This experiment indicates that the sulfide ions greatly enhance the Cd-binding capacity of the peptides.

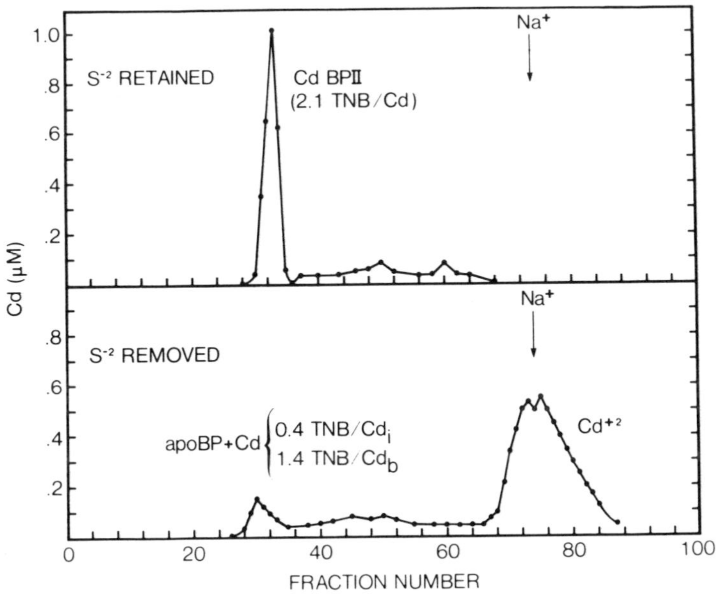

Figure 3. Effect of Sulfide Removal on Cadmium Binding Capacity of Cd-BP-II. Equal volumes (1.8 mL) of Cd-BP-II were acidified to pH 2. One was immediately readjusted to pH 8.6 and chromatographed over Sephadex G-25; the second was exhaustively evacuated to remove H_2S, readjusted to pH 8.6, and similarly chromatographed.

Previously published data (2) demonstrate that Cd-BP-II has a half-titration point of 5 for loss of bound cadmium ions. Data which will be published elsewhere establish 2 inflection points for the displacement of Cd^{+2} from BP-I. One is relatively high, ca. pH 8-9, and the major one is at pH ~5. This behavior indicates that under acidic conditions the BP-I and BP-II lose Cd^{+2} more readily than mammalian MTs.

Preliminary amino acid analyses were obtained on Cd-BP-I and -II purified by successive chromatography over Sephadex G-75, DEAE Sephadex A-25 and Sephadex G-50. As shown in Table II, BP-II contains almost exclusively cys, glx and gly in an approximately 3:3:1 ratio, as found in other metallochelatins. The small amounts of other amino acids (< 10%) are attributed to impurities. Cd-BP-I, in contrast, has an amino acid composition different from that of either the metallochelatins or the mammalian metallothioneins. The low cysteine and high asx + glx content of BP-I resemble the distribution found in the Cu-Cd-BP of *Datura innoxia* (6) and the Cu-Th of *Saccharomyces cerevisiae* (8).

TABLE 1

Amino Acid Analyses[a] of *Euglena gracilis* Cd-BP-I & -II

Amino Acid	BP-I	BP-II	Amino Acid	BP-I	BP-II
Cys	14	40	Met	0.5	nd
Glx	15	35	Ile	1.6[b]	0.7[b]
Gly	16	14	Leu	1.6[b]	---
Asx	21	2.1	Tyr	1.3	0.0
Ser	8.0	3.0	Phe	1.5	0.5
Thr	3.4	1.0	Lys	1.9	0.8
Pro	5.3	nd	His	1.3	0.4
Ala	5.5	1.4	Arg	3.0	0.8
Val	4.3	0.6			

[a] 22, 46 and 72 h HCl digestions and perfonmic acid/HCl digestion were used for each peptide.

[b] Tris buffer interference

Kinetics and Mechanisms of *E. gracilis* Cd-BP Reactions

As the composition and structures of the *Euglena gracilis* Cd-BPs are elucidated, it is desirable to simultaneously probe their chemical reactivities. The thermodynamic and kinetic properties of their reactions are required, in addition to their composition and static structures, in order to evaluate their putative functions in metal detoxification and their mechanisms of metal binding and release. To explore their chemical properties, we have examined the kinetics and mechanisms of their reactions with the following reagents:

a. 5,5'-dithiobis(2-nitrobenzoic acid) [DTNB = $\phi SS\phi$]
b. Pyridylazoresorcinol [PAR]
c. EDTA

DTNB reacts with both the thiolates and the sulfides of Cd-BP-II (3). The reaction can be convienently followed at 412 nm. The stoichiometry is complex, because thiolates generate one 5-thio-2-nitrobenzoate ion, ϕS^-; while sulfide ions generate two and cadmium interferes with the latter reaction:

$$RS_2^- + \phi SS\phi \longrightarrow \phi S^- + RSS\phi$$
$$S^{-2} + \phi SS\phi \longrightarrow 2\phi S^- + S^o$$
$$CdS + \phi SS\phi \longrightarrow \text{No Reaction}$$

The reactions with Cd-BP-I & -II are complex reactions with three distinct components. A rapid reaction, complete within the mixing time, accounts for about half the TNB generated. The remainder of the reaction occurs as two measurable steps, referred to as the fast and slow phases. The rate laws for the measurable portions of the reaction are identical in form to that previously reported for metallothionein reactions. The apparent rate constants for the fast (f) and slow (s) steps are linear functions of the DTNB concentrations and have non-zero intercepts indicating that each phase consists of a first-order and a second-order component:

$$k_f^{obs} = k_{f,1} + k_{f,2}[\phi SS\phi] \qquad k_s^{obs} = k_{s,1} + k_{s,2}[\phi SS\phi].$$

As shown in Table III, each rate constant for Cd-BP-I is similar in magnitude and slightly greater than the corresponding BP-II constant. Both sets of constants are approximately three- to five-fold larger than those for the reactions of typical mammalian MTs. Thus, a hierarchy of reactivity toward DTNB is established:

BP-I ≥ BP-II > mammalian MT

The first order reactions of methallothioneins and DTNB have been previously attributed to rate-limiting Cd-thiolate bond opening processes and the second order reactions to electrophilic attack of DTNB at the coordinated thiols (9). It is not yet clear that the same mechanistic conclusions can be drawn here.

TABLE 2

Rate Constants (SD)[a] for the DTNB Reactions of Cd-BP-I & -II

Peptide	fast step		slow step		TNB/Cd
	$k_1 \times 10^2$	k_2	$k_1 \times 10^3$	$k_2 \times 10^2$	
BP-I	1.0(1)	1.0(1)	1.8(2)	8.3(2)	3.0(5)
BP-II	0.82(5)	0.87(5)	1.2(2)	7.9(2)	2.0(3)
Cd-Th[b]	0.24	0.24	0.075	2.8	
Zn-Th[b]	0.20	0.3	0.14	4.8	

[a] units: k_1, s^{-1} and $k_2, M^{-1}s^{-1}$; (SD = Standard Deviations)
[b] ref 9.

Pyridylazoresorcinol is a tridentate, metallochromic reagent which undergoes a pronounced visible absorption shift from 413 to 495 nm upon complexation with Cd^{+2} or Zn^{+2}. [The absorbance of the resulting bis-chelate complex, $(HPAR)_2Cd$ is pH sensitive due to weakly acidic protons on the ligand (10), and the extinction coefficient must be determined for the specific reaction conditions.] PAR reacts slowly with metallothionein under these conditions, while the Cd-BPs react rapidly. The cadmium ions in BP-I react completely; while only ca. 20% of the BP-II cadmium ions react in the absence of other reagents. These reactions, like those of DTNB, occur in three phases: one complete within mixing time and measurable fast and slow steps. The rate laws for the measurable reaction phases are, however, very different. Both the slow and fast phases

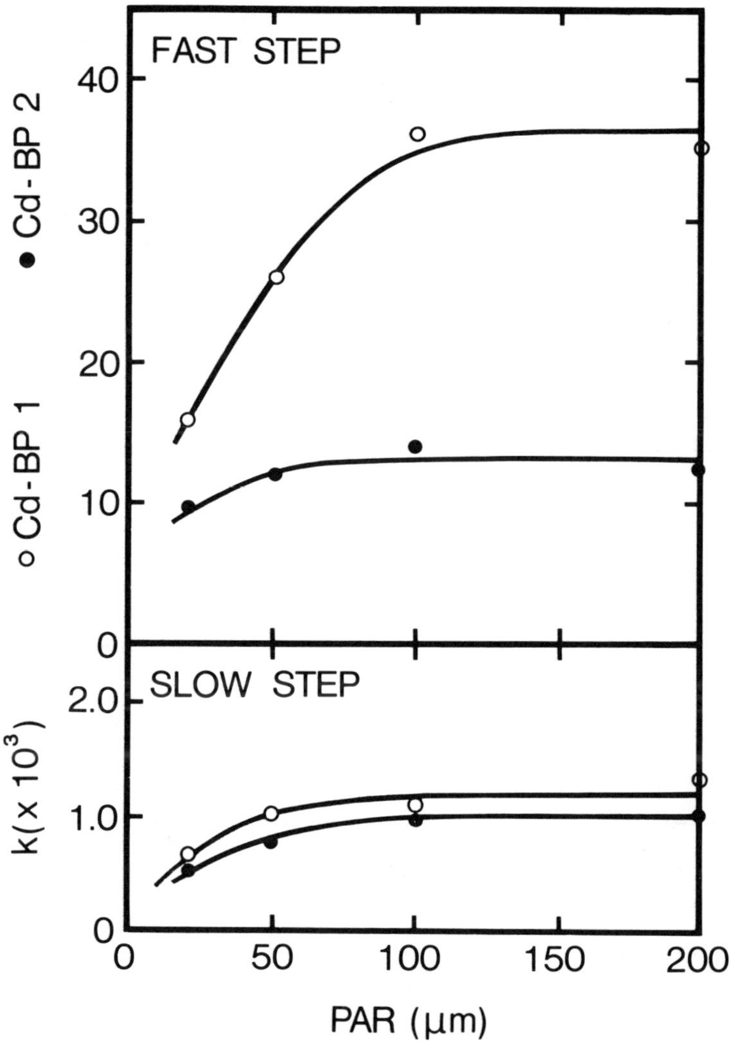

Figure 4. Saturation Kinetics Observed for the PAR Reactions of Cd-BP-I & -II. $k_{app,f}$ and $k_{app,s}$ were determined by conventional plots of $\ln(A_\infty - A_t)$ and are plotted versus [PAR]. The reactions were monitored at 485 nm and carried out in 50 mM Tris-HCl/100 mM $NaClO_4$ buffer, pH 7.5, [PAR] = 20-200 μM; [Cd] = 6 μM.

exhibit saturation kinetics, Figure 4, indicative of complex formation proceeding the rate determining step. These data can be treated by standard double-reciprocal plots (not shown) to yield the equilibrium complexation constants and rate constants defined in the equation below and presented in Table 3:

$$PAR + Cd\text{-}BP \underset{}{\overset{K_c}{\rightleftharpoons}} PAR\blacklozenge Cd\text{-}PAR \xrightarrow{k_d} BP + Cd(PAR)_2$$

Even the slowest of these reactions is more rapid than those of metallothionein with PAR. Since PAR is a tridentate N,N,O donor system, it is reasonable to postulate that the complex PAR♦Cd-BP involves bidentate or tridentate coordination of PAR to a cadmium bound to two or more thiolate or sulfide ligands. The kinetics data do not, however, provide any insight into the bonding or structure of the complex.

TABLE 3

Kinetic Constants for the PAR Reactions of Cd-BP-I and -II[a]

	fast step		slow step		
Peptd	$K_c \times 10^2$	$k_d \times 10^2$	$K_c \times 10^3$	$k_d \times 10^3$	$[Cd]_{PAR}/[Cd]_{BP}$
BP-I	35.9(1)	4.5(1)	2.3(1)	1.4(1)	1.0(1)
BP-II	3.0(1)	0.72(5)	1.5(1)	1.25(3)	0.20(2)

[a] as defined in the text; units: K_c, M^{-1}; and k_d, s^{-1}

Finally the reactions of a Sephadex G-50 Cd-BP mixture (BP-I & -II) with EDTA were measured using a pseudo-first order excess of the ligand (2) Under these conditions, the reaction was biphasic and can be modelled with two first order rate constants (conditions: 25°; 50 mM Tris-HCl, pH 8.6, 100 mM KCl; 0.4 mM EDTA):

$$k_f^{obs} = 4 \pm 3 \times 10^{-4} sec^{-1} \qquad k_s^{obs} = 7 \pm 2 \times 10^{-5} sec^{-1}$$

Both the slow and fast phases were faster than the analogous reaction of a cadmium-rich metallothionein and EDTA, which occurred over three days. The rate constant for this reaction also under pseudo-first order conditions (1mM EDTA; 10 mM Tris-HCl; pH 7.5; 100 mM KCl; $Cd_{5.1}Zn_{1.3}Cu_{0.6}MT$) is:

$$k^{obs} = 3.7 \times 10^{-6}$$

The EDTA reactions also fit the pattern that the BPs are more reactive than MT in reactions affecting the metal-sulfur linkages.

DISCUSSION

The amino acid analysis and additional physiochemical properties of Cd-BP-II presented in the first part of the text establish striking similarities between BP-II and the plant and yeast phyto- or metallo-chelatins described by others (5,6). The presence of sulfide ions (3), which are also found in the yeast γ-glutamylcysteinyl peptides called cadystins (4), is also suggestive of a common composition and structure for the *E. gracilis* system. We are presently seeking direct evidence for the presence of the γ-glu-cys bonds, which will unambiguously support this identification.

The *E. gracilis* Cd-BP-I and -II differ in a number of properties that have been examined, the most definitive being the amino acid compositions. The BP-I analysis is more like those of the Type-II Cu- and Cu,Cd-binding metallothioneins of yeast and other organisms than those of mammalian (Type-I) MTs (which have more cysteine and less glx and asx) or metallochelatins (which have only cys, gly and glu). These findings raise the interesting possibility that the *E. gracilis* may have both a plant-like metallochelatin and a type-II metallothionein. While our cells grown in the dark contain both types of BPs, suggests that *E. gracilis* cells grown in the dark may generate only the Type-II metallothionein in response to cadmium (11).

BP-I and -II are consistently more reactive to chemical probes which target either the sulfur donor ligands, cysteine and sulfide, or the cadmium ions of the complexes than are mammalian MTs. The subsititution of sulfide ions for cysteines and the presence of only smaller peptides with an average of three cysteines per unit (as inferred from the amino acid analysis) should render the BP-II structure more open than the clusters of MT which are somewhat protected by the thionein protein chain.

The ease and speed with which *E. gracilis* cells can be grown on controlled media and the large amounts of Cd-BPs that can be harvested suggest that this may be an excellent model system in which to examine the effects of environment on the composition, structure and reactivity of metallochelatins.

Note Added: *Euglena gracilis* "Z" strain (#753) was purchased from the Culture Collection of Algae, University of Texas-Austin.

REFERENCES

1. Gingrich DJ, Petering DH, Shaw III CF (1984). Mar. Env. Res. 14:89.
2. Gingrich DJ, Weber DN, Shaw III CF, Garvey JS, Petering DH (1986). Env. Health Res. 65:77.
3. Weber DN, Petering DH, Shaw III CF (1987). J. Biol. Chem. 262:6962.
4. Murasugi A, Wada C, Hayeshi Y (1983). J. Biochem. 93:661.
5. Grill E (1987). Experientia Suppl. 52:317; Bernhard WR, Kägi JHR (1987). Experientia Suppl. 52:309.
6. Robinson NJ, Barton K, Naranjo CM, Silland LO, Treshella J, Watt K, Jackson PJ (1987). Experientia Suppl. 52:323.
7. Phillips JL (1983). Biochem. Biophys. Res. Commun. 114:998
8. Butt TR, Sternberg EJ, Gorman JA, Clark P, Hamer D, Rosenberg M, Crooke ST (1984). PNAS 81:3332.
9. Li T-Y, Minkel DT, Shaw III CF, Petering DH (1981). Biochem. J. 193:441.
10. Hnilickova M, Sommer L (1961). Czech. Chem. Commun. 26: 2198; Sommer L, Kuban V, Langova M (1982). Freneius S. Anal. Chem. 310:51; Corsini A, Fernando Q, Freiser H (1963). 2:224.
11. Mazus B, Falchuk KH, Vallee BL (1988). 72nd FASEB Meeting, May 1988, Abstract 2737.

Metal Ion Homeostasis: Molecular Biology
and Chemistry, pages 325–336

PERSPECTIVES ON Cd AND Zn ACCUMULATION, ACCOMMODATION AND TOLERANCE IN PLANT CELLS: THE ROLE OF Cd-BINDING PEPTIDE VERSUS OTHER MECHANISMS

George J. Wagner and Rachel M. Krotz

Plant Physiology/Biochemistry/Molecular Biology Program
Department of Agronomy, University of Kentucky,
Lexington, KY 40546-0091

ABSTRACT Metal binding peptides containing (γglu-cys)$_n$, also called Cd-binding peptide, phytochelatin, etc., have received much attention in the recent literature. Their unique structure and apparent mode of biosynthesis make them very interesting subjects for study. Since existing reports concerning formation of these peptides in plants and plant cultured cells have utilized high or very high levels of metals and have focused on the basis of tolerance to high metal challenge, the importance of these peptides in the transfer of Cd from vegetable foods - produced under low level exposure conditions - to the human diet is not known.

In an attempt to study responses of plant cells to Cd and Zn in the absence of Cd-peptide and to approach the low-level exposure case, we have studied tobacco cultured cells exposed to subtoxic levels of metal before Cd-peptide has accumulated to any degree. Responses under toxic level exposure were also studied in the presence of buthionine sulfoximine, an inhibitor of peptide formation. The possible role of vacuolar organic acids as complexing agents for Cd and Zn, and of vacuolar compartmentation in metal accommodation and accumulation has also been assessed. Vacuolar accumulation of Cd and Zn as organic acid salts and cell wall binding may be principal mechanisms for metal accumulation under low level exposure conditions. In the system studied, even high, toxic levels of Zn do not induce significant formation of ligand of Cd-induced peptide.

INTRODUCTION

To this author, one might consider studying (γglu-cys)$_n$ metal binding peptides for several reasons:

1) In a fundamental understanding sense, because they form the basis for a unique, metal-stimulated, metal binding peptide system. This system appears to be composed of peptide, metal and acid labile sulfide (Winge et al., this volume).

2) In an applied sense, in that one may ask if these peptides can form the basis of a "super-tolerant" plant which could be used to detoxify effluents, soils or to "mine" valuable metals.

3) In a human health sense. Since ≥70% of dietary Cd in man comes from vegetable components of the diet, one may ask if (γglu-cys)$_n$ peptides or other mechanisms are important in Cd accumulation in agriculture. Agricultural crops contribute to intake of other toxic metals in man as well. Also, one may ask if manipulation of plant metal accumulation mechanisms might lead to the desired condition of lowered toxic metal in the diet.

Fundamental Understanding of the Chemistry and Biology of (γglu-cys)$_n$ Metal Binding Peptides.

It should be emphasized that these peptides are fundamentally interesting in their own right. Their unique structure (1,2), their high affinity for Cd (K_3 Cd=10^{19}, 3) and their apparent contribution to Cd tolerance after high level Cd exposure (4-8) make them interesting to study. We have yet to understand if most higher and lower plants are capable of forming these peptides, though E. Grill's data (this volume) suggest that the capacity to form peptides may be ubiquitous in higher plants. We do not as yet understand the specificity of metal binding *in vivo*, the mode of peptide biosynthesis, where peptide occurs in plant cells, or if non-metal inducers can cause their formation. We also have little information regarding turnover, a possible role in metal uptake, if peptide can be the sole factor in metal tolerance or accumulation, and the relevance of peptide to the low level metal exposure condition. It is premature to suggest complex roles based on the present limited information. One is reminded that the issues of the tolerance and nutrient-ion-homeostasis roles for metallothionein (MT) are still debated, twenty years after extensive investigation of this metalloprotein. The importance of metallothionein in total metal accumulation is still a question for the yeast case and also for animal cells.

It can be argued that the nutrient-ion homeostasis role for MT is consistent with the occurrence of a constitutive component of MT in yeast and animals and the requirement for only low levels of metal (1μM Cd and 25μM Zn) to induce this protein. In contrast, the constituent nature of (γglu-cys)$_n$ peptide is not established, indeed most reported results suggest that it does not occur constitutively. Also, most reports indicate a need for high level metal exposure to induce significant peptide accumulation. The work of Winge's group demonstrates the involvement of acid labile sulfur in metal peptide complexes, at least under certain circumstances (this volume). Thus, we are just beginning to understand the biology of this interesting metal binding peptide.

(γglu-cys)$_n$ Metal Peptide as a Potential Means for Soil and Effluent Detoxification.

If one wished to consider the possible exploitation of this metal chelator in an applied sense, one may be encouraged by the demonstration in at least two systems that tolerance (lack of growth inhibition) is correlated with occurrence of Cd-peptide (4,8). One must keep in mind however that these systems are cultured cells. Tolerance has not been correlated with peptide in an intact plant system. But, Rauser has suggested that the rate of Cd and Cu binding peptide(s) formation may be associated with tolerance in *A. gigantea* (5). Indeed, at this meeting, E. Grill described the occurrence of low levels of peptide in plants growing on soil containing 2% Zn. Peptide was said to account for only a few percent of total Zn in these plants, the majority being associated with a soluble, low mol wt constituent. In cultured tobacco cells not acclimated to growth in the presence of high levels of Cd or Zn ("non-tolerant"), lack of growth inhibition was positively correlated with the amount of Cd-peptide and inversely correlated with the amount of low mol wt non-peptide Cd (under buthionine sulfoximine inhibition conditions, 8). Since cultured cells can be easily acclimated to growth with high levels of metal, it may be possible to select tolerant variants. However, to this author's knowledge, these tolerant lines have not been shown to be stable variants which can be regenerated to form tolerant plants which possess a seed-transmissible tolerance trait. It is noteworthy that Huang et al. reported the selection of Cd but not Pb, Zn, Hg or Ag resistant tomato cultured cells which retained resistance after growth without selection pressure (9). Also, Kishinami and Widholm selected Cu and Zn resistant *Nicotiana* cells said to retain resistance (10).

It is apparently not difficult to select intact plants for Cd tolerance. We have selected variants (5 to 8% of a population) among populations of tobacco variety KY 14 which are survivors when grown for extended periods of time in 1/2 strength Hoagland's solution containing 90μM Cd. If these plants are rescued after death of less tolerant plants by transfer to Cd-free solution, they resume vigorous growth. After recovery, they can be transferred to 90μM Cd and grow as though tolerant. Cd-peptide appears to be a principal form of soluble Cd in these plants. We have not yet examined progeny of these plants for inheritable tolerance (unpublished). Therefore, available evidence would suggest that peptide may be useful in increasing the capacity of plants to accumulate Cd. It is noteworthy in this context that the water hyacinth *Eichhornia crassipes* has been shown to efficiently concentrate metals from water and was recently shown to accumulate Cd as Cd peptide (11).

A role for peptide in accumulation of valuable metals is suggested by experiments of Grill et al. using silver (7); however, little data is available on the ability of various metals other than Cd and Cu to specifically induce and bind to peptides in plants. The work of Grill et al. suggests that various metals can cause varying degrees of peptide induction in *Rauvolfia serpentina* cultured cells (7) and *Schizosaccharomyces pombe* cells (12). However, experiments were limited to single high concentrations and single metal ion salts. Also, only ligand of peptide was monitored and it has not been shown if many of the ions which caused formation of peptide can actually bind *in vitro* or *in vivo*. Therefore, indirect, perhaps stress-related formation has not been ruled out. In the *Rauvolfia* system, Zn is said to be an efficient inducer of peptides while in tobacco cultured cells (8) and plants, and in *S. pombe* cells (Winge et al., this volume), Zn does not cause formation of substantial peptide, even after high level Zn exposure. We conclude that the literature is incomplete and inconsistent regarding induction of peptide by various metals in a "metallothionein-like" fashion. Since metallothionein is induced at the transcriptional level and peptide is not, it would not be surprising to find different metal specificities in mechanisms of metal induction.

It should be pointed out here, and will be more thoroughly discussed below, that metal tolerance has been correlated with increased levels of organic acid in several plants which have evolved tolerance to enable growth on high level metal containing soils. In this group is the *Silene* plant described by E. Grill as producing only low levels of peptide after high level Zn accumulation from soil.

Cadmium, Crop Plants and the Human Diet.

Consideration of the mechanisms of Cd accumulation as pertains to transfer of Cd from crop plants to the human diet must include a realization that agricultural plants are normally exposed to very low levels of metal (13, 14) (perhaps $\leq 0.04\mu M$ Cd and $\leq 25\mu M$ Zn) in unpolluted to moderately polluted soil solutions. It is difficult to estimate Cd and Zn content of a soil solution to which a plant root is exposed since many factors play in such complex systems (15). However, mean Cd and Zn concentration of soils have been estimated to be 0.06 and $50\mu g$/gm dry wt, respectively (16), and if one assumes a mean soil density of 1.3 g/cc and that under near neutral pH conditions, plants can access only about 5% of soil content, the available Cd and Zn would be about 0.04 and $50\mu M$, respectively. However, it should be noted that soil-to-plant concentration factors for Cd after low level exposure are generally found to be several fold higher than those for Zn (15,16). Thus Zn availability may be lower than that predicted above.

The above estimated values for soil solutions are in contrast to ranges of 45 to $1000\mu M$ Cd and 300 to $2000\mu M$ Zn used in studies of cultured cells. Levels often used in studies of solution cultured plants range from 0.1 to $6\mu M$ Cd and 2 to $150\mu M$ Zn. Thus, often for reasons of expedience, most studies are conducted far above levels expected to be encountered by most plants in nature. One may therefore question whether responses of plants and plant cells to high concentrations of metals reflect the low level exposure case.

Plants may encounter high levels of certain metals in nature due to high natural soil abundance, acid soils, application of metal contains sludge or excessive use of high Cd containing phosphate rock fertilizer, but fortunately high exposure is not the norm (17).

Other factors affecting Cd in food plants which are poorly understood are mechanisms of uptake at the root level and mechanisms of metal partitioning among plant tissues. We do not know if siderphore mechanisms exist for this metal or if rhizosphere organisms may concentrate the metal and present the root surface with a higher concentration of the metal than that expected from soil content. We also do not understand mechanisms of partitioning of metals between roots, shoots, etc. These gaps in knowledge also exist for the related metals, Zn, Cu, Hg as well. Partitioning of metals between roots, leaves, fruits, etc. varies from plant to plant and has not been studied systematically (14). The question of possible remobilization of Cd during fruit and grain development is not resolved.

Most plants are relatively tolerant to growth in the presence of low to moderate levels of Cd and Zn. Indeed, low levels of Cd appear to stimulate growth of tobacco seedlings (unpublished). Because of this relative tolerance, healthy plants can contribute substantial metal to the human diet.

RESULTS AND DISCUSSION

Fig. 1 shows the accumulation of soluble Cd in tobacco cultured cells exposed to subtoxic (8) 45μM Cd. As shown, a high level of Cd penetration into cells has occurred before cells can be washed with 10mM $CaSO_4$ (to remove free space Cd) and recovered (~8 minutes). Washed cells were homogenized in 25mM KPO_4 buffer, pH 8 as previously described (8). A low level of accumulation or equilibration with the medium occurs over a 6 hour period. Cd-peptide is shown to occur under these conditions, but only after about 2 hours and after 36 hours Cd peptide only accounts for 30% of soluble Cd. The remainder occurs as $\geq$300 dalton soluble metal. Peptide bound and non-peptide soluble Cd were

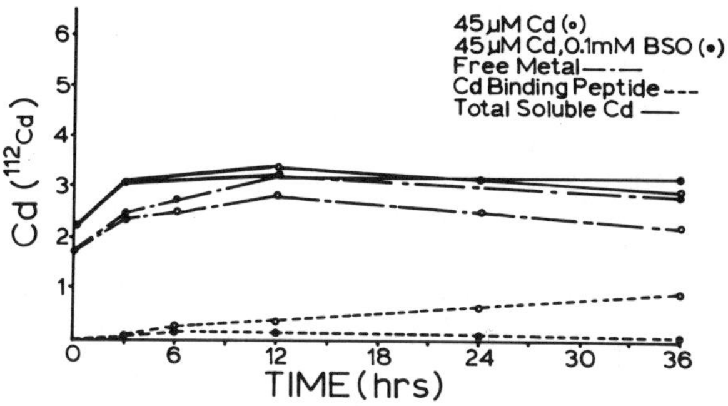

FIGURE 1. Time course of Cd accumulation in tobacco cultured cells in presence and absence of Cd-peptide-subtoxic exposure.

distinguished using G-15 desalting as previously described (8). The level of insoluble Cd (assumed to be cell wall and/or membrane bound metal) under these conditions is <10% of total soluble metal and is little changed from 0 time to 36 hours (not shown). Thus, cells take up Cd under these non-toxic exposure conditions in the absence of Cd peptide. Similarly, Zn is taken up rapidly in cells exposed to 300µM Zn (not shown) but no ligand of Cd-peptide is formed (8). One may then ask what the form of intracellular, soluble Cd and Zn is in the absence of peptide.

Correlations have been reported between organic acids and Cu and Zn levels in plants tolerant to these metals. Ernst presented a model in which organic acids serve to shuttle Zn from the cytoplasm to the vacuole and then to complex metals within the vacuole (see 18). A number of comparative studies of tolerant versus non-tolerant plants have demonstrated increased acid levels in tolerant clones (i.e., 10,19). However, correlations between metal accumulation and increased organic acid are not positive in all cases and alternative models for metal accumulation have been proposed (i.e., see 18,20).

We compared organic acid and metal levels in tobacco cultured cells exposed to Cd and Zn to determine if sufficient endogenous organic acid occurs in cells to accommodate the immediate presence of soluble metal after exposure and also if organic acid biosynthesis is stimulated by metal exposure. Substantial Cd and Zn are found within cells, but concentrations of organic acids (principally malic and citric acids) were such that even after exposure to very high Cd (600μM) and Zn (2000μM) there is still excess organic acid to accommodate the metal - assuming one mole of metal is complexed by one mole of acid (not shown). We have determined the cation content of tobacco cultured cell lysates and assuming that the bulk of ions is contained in the vacuole, we calculate that exposed and unexposed cells contain 60mM K, 1mM Ca, 2mM Mg, 3mM nitrate, and low levels of protein. The pH of the vacuole is estimated to be pH 5. Under these conditions charge balance is realized. Given that log conditional stability constants for malate and citrate, Cd complexes are reported to be (3.8 est.) and 3.9 and for Zn complexes 2.8 and 4.9, respectively and that similar constants for Ca and Mg and protons are at least 1 order of magnitude lower (21), Cd and Zn are expected to easily displace Mg and Ca from these ligands (Krotz and Wagner, in preparation).

The bulk of organic acids is expected to occur in vacuoles and the above considerations assume metals also accumulate in the vacuole. We have directly examined the localization of Cd and Zn in protoplasts prepared from leaves of Cd exposed wheat seedlings and those prepared from *Datura* cultured cells. We were unable to prepare vacuoles from tobacco cultured cell protoplasts which were sufficiently pure for meaningful analysis. In both wheat leaves and *Datura* cells we calculate that the bulk of Cd and Zn occurs in vacuoles of tissue/cells exposed to 45μM Cd or 300 μM Zn (Krotz and Wagner, in prparation). Thus, under these relatively high but subtoxic levels of exposure, the metals accumulate in the vacuole.

Organic acid levels appeared to increase with time of culture. This, however, reflected cell growth since the rate of acid increases the same as the rate of cell growth. We used radiolabeled acetate to determine if label in cationic, anionic or neutral components of extracts varied between control vs 45μM Cd vs 300μM Zn treated cells. No stimulation of incorporation due to metal occurred up to 14 hours. As expected, the anionic or organic acid containing fraction was principally labeled. Analysis of specific acids by HPLC showed malic and citric acids to be the principal labeled components.

These results are consistent with the rapid uptake of Cd and Zn under the exposure conditions tested and accumulation of metal in the vacuole as complexes with substantial endogenous organic acid. With time, Cd - but in this system, not Zn - stimulates formation of Cd peptide which may enhance metal accommodation within cells.

The role of cell walls and cellular insoluble components in complexing Cd and Zn has not been discussed here. In certain whole plants, this fraction is known to be significant in metal accumulation (14,18).

To test the influence of Cd peptide on total Cd accumulation, cells were exposed to 250μM Cd in the presence and absence of buthionine sulfoximine. As shown in Fig. 2, Cd accumulation is only partially reduced in the presence of 0.2mM inhibitor, which inhibits Cd-peptide formation by ≥85%. Huang et al. have observed similar results in Cd-tolerant tomato cells (9). Thus, peptide does not appear to be principally responsible for Cd accumulation under the conditions tested.

The above results suggest that in cultured tobacco cells accommodation to subtoxic levels of Cd and toxic as well as subtoxic levels of Zn may occur as a result of complexation in the vacuole with endogenous organic acid.

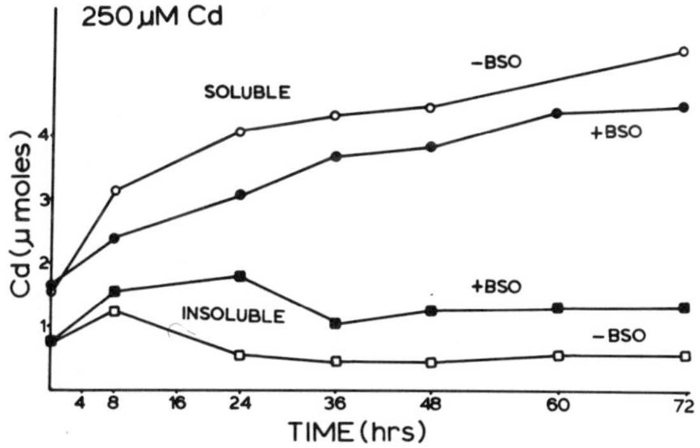

FIGURE 2. Time course of Cd accumulation in tobacco cultured cells in the presence and absence of Cd-peptide-toxic level exposure.

SUMMARY

Heavy metal uptake by plants and the significance of (γglu-cys)$_n$ peptides may be viewed from several perspectives. The unique peptides are very interesting subjects for study from both a biochemical and physiological point of view. More study is needed to understand their roles in metal tolerance, accommodation and accumulation, which metals and factors stimulate their formation, etc. If one is interested in modifying plants to prepare "super-accumulators" for detoxifying soils or effluents, one may be encouraged that peptide may serve in this capacity. However, available data suggest that even in "metal-tolerant" systems, peptide is not solely responsible for accumulation and the stability of metal tolerance in variants can be questioned. If one is interested in heavy metal accumulation in plants as it relates to the human diet and human health, one must realize that vegetable components of the diet are acquired under low level exposure conditions. Until we understand the degree of involvement of (γglu-cys)$_n$ metal binding peptides in plants exposed to levels of metal which occur in agricultural soils, we should avoid discussing peptides in a human-diet, human-health context.

We are currently exploring alternative means for modifying metal partitioning through introduction of the mouse metallothionein gene into plants (22).

Finally, mechanisms other than peptide sequestration have been suggested for heavy metal accumulation, accommodation and tolerance in plants. These include vacuolar compartmentation as organic acid complexes, binding to cell wall exchange sites, aggregation with sulfide, and others. Keeping in mind the diversity which exists among terrestrial plants, we may expect to see different mechanisms predominate, depending on the metal and the plant species.

REFERENCES

1. Kondo N, Imai K, Isobe M, Goto T (1984). Cadystin A and B, major unit peptides comprising cadmium binding peptides induced in a fission yeast. Tetrahedron Lett 25:3869.
2. Grill E, Winnacker EL, Zenk MH (1985). Phytochelatins: The principal heavy-metal

complexing peptides of higher plants. Science 230:674.
3. Reese NR, Wagner GJ (1987). Properties of tobacco cadmium binding peptide(s). Biochem J 241:641.
4. Jackson PJ, Roth EJ, McClure PR, Navanjo CM (1984). Selection, isolation and characterization of cadmium resistant *Datura* inoxia suspension cultures. Plant Physiol 75:914.
5. Rauser WE (1986). The amount of Cd associated with Cd-binding protein in roots of *Agrostis gigantea*, maize and tomato. Plant Sci 43:85.
6. Steffens JC, Hunt DF, Williams BG (1986). Accumulation of non-protein metal binding polypeptides in selected cadmium resistant tomato cells. J Biol Chem 261:13879.
7. Grill E, Winnacker EL, Zenk MH (1987). Phytochelatins, a class of heavy metal binding peptides from plants are functionally analogous to metallothioneins. Proc Natl Acad Sci USA 84:439.
8. Reese NR, Wagner GJ (1987). Effects of buthionine sulfoximine on Cd-binding peptide levels in suspension-cultured tobacco cells treated with Cd, Zn or Cu. Plant Physiol 84:574.
9. Huang B, Hatch E, Goldsborough PB (1987). Selection and characterization of cadmium tolerant cells in tomato. Plant Sci 52:211.
10. Kishinami I, Widholm J (1986). Selection of copper and zinc resistant *Nicotiana plumbaginifolia* cell suspension cultures. Plant Cell Physiol 27:1263.
11. Fujita M (1985). The presence of two Cd-binding components in the roots of water hyacinth cultivated in Cd-containing medium. Plant Cell Physiol 26:295.
12. Grill E, Winnacker EL, Zenk HH (1986). Synthesis of seven different homologous phytochelatins in metal-exposed *Schizosaccharomyces pombe* cells. FEBS Lett 197:115.
13. Wagner GJ, Nulty E, LeFevre M (1984). Cadmium in wheat grain: Its nature and fate after ingestion. J Tox Environ Health 13:979.
14. Wagner GJ, Yeargan RY (1986). Variation in Cd accumulation potential and tissue distribution of Cd in tobacco. Plant Physiol 82:274.
15. Cataldo DA, Wildung RE (1978). Soil and plant factors influencing the accumulation of heavy metals by plants. Environ Health Persp 27:149.

16. Hopps H (1974). Overview. In "Geochemistry and the environment Vol 1, The Relation of Selected Trace Elements to Health and Disease," Washington, D.C.: National Academy of Sciences, p 3.
17. Ryan JA, Pahren HR, Lucas JB (1982). Controlling Cd in the human food chain: A review and rational based on health effects. Environ Res 28:251.
18. Woolhouse HW (1983). Toxicity and tolerance in the responses of plants to metals. In Lange OL, Nobel PS, Osmond CB, Ziegler H (eds): "Encyclopedia of Plant Physiology," New Series, Vol 12c, p 245.
19. Godbold DL, Horst WJ, Collins JC, Thurman DA, Marschner H (1984). Accumulation of Zn and organic acids in roots of Zn tolerant and non-tolerant ecotypes of *Deschampsia caespitosa*. J Plant Physiol 116:59.
20. Van Steveninck RFM, Van Steveninck ME, Fernando DR, Horst WJ, Marschner H (1987). Deposition of zinc phytate in globular bodies in roots of *Deschampsia* ecotypes; a detoxification mechanism. J Plant Physiol 131:247.
21. Sillen LG, Martell AE (1964). Stability constants special publication No. 17. The Chemical Society of London, Metcalfe and Cooper, England.
22. Maiti IB, Hunt AG, Wagner GJ (1988). Seed-transmissable experssion of mammalian metallothionein in transgenic tobacco. Biochem Biophys Res Comm 150:640.

Metal Ion Homeostasis: Molecular Biology and Chemistry, pages 337–346

Cd-TOLERANCE IN PLANT CELLS: A COMPARISON OF BIOCHEMICAL AND MOLECULAR PROPERTIES OF TOLERANT AND SENSITIVE CELLS[1]

Paul J. Jackson, Nigel J. Robinson[2], and Emmanuel Delhaize

Genetics Group, Life Sciences Division, Los Alamos National Laboratory, Los Alamos, NM 87545

ABSTRACT Plants and plant cells can be selected for the ability to grow in the presence of normally toxic concentrations of certain trace metal ions. Metal-tolerance is often associated with the ability to produce large amounts of certain metal-binding polypeptides, poly(γ-glutamylcysteinyl)glycines. The ability to produce these polypeptides plays an important role in Cd-tolerance since inhibition of their synthesis results in rapid cell death in the presence of metal ions. However, Cd-sensitive cells are also capable of synthesizing equivalent amounts of these compounds. Therefore, some other biochemical or physiological mechanism must also contribute to tolerance. Molecular and biochemical properties of Cd-tolerant *Datura innoxia* cells grown in the presence and absence of Cd were compared to those of Cd-sensitive cells grown under the same conditions. Certain biochemical and molecular differences which may contribute to tolerance were apparent.

INTRODUCTION

Angiosperms have rapidly colonized environments containing high concentrations of certain metal ions (1).

---

[1] This work was supported by the U.S. Department of Energy, Ecological Research Division and the Southwest Consortium on Plant Genetics and Water Resources.

[2] Present address: Department of Botany, University of Durham Science Laboratory, South Road, Durham, DH1 3LE, England.

While much is known about the genetics and evolution of metal tolerance in plants, the physiological and biochemical basis of tolerance is not as well understood. Poly-(γ-glutamylcysteinyl)glycines [$(\gamma EC)_nG$, n=2-5] have been implicated in metal-detoxification in a number of plant species (2,3,4). These compounds are not directly encoded by structural genes but are products of a biosynthetic pathway which consumes glutathione (5). They are clearly involved in metal-tolerance, since inhibition of their synthesis results in rapid cell death (6). However, Cd-sensitive cells are also able to synthesize similar amounts of these compounds in the presence of Cd. This suggests that some other factors must also contribute to tolerance. This paper describes the results of experiments designed to determine differences between Cd-tolerant and sensitive cells derived from a common cell line.

METHODS

Maintenance of Plant Suspension Cultures and Measurement of Growth Rates.

Suspension cultures of Cd-tolerant and sensitive *D. innoxia* were maintained as described (7). Packed cell volumes of 5 ml of culture were used to determine growth rates of cultures (8).

Radioisotope Labeling of Cells.

Carrier-free [^{109}Cd]Cl_2, L-[^{35}S]cysteine (>300 Ci $mmol^{-1}$), L-[4,5-^{3}H(N)]leucine (50 Ci $mmol^{-1}$) and [6-^{3}H]uridine (25-30 Ci $mmol^{-1}$) were purchased from E.I. du Pont de Nemours & Co., Inc., Boston, MA. Labeling was as described previously (2,5,7).

Determination of RNA and Protein Synthesis.

Cells were pulse-labeled with either 10 μCi/1.0 μg/ml [^{3}H]leucine or 5 μCi/0.1 μg/ml [^{3}H]uridine for 30 minutes. To determine uridine incorporation, four 5 ml aliquots were centrifuged at 800 g for 2 minutes to collect the cells. Pellets were then resuspended in 5 ml ice-cold 20% trichloroacetic acid (TCA) for 30 minutes. Cells were collected by centrifugation as before and the pellet was washed three times with ice-cold 5% TCA. Cells were then collected on glass fiber filters by vacuum filtration. Filters were washed five times with 5 ml ice-cold 5% TCA and twice with

95% ethanol, dried and assayed by scintillation spectrophotometry. Leucine incorporation was measured as described previously (5).

Cell Extraction and Analysis of Polypeptides.

Cells (10 ml of culture) were collected by centrifugation at 800 g for 1 minute, resuspended and washed twice in ice-cold buffer containing 10 mM Tris-HCl, pH 7.4, 10 mM KCl, and 1.5 mM $MgCl_2$. Pellets were resuspended in 0.7 ml washing buffer containing 50 mM 2-mercaptoethanol and were homogenized in an Elvehjem tissue grinder. Samples were centrifuged (20,000 g for 30 minutes) prior to analysis by gel filtration HPLC. Extracts to be analyzed by reverse phase HPLC were acidified by addition of 0.6 ml of 1 N HCl then centrifuged for 10 minutes at 10,000 rpm in a microcentrifuge prior to passage through Centricon filters (Amicon Corp., >30,000 d exclusion) and analysis.

In vitro Translation.

Total RNA and (A^+)RNA were purified as described previously (5). (A^+)RNA was translated in a reticulocyte lysate kit (Amersham Corp.) using $[^{35}S]$methionine as the labeled amino acid. The procedures that accompanied the kit were followed for the translation of 1 μg (A^+)RNA. Translation products were separated by 2-dimensional gel electrophoresis using the method of O'Farrell (9) with the following modification; CHAPS was used instead of NP-40 in the first dimension and the first dimension rod was not equilibrated prior to electrophoresis through a 7.5-15% SDS gradient gel. Gels were impregnated with ENHANCE (E.I. du Pont de Nemours & Co., Inc.) prior to fluorography at -70°C.

HPLC Analysis.

Reverse phase HPLC. Two hundred μl aliquots of acidified cell extracts were applied to a 250 x 4.6 mm column of nucleosil C-18 (BioRad) equipped with a 30 x 4.6 mm precolumn. Samples were eluted with a 20 ml linear gradient of 0.1% (v/v) trifluoroacetic acid to 0.1% trifluoroacetic acid containing 20% (v/v) acetonitrile, at a flow rate of 2 ml min^{-1}. One ml fractions were collected and assayed for the presence of thiols.

Gel filtration HPLC. Twenty μl aliquots of cell extracts were applied to two 300 x 7.5 mm columns (in series) of Spherogel-TSK 3000SW (Beckman Instruments) equilibrated with 50 mM Tris-HCl, pH 7.0, 150 mM NaCl. Samples were eluted with the same buffer (0.5 ml min^{-1}) and 250 µl fractions were collected and assayed for the presence of ^{109}Cd.

Thiol Assay.

The total number of thiol groups was determined by the method of Ellman (10).

RESULTS

Growth of cells in the presence or absence of Cd. Cultures of Cd-sensitive *Datura innoxia* cells and cells tolerant of long-term exposure to 250 µM Cd divide with similar frequencies in the absence of the toxic metal ion. Exposure to 250 µM Cd results in an initial decrease in the viability of both cultures as determined by the exclusion of trypan blue (Fig. 1). However, while the viability of the tolerant culture stabilizes at approximately 80% and eventually recovers, the viability of the sensitive culture is reduced to approximately 20% within 48 hours after exposure to Cd and continues to decline.

Protein and RNA synthesis. Cd-tolerant and sensitive cells were exposed to 250 µM $CdCl_2$ for different periods. Cells were then incubated for 30 minutes in either [^{3}H]-leucine or [^{3}H]uridine to determine the rate of protein and RNA synthesis occurring in the cells. Figure 2 shows a rapid decrease in RNA synthesis in sensitive cells within eight hours after exposure to Cd. This is followed by a rapid decrease in protein synthesis. Both RNA and protein synthesis decrease slightly in Cd-tolerant cells but recover to previous levels.

Synthesis and assembly of Cd-binding complexes. Initially, Cd-tolerant and sensitive cells synthesize equivalent amounts of $(\gamma EC)_nG$ upon exposure to 250 µM Cd (Table 1). The amount of synthesis is similar for 24 hours. After this time there is a decrease in the amount of these compounds accumulated in the sensitive cells. However, the viability of the sensitive culture has decreased by this time. This suggests that the ability to synthesize metal-binding polypeptides is only part of the tolerance mechanism. Analysis of metal-binding complexes in the two cell cultures demonstrates a difference in the amount of Cd bound shortly after exposure of the cells to Cd (Figure 3). In the first 4 hours (Fig. 3a), most of the Cd found in extracts from the sensitive cells elutes at a Kd similar to a

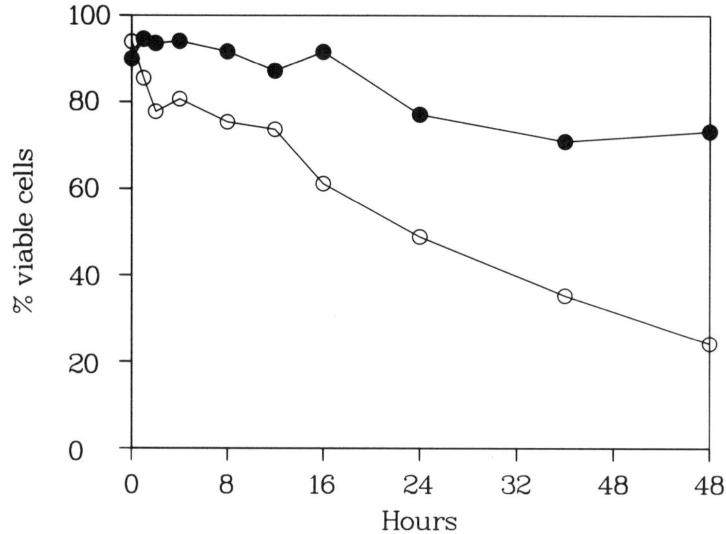

FIGURE 1. Determination of cell culture viability by exclusion of a vital stain. A small sample of culture was placed on a microscope slide and stained with a solution containing 2% (w/v) trypan blue. After five minutes cells were viewed microscopically and at least 800 cells were scored for the absence or presence of the dye within the cytoplasm. (●), Cd-tolerant cells; (o), sensitive cells.

2-mercaptoethanol:Cd complex (fraction 41) while a portion of the Cd found in extracts from Cd-tolerant cells is associated with metal complexes known to contain $(\gamma EC)_nG$. After 8 hours (Fig. 3b), the majority of the Cd in extracts from tolerant cells is associated with these complexes. After 48 hours (Fig. 3c), sensitive cells have formed larger Cd:$(\gamma EC)_nG$ complexes but, by this time, cell death has occurred.

Analysis of in vitro translation products by two-dimensional gel electrophoresis. Total mRNA was isolated from Cd-tolerant and sensitive cells grown in the absence of Cd and from cells exposed to 250 μM Cd for four hours. Figure 4 shows the results of 2-dimensional gel electrophoresis of the products of *in vitro* translation of this mRNA. Results demonstrate that some mRNA sequences are synthesized constitutively by tolerant but not by sen-

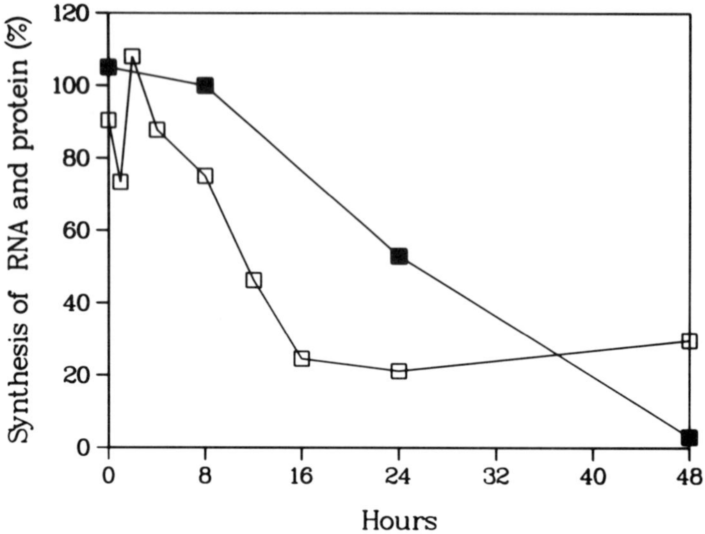

FIGURE 2. RNA and protein synthesis in Cd-tolerant and sensitive cells exposed to Cd for different periods. Cd-tolerant and sensitive D. innoxia cells growing in the presence or absence of Cd for different periods were pulse-labeled for 30 minutes with either [^{3}H]leucine or [^{3}H]-uridine. Total radioactivity incorporated during this time was then determined. The rates of RNA (□) and protein (■) synthesis for sensitive cells are shown as a percentage of the corresponding rates for tolerant cells.

sitive cells in the absence and presence of Cd (panels a and b, arrows for example). In addition, some mRNA sequences increase in the presence of Cd in both tolerant and sensitive cells (panels b and d, arrows for example) while others decrease. The activities of several genes are therefore modulated, either directly or indirectly, by exposure of the cells to Cd.

DISCUSSION

The ability to synthesize large amounts of $(\gamma EC)_n G$ has been correlated with the ability of plants and plant cells to grow in the presence of normally toxic concentrations of certain trace metal ions (2,3,4,6,7). These compounds play an important role in metal-tolerance since the inhibition of their synthesis results in rapid cell death (6 and our

TABLE 1.
SYNTHESIS OF $(\gamma EC)_nG$ IN Cd-TOLERANT AND SENSITIVE D. INNOXIA CELLS AFTER EXPOSURE TO Cd[a].

	μg GSH equivalents/mg protein	
	Cd-tolerant cells	Cd-sensitive cells
Hours		
0	0	0
8	0.17	0.23
24	0.77	0.86
48	1.32	0.52

[a] Extracts from cells were separated by reverse phase HPLC and the amounts of thiols present in each fraction determined. Fractions containing material with retention times equal to the different $(\gamma EC)_nG$ were used to determine the total number of thiols, expressed as μM GSH equivalents/mg protein, and, therefore, the total amount of these polypeptides present. Time is the number of hours after addition of Cd.

own unpublished results). The enzyme(s) responsible for the synthesis of these compounds are present constitutively in both Cd-tolerant and sensitive cells, even in the absence of toxic metal ions (5). This suggests that either the metal-binding polypeptides or the enzymes perform some other metabolic function in the absence of metal ions. While $(\gamma EC)_nG$'s play a role in metal tolerance, the mechanism of resistance is more complex. Both Cd-tolerant and sensitive cells can synthesize these compounds upon exposure to Cd. However, only the tolerant cells survive. There are clearly other factors which contribute to tolerance.

Results presented here demonstrate a clear difference in viability between Cd-tolerant and sensitive cells exposed to Cd. This is first manifested in sensitive cells as a reduction in rates of RNA and protein synthesis and is followed by a rapid reduction in viability of the culture. However, these events precede any effects of Cd on $(\gamma EC)_nG$ accumulation in sensitive cells. Therefore, some other factor(s) must be involved in tolerance. Equivalent synthesis of $(\gamma EC)_nG$ is not accompanied by equivalent accumu-

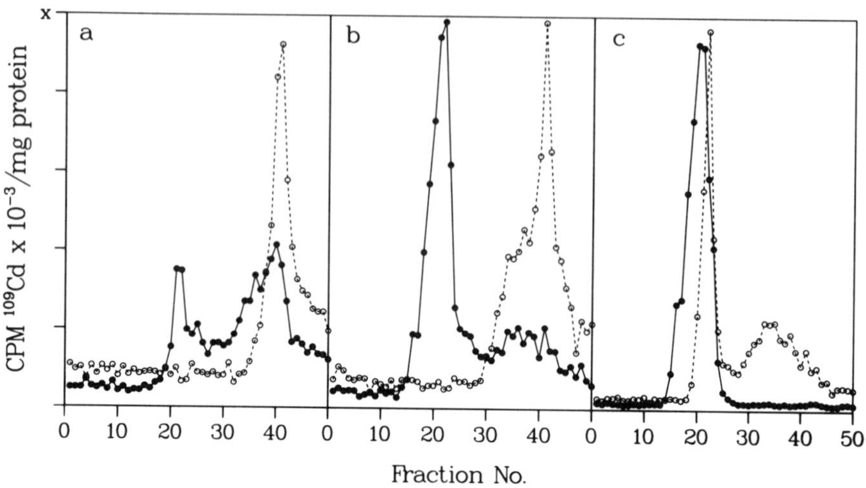

FIGURE 3. The accumulation of Cd-binding complexes in Cd-tolerant and sensitive cells exposed to $CdCl_2$ for different periods. Cd-tolerant and sensitive D. innoxia cells were grown in media containing 250 μM $CdCl_2$ for 4, 8, and 24 hours (panels a, b, and c, x = 50, 70, and 325 respectively). They were then extracted and analyzed by gel filtration HPLC. (●), Cd-tolerant cells; (o), sensitive cells.

lation of Cd^{2+} in metal-polypeptide complexes in the first four to eight hours after exposure to Cd. In some mammalian cells, tolerance to different concentrations of Cd is tightly correlated with the ability to produce specific amounts of metallothionein immediately following exposure to Cd (11). The ability to form Cd:$(\gamma EC)_n G$ complexes during initial exposure to Cd may determine whether or not cells can survive long-term exposure to this metal ion. The results presented here suggest that Cd-sensitive cells are unable to form $(\gamma EC)_n G$:Cd complexes during early periods of Cd exposure. This may result because the polypeptides are somehow separated from the Cd in the cells. Alternatively, the size or structure of the polypeptides may differ between the tolerant and sensitive cells. Reese and Winge have reported that sulfide is a component of polypeptide:Cd complexes isolated from Schizosaccharomyces pombe (12). It is possible that tolerant cells are able to syn-

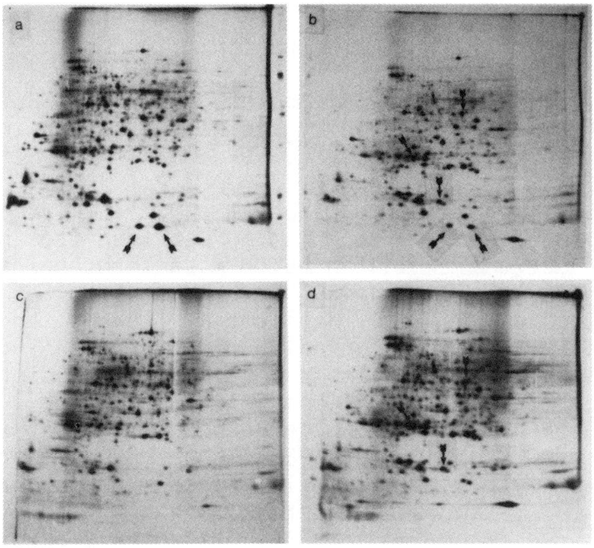

FIGURE 4. Two-dimensional gel electrophyoresis of *in vitro* translation products of mRNA isolated from (a) Cd-tolerant cells growing in the absence of Cd; (b) Cd-tolerant cells growing for 4 hours in 250 μM $CdCl_2$; (c) Cd-sensitive cells growing in the absence of Cd; and (d) Cd-sensitive cells growing for 4 hours in 250 μM $CdCl_2$. Arrows common to panels a and b represent products encoded by mRNA sequences which are constitutively expressed in Cd-tolerant cells growing in the presence and absence of Cd. Arrows common to panels b and d represent products encoded by mRNA sequences which are induced by Cd.

thesize sufficient sulfide during early exposure to Cd to form stable Cd complexes while sensitive cells lack this ability.

An analysis of the products of *in vitro* translations of mRNA demonstrates several differences in gene expression between tolerant and sensitive cells. Several genes which are not expressed in sensitive cells are constitutively expressed in their tolerant counterparts in the presence and absence of Cd and might encode "tolerance" proteins. The mRNAs induced by Cd in both cell lines might encode enzymes of the $(\gamma EC)_nG$ biosynthetic pathway. Although the initial synthesis of $(\gamma EC)_nG$ is cycloheximide insensitive (5), continued synthesis may require induction of the synthesis of these enzymes.

REFERENCES

1. Gemmel RP (1977). Colonization of Industrial Wasteland. London: Arnold.
2. Jackson PJ, Unkefer CJ, Doolen JA, Watt K, Robinson NJ Poly(γ-glutamylcysteinyl)glycine: Its role in cadmium resistance in plant cells. Proc Natl Acad Sci USA 84:6619.
3. Grill E, Winnacker E-L, Zenk M (1987). Phytochelatins, a class of heavy-metal-binding peptides from plants, are functionally analogous to metallothioneins. Proc Natl Acad Sci USA 84:439.
4. Steffens JC, Hunt DF, Williams BG (1986). Accumulation of non-protein metal-binding polypeptides (γ-glutamylcysteinyl)$_n$-glycine in selected cadmium-resistant tomato cells. J Biol Chem 261:13879.
5. Robinson NJ, Ratliff RL, Anderson PJ, Delhaize E, Berger JM, Jackson PJ (1988). Biosynthesis of poly-(γ-glutamylcysteinyl)glycine in cadmium-tolerant *Datura innoxia*. Plant Sci, In press.
6. Reese, RN, Wagner GJ (1987). Effects of buthionine sulfoximine on Cd-binding peptide levels in suspension-cultured tobacco cells treated with Cd, Zn, or Cu. Plant Physiol 84:574.
7. Jackson PJ, Roth EJ, McClure PR, Naranjo CM (1984). Selection, isolation, and characterization of cadmium-resistant *Datura innoxia* suspension cultures. Plant Physiol 75:914.
8. Chu YE, Lark KG (1976). Cell-cycle parameters of soybean (*Glycine max* L.) cells growing in suspension culture: Suitability of the system for genetic studies. Planta 132:259.
9. O'Farrell PH (1975). High resolution two-dimensional electrophoresis of proteins. J Biol Chem 250:4007.
10. Ellman GL (1959). Tissue sulfhydryl groups. Arch Biochem Biophys 82:70.
11. Hildebrand CE, Enger MD, Tobey RA (1980). Comparative studies of zinc metabolism in cultured Chinese hamster cells with differing metallothionein-induction capacities. Biol Trace Element Res 2:235.
12. Reese RN, Mehra RK, Tarbet EB, Winge DR (1988). Studies of the γ-glutamyl Cu-binding peptide from *Schizosaccharomyces pombe*. J Biol Chem 263:4186.

Metal Ion Homeostasis: Molecular Biology and Chemistry, pages 347–357

THE ROLE OF METAL-BINDING COMPOUNDS IN THE COPPER TOLERANCE MECHANISM OF SILENE CUCUBALUS

Jos A.C. Verkleij[1], Paul Koevoets[1], Jan van 't Riet[2], Mary C. van Rossenberg[2], Ruud Bank[3] and Wilfried H.O. Ernst[1]

[1] Department of Ecology and Ecotoxicology, Faculty of Biology
[2] Biochemical Laboratory, Faculty of Chemistry
[3] Institute of Human Genetics, Medical Faculty, Free University, P.O. Box 7161, 1007 MC Amsterdam, The Netherlands

ABSTRACT Various mechanisms have been proposed to explain the metal tolerance in higher plants. These include: exclusion of metals from the plant, compartmentation of metals in vacuoles and cell walls, evolution of metal-tolerant enzymes and specific metal-binding proteins. According to Grill and co-workers (1) only metal-binding peptides (phytochelatins) are involved in the detoxification of heavy metals in higher plants. However, the existence of higher Mol. Wt. metal-binding proteins cannot be disregarded although the described high Mol. Wt. forms may be association artefacts. In *Silene cucubalus*, a species occurring on heavy metal contaminated soils, a number of tolerant and non-tolerant strains has evolved. In our study of the molecular basis of copper-tolerance we characterized the metal binding compounds, synthesized in tolerant and non-tolerant plants after 20 days of copper-treatment. We could demonstrate by gelfiltration on FPLC-Superose-12 in root extracts from both types of plants, grown on 40 µM copper, two compounds: one with an apparent Mol.Wt. of 12 kD and containing more than 60% of the copper, and another one with a Mol.Wt. of 5.3 kD. After further purification by ion-exchange chromatography and gelfiltration of FPLC, the same copper-containing compounds appeared. Determination of the amino-acid content of the purified 12 kD and 5.3 kD peaks from both types of plants revealed a striking similarity with the amino-acid composition of phytochelatins. These results indicate that both tolerant and sensitive plants induce the synthesis of copper-binding compounds with apparently the same Mol.Wt. (most probably phytochelatins) upon growth in the presence of high amounts of copper. Although in both types of plants these compounds may play a role in copper-detoxification and homeostasis, further research is necessary to elucidate their significance in the copper-tolerance mechanism.

INTRODUCTION

Excesses or deficiences of heavy metal ions have effects on plant growth, morphology and physiology (2). A variety of mechanism have evolved in plants for coping with the toxic effects of a surplus of heavy metals. One of the mechanism could be the synthesis of large amounts of metal-binding compounds, which are produced especially after exposure to copper and cadmium (3). Two classes of compounds have been described. Most reports uptill 1985 described the presence of metallothionein like metal-binding compounds (MT) in plants. However, since 1985, another class of metal-binding peptides, the so-called phytochelatins came into attention (1). These peptides have the general structure (γ-glutamyl-cysteinyl)n-glycin, in which n is mostly 2-4. Some yeast strains also synthesize cadmium-binding peptides after Cd-treatment (cadystins) which have the same composition as phytochelatins (4). However, other yeast strains produce MT-like proteins (5) or both compounds (6).

In plants phytochelatins have been found in Cd-treated plants and plant cells. For Cu and also for Cd however, the existence of higher Mol Wt metal-binding proteins cannot be disregarded although the described high Mol Wt forms may be association artefacts.

In Silene cucubalus, a pioneer species of soils, with various degrees of heavy metal contamination a number of tolerant and non-tolerant populations has evolved (7). These populations are excellent objects to investigate the role of metal-binding compounds in the tolerance mechanism.

In continuation of earlier work (8) we are studying the molecular basis for Cu-tolerance in Silene cucubalus.

MATERIAL AND METHODS

Plant Material

Seeds were collected from two populations of Silene cucubalus (L.) Wib, a copper-sensitive one from the Botanical Garden of the Free University, Amsterdam, The Netherlands and a copper-tolerant one from a copper mine waste near Marsberg, F.R.G. (7).

Seeds of each population were sown on garden soils. Four weeks after germination seedlings were randomly placed in polyethylene tanks filled with 340 l permanently aerated 1/4 strength Hoagland solution, pH 5.5. After an initial growth period of seven days, copper was added to final concentrations of 1 and 40 µM. The climatic conditions in the greenhouse were: 21°/15°C and 16 h/8 h, respectively, day/night regime, 300 $\mu E\ m^{-2}s^{-1}$ and 70 ± 5% R.H. After 3 weeks growth the experiment was terminated. Roots were blotted between paper towels prior to weighing.

Extraction of Roots and Chromatography

Roots (5 g) from both tolerant and sensitive plants from the experiments with a copper concentration of 1 µM (control treatment) and 40 µM (copper treatment) were harvested, washed in deïonized water and homogenized in 20 mM Tris-HCl, 5 mM β-mercaptoethanol and 1 µM phenylmethansulfonyl fluoride, pH 7.9 at 4°C using a blender.

The slurry was strained through four layers of cheese-cloth and the filtrate was centrifuged at 100.000 g for 90 min at 4°C. Supernatants were applied on an FPLC-Superose 12 column (HR 10/30, Pharmacia Fine Chemicals, Uppsala, Sweden). Elution (24 ml h^{-1}) took place with NH_4HCO_3 (100 mM, pH 8.0) at room temperature. Fractions (5 ml) were collected with a Pharmacia Frac 100 fraction collector and the effluents monitored at 254 nm. Copper in the fractions was determined by atomic absorption spectrophotometry (Perkin-Elmer 4000, Überlingen FRG).

In order to purify and characterize the copper-binding compounds supernatants were applied on FPLC-mono Q anion exchange chromotography columns (HR 5/5) and eluted by a linear salt gradient (0.0-1.5 M NaCl) in 20 mM Tris-HCl, pH 8.0. Top fractions obtained by anion exchange chromatography on FPLC-Mono Q (fractions from peak I, II en III, Fig 3) were chromatographed on an FPLC superose 12 column (HR 10/30) equilibrated with 100 mM NH_4HCO_3, pH 8.0. Columns were calibrated with BSA, α-chymotrypsinogen, trypsin inhibitor, myoglobin III, cyt c, aprotinine, insuline, α and β chain having Mol.Wts. of 67, 23.2, 20.0, 18.8, 12.5, 6.5, 3.4 and 2.4 kD, respectively.

Aminoacid analysis

1 ml fractions of 12 kD Cu-containing peaks from a FPLC-Superose-12 column, following FPLC-Mono Q anion exchange chromatography were freeze dried, oxidized with performic acid (9), freeze dried and hydrolyzed in 6 N HCL with 0.01% phenol, for 1 h at 150°C in evacuated sealed tubes.

Phenylisothiocyanate (PITC) derivatization and separation by reversed phase (RP) HPLC were according to (10).

RESULTS

After 3 weeks of growth on a high copper concentration (40 µM) copper-tolerant plants exhibit a similar biomass production as those grown on the control solution (1 µM Cu). However, growth of the sensitive plants was strongly inhibited by the high copper concentration (7, 8). High and comparable amounts of copper, accounting for about 90% of the total copper content on a whole plant basis, were found in the roots of both tolerant and non-tolerant plants (Table 1).

TABLE 1

The (total) amount of copper in roots and cell-free extracts of roots of <u>Silene cucubalus</u> (± sd), grown for 3 weeks on 40 µM Cu in a nutrient solution.

	non-tolerant	Cu-tolerant
Cu-treatment (µM)	40	40
total amount of copper in the roots (µmol/g.fr.w.)	1.54±0.22	1.74±0.27
amount of copper in cell-free extracts of roots (µmol/g.fr.w.)	0.64±0.05	0.56±0.04
amount of copper in cell-free extracts of roots (nmol/µg protein)	0.35±0.04	0.40±0.03

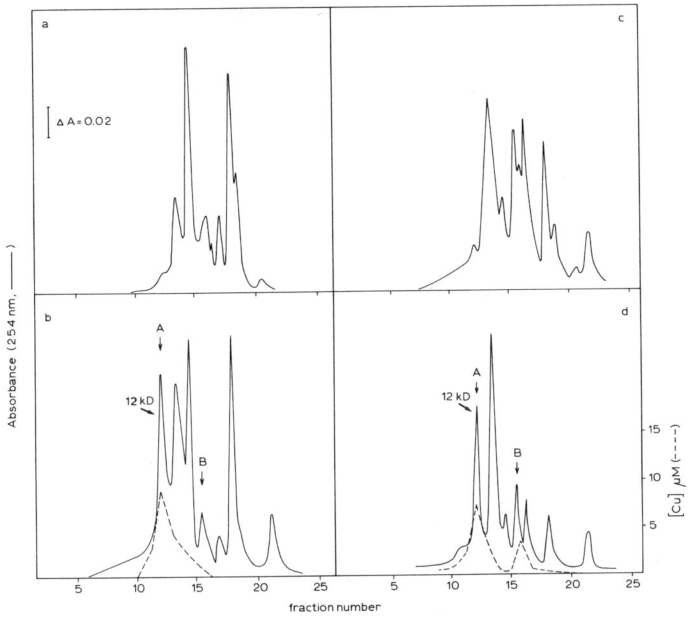

FIGURE 1. a-d. FPLC-Superose 12 gel filtration profiles of root extracts of a copper-tolerant (a,b) and a -sensitive (c,d) plant of Silene cucubalus after 3 weeks growth on 1 μM (a,c) and 40 μM copper (b,d).

Between 35-40% of the total copper present in the roots was recovered in the cell-free extracts. The extracts from both types of plants, grown on 1 μM and 40 μM copper respectively, were analyzed by gelfiltration on FPLC-Superose 12 (Fig. 1). In either extract of plants grown on 40 μM copper a compound with an apparent Mol.Wt. of 12 kD, and containing more than 60% of the copper, is present (Fig. 1, peak A). In extracts from tolerant and sensitive plants, grown on 40 μM copper, another copper-containing compound with a Mol.Wt. of 5.3 kD is also present (Fig. 1, peak B).

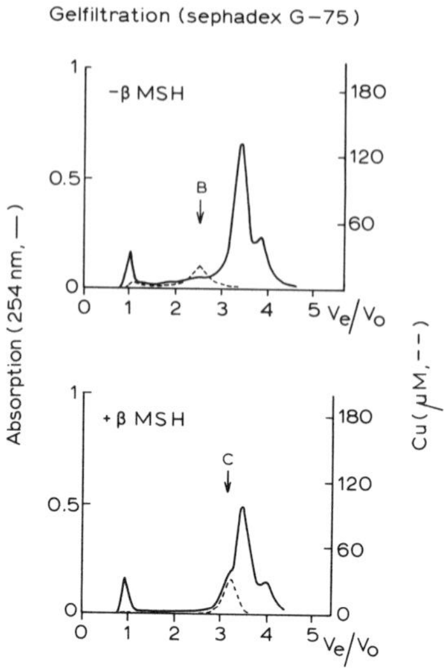

FIGURE 2. Analysis of G-75 gel filtration of root extracts of copper tolerant plants of Silene cucubalus after 3 weeks growth on 40 μM copper (with or without β-mercaptoethanol).

In both peaks the A_{254} nm/A_{280} nm ratio was about 1.5, suggesting the presence of mercaptide bonds. Unless copper was added as $CuSO_4$ to the extracts, copper was always present in larger Mol.Wt. complexes (Mol.Wt. 12 and 5.3 kD respectively) and never eluted with the total included volume. Both tolerant and sensitive plants induce the synthesis of copper binding compounds upon growth in the presence of high amounts of copper. The amount of copper, bound to this 'putative' 12 kD compound, calculated on protein basis (11) did not differ between the tolerant and non-tolerant plants.

Upon gelfiltration on Sephadex G-75 of the cell free extracts of both types of plants, copper was found in material eluting with a high and a low Mol.Wt., respectively (Fig. 2). If elution took place in the presence of β-mercaptoethanol, all the copper eluted as the low Mol.Wt. material. This result however was also obtained if both extraction and chromatography were performed in the absence of β-mercaptoethanol. Thus either complex reduction or oxidation gave rise to lower Mol.Wt. copper-complexes.

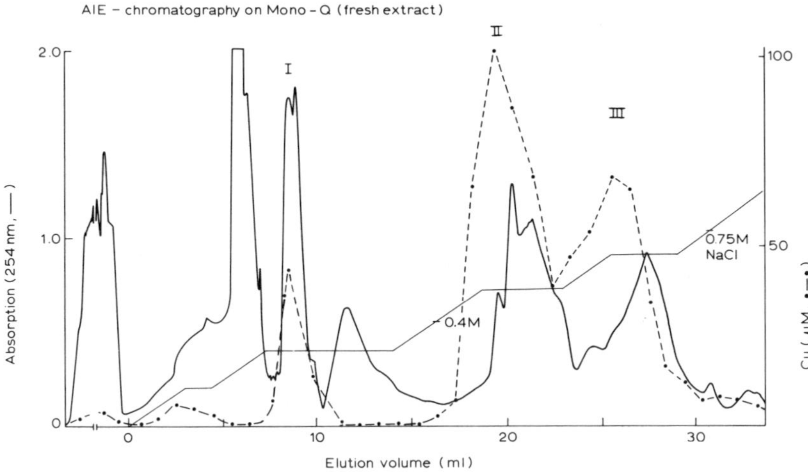

FIGURE 3. Analysis of FPLC-Mono Q exchange chromatography of root extracts of Silene cucubalus plants after 3 weeks growth on 40 μM copper.

After anion-exchange chromatography of the cell-free extracts on FPLC-Mono Q, three major copper-containing peaks were observed eluting from the column at different salt concentrations (Fig. 3). The highest copper-content was found in fraction II, which accounted for about 50% of the recovered metal. Various peaks of the fractions I, II en III were further purified on FPLC-Superose 12 gelfiltration. Then the three complexes of the AIE chromatography on Mono Q appeared to consist of high and a low Mol.Wt. compounds in varying ratio's. The less electro-negative complex contained the highest amount of the 5.3 kD compound (Fig. 4). Determination of the amino-acid composition of the purified copper-containing peaks from both types of plants revealed a striking similarity with the amino-acid composition of the so-called phytochelatins. Glx, Cys and Gly accounted for about 80% of all amino acids in a ratio of 2:2:1, respectively.

Also after SDS-Polyacrylamide gelelectrophoresis we could detect low Mol.Wt. silver stainable compounds migrating with almost the same mobility as the trailing dye (Bromo Phenol Blue) and only detectable in roots extracts of both types of plants after exposure to high concentration of copper.

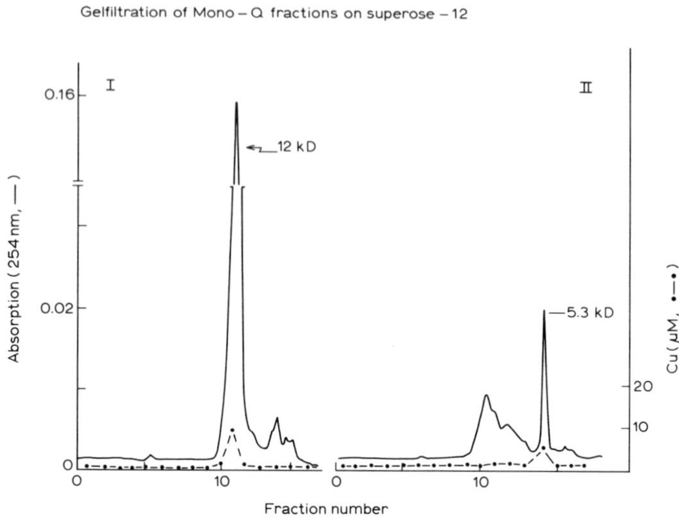

FIGURE 4. Analysis of FPLC Superose 12 gel filtration of a copper containing peak fraction following FPLC-Mono Q anion exchange chromatography.

DISCUSSION

No difference in copper uptake could be observed between the tolerant and non-tolerant strain of Silene cucubalus (Table 1). By contrast, another copper-tolerant strain collected from a copper mine waste near Imsbach FRG, showed a diminished uptake of copper (8). So partial exclusion of the metal seems not to be an exclusive physiological basis of copper-tolerance. In both types of plants copper-binding compounds are present and induced by increased copper in the roots (Fig. 1). Although in both tolerant and sensitive plants these compounds may play a role in copper-detoxification and homeostasis, their presence as such seems not be essential for the genetically defined "copper and cadmium tolerance". So tolerance is not simply due to the (enhanced) expression of gene(s), that code for these metal-binding compounds or for enzymes involved in their synthesis.

The amino-acid composition of the high and low Mol.Wt.

purified compounds revealed a striking similarity with that of phytochelatins (n=2). Similar compounds with a comparable structure were found in roots of cadmium-tolerant and sensitive plants of Silene cucubalus after exposure to high amounts of cadmium (12). The conversion of the high Mol.Wt. compounds into the low Mol.Wt. compounds in the presence of β-mercaptoethanol or high salt during the gelfiltration step could be the result of complex reduction or oxidation of the copper-binding compounds (13). The existance of a high Mol.Wt. form could also be caused by binding of acid labile sulfide as was found in Euglena after cadmium exposure (14). We have no indication as yet that acid-labile sulfide is involved in the sequestration of copper to these compounds.

Since at most 30-35% of the total copper present in the roots was found to be bound to these compounds, the mechanism of tolerance does not simply involve the sequestration of all excess copper in this way. Also we could not observe a higher amount of copper bound to the observed metal binding compounds in the copper-tolerant strain. However, an increased production of the metal binding compound resulting in an initially higher binding capacity for copper could be a physiological basis of copper tolerance.

The present study indicates that the roots of tolerant and sensitive plants of Silene cucubalus synthesize phytochelatins (which we prefer to mention metallothiopeptides) in response to treatment of copper. However, in another copper-tolerant strain, which is also cadmium and zinc-tolerant (15), a metallothionein-like copper complex still may be involved in the copper-detoxification as was also observed in Euglena gracilis (14). If phytochelatins are implicated in the tolerance mechanism, glutathione and enzymes, which synthesize precursors of these compounds, must be studied more in detail. Furthermore the role of copper in producing hydroxyl radicals causing lipid peroxidation and damage of the cell membranes, seems to be important. In a preliminary experiment, De Vos et al. (16) could demonstrate a significant difference in sensitivity of changes in permeability and lipid peroxidation of membranes induced by copper of the roots of copper-tolerant and sensitive plants of Silene cucubalus.

In de mechanisms of copper-tolerance a variety of factors seems to be crucial, in which the role of metal binding compounds still has to be investigated.

REFERENCES

1. Grill E, Winnacker EL, Zenk MH (1987). Phytochelatins, a class of heavy-metal-binding peptides from plants, are functionally analogous to metallothioneins. Proc Natl Acad Sci USA 84: 439.
2. Woolhouse HW (1983). Toxicity and tolerance in the responses of plants to metals. In Lange OL, Nobel PS, Osmond CB, Ziegler H (eds): "Encyclopedia of Plant Physiology, New Series, Vol 12B" New York: Springer Verlag, p 195.
3. Robinson NJ, Jackson PJ (1986). "Metallothionein-like" metal complexes in angiosperms; their structure and function. Physiol Plant 67: 499.
4. Kondo N, Isobe M, Imai K, Goto T (1985). Synthesis of metallothionein-like peptides cadystin A and B occurring in fusion yeast, and their isomers. Agric Biol Chem 49: 71.
5. Hamer DH (1986). Metallothionein. Ann Rev Biochem 55: 913.
6. Mehra RK, Tarbet EB, Gray WR, Winge DR (1988). Distinct mechanism of cadmium and copper resistance in Torulopsis glabrata. J Cell Biochem Suppl 12D p. 346.
7. Verkleij JAC, Bast-Cramer WB (1985). Cotolerance and multiple heavy metal tolerance in Silene cucubalus from different heavy metal sites. In Lekkas TD (ed): "Heavy Metals in the Environment, Vol 2" Edinburgh: CEP Consultants p 174.
8. Lolkema, PC, Donker MH, Schouten AJ, Ernst WHO (1984). The possible role of metallothioneins in copper tolerance of Silene cucubalus. Planta 162: 174.
9. Moore S (1963). On the determination of cystine as cysteic acid. J Biol Chem 238: 235.
10. Bidlingmeyer BA, Cohen SA, Tarvin TL (1984). Rapid analysis of amino-acids using pre column derivation. J Chromat 336: 93.
11. Lowry OH, Rosebrough NJ, Farr AL, Randall RJ (1951). Protein measurement with the Folin phenol reagent. J Biol Chem 193: 265
12. Verkleij JAC, Koevoets P, van 't Riet J, van Rossenberg MC, Bank R, Ernst WHO (1988). Cadmium binding compounds in metal-tolerant and non-tolerant Silene cucubalus (submitted).
13. Lue-Kim H, Rauser WE (1986). Partial characterization of cadmium-binding protein from roots of tomato. Plant Physiol 81: 896.

14. Taylor P, Weber D, Gingrich D, Shaw III CF, Petering DH (1987). Cd, Zn speciation and metabolism in Euglena gracilis: metal-metal interactions and cell proliferation. In Lindberg SE, Hutchinson TC (Eds). "Heavy Metals in the Environment" Vol 2. Edinburgh: CEP Consultants p 250.
15. Verkleij JAC, Prast JE (1988). Cadmium tolerance and cotolerance in Silene cucubalus. New Phytol (submitted).
16. De Vos CR, Schat H, Ernst WHO (1988). Copper mediated changes in membrane permeability. J Cell Biochem Suppl 12A p 49.

Metal Ion Homeostasis: Molecular Biology and Chemistry, pages 359–366

INCREASED ACTIVITY OF γ-GLUTAMYLCYSTEINE SYNTHETASE IN DMSO-PERMEABILIZED Cd^R PLANT CELLS

John C. Steffens[1] and Bill G. Williams[2]

[1]Department of Plant Breeding, Cornell University, Ithaca, NY 14853
[2] BioVision, Inc., San Jose, CA.

ABSTRACT Cd^R tomato cells overproduce the metal binding polypeptides (γ-Glu-Cys) $_n$-Gly (phytochelatins) and are resistant to growth inhibition by buthionine sulfoximine, a specific inhibitor of γ-glutamylcysteine synthetase. Assay of this enzyme using DMSO-permeabilized Cd^R and Cd^S cells shows that increased activity of γ-glutamylcysteine synthetase may be an important component of the Cd^R phenotype.

INTRODUCTION

Plant cells respond to the presence of excess heavy metal ions by synthesizing non-protein metal binding polypeptides. These peptides are composed of a series of peptides related to glutathione, γ-Glu-Cys-Gly, but differ in the number of (γ-Glu-Cys) moieties (1,2,3,4). Thus the primary forms of heavy metal binding polypeptides found in tomato cell suspension cultures are $(\gamma\text{-Glu-Cys})_3$-Gly and $(\gamma\text{-Glu-Cys})_4$-Gly (3). In other species, however, these metal-binding complexes may be isolated in the forms $(\gamma\text{-Glu-Cys})_n$-Gly, where n= 2-7 (2). These peptides are believed to be functional analogues of the metallothioneins utilized by animal cells in the sequestration and detoxification of heavy metals. The ubiquity of the metal-binding phytochelatins across the major phyla of the plant kingdom is paralleled by the absence of metallothionein-like proteins in plants (5).

Recently we characterized a cadmium-resistant (Cd^R) tomato cell suspension line derived from *Lycopersicon esculentum* Mills. cv VFNT Cherry (Cd^S). Growth of the Cd^R line is unaffected at concentrations of Cd ion lethal to Cd^S cells. Uptake of cadmium ion by both the Cd^S and Cd^R cell lines is similar, and both lines sequester this ion in thiolate complexes predominantly composed of (γ-Glu-Cys)$_3$-Gly and (γ-Glu-Cys)$_4$-Gly. The Cd^R line is distinguished biochemically by its accumulation of high amounts of these peptides in response to added Cd ion (3,6). This increased accumulation was shown to be necessary for achieving tolerance to toxic levels of this ion.

The structural similarity of phytochelatins to glutathione suggests that their biosynthesis shares the first step of glutathione synthesis, *i.e.*, the condensation of glutamate and cysteine by γ-glutamylcysteine synthetase (EC 6.3.2.2).Specific inhibition of this enzyme is accomplished by the transition-state analog inhibitor, buthionine sulfoximine (BSO) (7). Buthionine sulfoximine, in the presence of cadmium ion, prevents synthesis of the phytochelatins and abolishes resistance of Cd^R cells to high concentrations of Cd ion (3). Similarly, low concentrations of BSO sensitize Cd^S cells to extremely low concentrations of cadmium ion (6). Thus the ability to overproduce phytochelatins and the biosynthesis of glutathione are tightly linked. The observation that Cd^R cells are able to overproduce phytochelatins and are resistant to growth inhibition by buthionine sulfoximine led us to ask whether the Cd^R phenotype can be explained on the basis of increased activity of γ-glutamylcysteine synthetase. Cells permeabilized by treatment with dimethylsulfoxide (DMSO) were used to develop conditions for assay of γ–glutamylcysteine synthetase.

MATERIALS AND METHODS

Cell Cultures. Tomato cell suspensions were maintained, and growth was measured in sidearm flasks and expressed as the settled cell-to-media ratio (SCM) after fifteen minutes of settling (3).

DMSO Permeabilization. Each assay was run in triplicate. Cells at an SCM of 0.3 were collected by vacuum filtration on a chilled Buchner funnel and 1 gfw was transferred to a chilled 30 ml Corex tube and centrifuged at 0^0 C, 2000 x g for 2 min. The loose pellet was resuspended in 15 ml ice-cold 50 mM MOPS-KOH pH 7.4, 10 mM DTT, 7.5% DMSO and centrifuged as previously. The supernatant was pipetted off and the pellet resuspended in about 1 ml of the buffer to make a slurry. 750 μl was removed using a pipet tip enlarged to accomodate the slurry, and added to a 1.5 ml polypropylene tube containing final concentrations of 5mM ATP, 6mM $MgCl_2$, and 6mM cysteine. 10 μCi ^{3}H-glutamic acid (52 Ci/mmol) was added for a final volume of 800 μl. The reaction was mixed by gentle inversion and incubated at 30^0 C for varying lengths of time

Incorporation of ^{3}H-glutamate was assayed either by precipitation of ^{3}H-GluCys as a cadmium mercaptide, or by HPLC radioactivity analysis of the monobromobimane derivative. Analysis of incorporation into mercaptides followed the procedure of Minnich et al. (8) with the following modifications. The reaction was stopped by vortexing with 25 μl 100% TCA. After centrifugation the supernatant was transferred to a fresh tube with 125 μl of 125 mM GSH, 40 mM $CdCl_2$, and then neutralized with approximately 200 μl 1 M KOH. The precipitated Cd-mercaptides were centrifuged for 5 min, the supernatant aspirated, and the pellet dissolved in 400 μl 7% TCA, and reprecipitated with 800 μl 1M MOPS-KOH pH 7.6. The insoluble mercaptides were pelleted, dissolved in TCA, and reprecipitated, and the final washed pellet was taken up in 400 μl 7% TCA and counted in aqueous scintillation cocktail.

Assays for HPLC analysis of specific incorporation of ^{3}H-glutamic acid into γ-glutamylcysteine was carried out similarly, except that the final pellet was dissolved in 100 μl 0.1 M HCl, to which was added 4 μl of 0.1 M monobromobimane (in acetonitrile), followed by 25 μl 34% (v/v) N-ethylmorpholine in reduced light. The reaction was incubated for 30 min at room temperature in reduced light. The labelling reaction was stopped with 90 µl 1M NaH_2PO_4, pH3 and 5 µl 3.6 N HCL, centrifuged 10 min and the supernatant transferred to an autosampler vial for HPLC analysis. Samples (50 µl) were injected onto HPLC via a Kortec autosampler onto a Waters 5 µm C18 radial compression column eluted with a 1-50% methanol gradient in 40 mM $NaOPO_4$, pH 3.5 at 1 ml/min. The thiol derivatives were identified using fluorescence detection. A Radiomatic Instruments HPLC radioactivity flow detector downline from the fluorescence detector was used to measure specific incorporation of ^{3}H-glutamic acid into γ-glutamylcysteine.

RESULTS

The activity of glutamyl cysteine synthetase was dependent, in both cell types, on Mg^{2+}, ATP, cysteine, and glutamate (Table 1). Larger amounts of carrier GSH during precipitations did not increase the yield of radioactivity. An additional precipitation step to further remove background unincorporated ^{3}H-Glu was successful in decreasing the background by a factor of two, but also decreased the recovery of ^{3}H-γ-glutamylcysteine about twenty-five percent.

Specificity of the assay was checked by derivativatizing the final pellet with monobromobimane, which permitted direct analysis of the pellet's radioactive components. In addition ^{3}H-γ-Glu-Cys, the radioactive product was identified by both its HPLC retention time and the dependence of its retention time on the bimane label. Approximately 10-15% of the total radioactivity in the mercaptide pellet was composed of unincorporated ^{3}H-Glu. In addition, about 5% of the radioactivity was found in a late-eluting peak whose retention time was dependent on glutamate, Mg-ATP, monobromobimane, but not cysteine. Reproducibility of the assay was good if replicated in triplicate.

Comparison of γ-glutamylcysteine synthetase activity (Table 2) in DMSO permeabilized Cd^R and Cd^S cells grown in the absence of Cd shows a two-fold increased activity of this enzyme in Cd^R cells.

TABLE 1

SUBSTRATE DEPENDENCE OF γ-GLUTAMYLCYSTEINE SYNTHETASE

Reaction	CPM
Full reaction	15,400
-ATP	1350
-$MgCl_2$	2300
-cysteine	580

Time-course analysis revealed that the reaction in both cell types had gone to completion within about fifteen minutes. Additional spikes of Mg^{2+}-ATP, cysteine, or dithiothreitol did not boost the reaction after this time. Addition of an ATP-generating system (phosphoenolpyruvate, pyruvate kinase) had no effect on the reaction (data not shown).

DISCUSSION

Initial efforts directed at detection of γ-glutamylcysteine synthetase activity in this cell line were unsuccessful. High levels of nonspecific ATPase activity in unpurified plant cell homogenates preclude use coupled assays involving cysteine and glutamate dependent ATP hydrolysis (9). In addition, the assay using conversion of γ-glutamylcysteine by bovine brain γ-glutamyl cyclotransferase (9) was inconvenient due to the inavailability of the cyclotranferase. The precipitation of radioactive γ-glutamylcysteine as a cadmium mercaptide became an attractive possibility for initial characterization of this enzyme. The repetitive precipitations, although unwieldy with a large number of samples, was made easier with a multi-sample vortexer. This assay, although crude, permitted the rapid development and optimization of reaction conditions for γ-glutamylcysteine synthetase.

Initially, we optimized the reaction conditions for γ-glutamylcysteine synthetase using fresh human erythrocytes. However, we were unable to use these reaction conditions to demonstrate activity of this enzyme in cell-free homogenates of Cd^R and Cd^S tomato cells, even after repeated efforts to modify homogenization procedures, buffers, etc. Extremely high rates of enzymatic activity have been reported in plant cells permeabilized with dimethylsulfoxide (DMSO). Such permeabilization has been demonstrated as an effective means to permeabilize the plasma membrane to the passage of small molecules while minimizing distruption of both the tonoplast and other subcellular structures (10). Thus, this semi-*in vitro* assay procedure represents an ideal means to characterize enzymatic activities under conditions causing minimal disruption of the cell, and provided a convenient means to assay γ-glutamylcysteine synthetase in these cell lines.

Given the increase in accumulation of $(\gamma\text{-Glu-Cys})_n$ upon exposure of both Cd^R and Cd^S cell to cadmium, we expected an increase of γ-glutamylcysteine synthetase activity in cells exposed to cadmium. However, in both cell lines, the synthetase activity of cells growing on cadmium was consistently lower than from cells growing in the absence of metal. However, the recoverable activity of γ-

TABLE 2

ACTIVITY OF γ-GLUTAMYLCYSTEINE SYNTHETASE IN Cd^R AND Cd^S TOMATO CELLS

Cell Line	$CdCl_2$, μM	CPM
CdR	0	16,740
CdS	0	7,780
CdR	600	11,390
CdS	600	4,550

glutamylcysteine synthetase was always higher in Cd^R than in Cd^S cell lines, and in the presence of cadmium, the difference in activity increased to about three-fold. We assume that the greater amount of phytochelatins in the Cd^R cells was responsible for decreasing the cytosolic Cd to levels which resulted in less inactivation of the enzyme.

In the absence of added Cd, the activity of γ-glutamylcysteine synthetase in Cd^R cell was approximately two fold higher than in the Cd^S cell line. Increased activity of γ-glutamylcysteine synthetase may be an important component in the development of the Cd^R phenotype. As phytochelatin biosynthesis, as well as glutathione regulation, is a function of the γ-glutamyl pathway, modulation of the rate-controlling enzyme would appear to be an efficient means of achieving metal tolerance via phytochelatin overproduction. Such a mechanism for overproducing phytochelatins would therefore be expected to also result in the ability to overproduce glutathione or any other γ-glutamyl product downstream from γ-Glu-Cys. Cd^R cells appear less sensitive to oxidative stresses typically dealt with through glutathione metabolism, such as stresses induced caused by H_2O_2 and thiocarbamate herbicides which typically are detoxified by GSH conjugation (Steffens and Williams, 1987).

The use of DMSO permeabilization to recover high enzymatic activity from plant cells provides a convenient starting point to optimize detection, characterization and purification of enzymes which are unstable or assayed only with difficulty. We are currently applying this approach to investigate the later steps in the biosynthesis of phytochelatins. The activity of γ-glutamylcysteine synthetase in cell-free preparations of Cd^R and Cd^S cells will be the focus of a later communication.

REFERENCES

1. Kondo, N.,Imai, K., Isobe, M., Goto, T. (1984). Cadystin A and B, major unit peptides comprising cadmium binding peptides induced in fission yeast--separation, revision of structures and

synthesis. Tetrahedron Letters 25: 3369.

2. Grill, E., Winnacker, E.-L., Zenk, M.H. (1985). Phytochelatins: the principal heavy metal complexing peptides of higher plants. Science 230:674.
3. Steffens, J.C., Hunt, D.F., Williams, B.G. (1986). Accumulation of non-protein metal-binding polypeptides (γ-glutamyl-cysteinyl)n-glycine in selected cadmium- resistant tomato cells. J. Biol. Chem. 261:13879.
4. Jackson, P.J., Unkefer, C.J., Doolen, J.A., Watt, K. Robinson, N.J. Poly(γ-glutamylcysteinyl)glycine: its role in cadmium resistance in plant cells. Proc. Natl. Acad. Sci. USA 84:6619.
5. Grill, E., Winnacker, E.-L., Zenk, M.H. (1987). Phytochelatins, a class of heavy metal binding polypeptides from plants, are functionally analogous to metallothioneins. Proc. Natl. Acad. Sci. USA 84:439.
6. Steffens, J.C., Williams, B.G. (1987). Heavy metal tolerance in tomato. In Nevins D.J.Jones R.A., (eds): "Plant Biology" Vol. 4, New York: Alan R. Liss, p 109.
7. Griffith, O.W., Meister, A. (1979). Buthionine sulfoximine, a potent and specific inhibitor of γ-glutamylcysteine synthetase. J. Biol. Chem. 254:7558.
8. Minnich, V., Smith, M.B., Brauner, M.J., Majerus, P.W. (1971). Glutathione biosynthesis in human erythrocytes. I. Identification of the enzymes of glutathione synthesis in hemolysates. J. Clin. Invest. 50:507.
9. Meister, A., Anderson, M.E. (1983). Glutathione and related γ-glutamyl compounds: biosynthesis and utilization. Annu. Rev. Biochem. 52:711.
10. Delmer, D. (1979). Dimethylsulfoxide as a potential tool for analysis of compartmentation in living plant cells. Plant Physiol. 64: 623.

Metal Ion Homeostasis: Molecular Biology and Chemistry, pages 367–374

COPPER PHYTOCHELATINS OF MIMULUS GUTTATUS[1]

David A. Thurman, David E. Salt and A. Brian Tomsett

Departments of Botany and Genetics,
University of Liverpool, Liverpool, L69 3BX, U.K.

ABSTRACT A copper-containing peptide has been isolated from the roots of copper-tolerant *Mimulus guttatus* after growth in copper. The behaviour of the peptide during gel filtration and ion exchange chromatography was monitored by estimating copper. The peptide contained 39.1% glutamic acid/glutamine, cysteine 39.1% and 13.4% glycine which suggests that this copper-binding peptide is a phytochelatin with a ratio of glx:cys:gly of 3:3:1. Its apparent molecular weight estimated under dissociating conditions was approximately 1600. A second copper-containing peptide, which was not purified to such an extent, contained 29.7% glutamic acid/glutamine, 29.7% cysteine and 16.2% glycine; this peptide may also be a phytochelatin with a ratio of glx:cys:gly 2:2:1.

[1]This work was supported by The Royal Society and S.E.R.C.

INTRODUCTION

A number of plants and plant cell cultures are known to contain high levels of poly (γ-glutamycysteinyl) glycines (phytochelatins, PCs) when grown in the presence of cadmium (1). This metal is the most effective one known so far in its ability to increase the PC content of plant cells. Thus *Rauvolfia serpentina* suspension cultures when grown in cadmium contained higher levels of PCs than cultures grown in a range of other metals, however not all metals tested were able to increase the PC contents of these cultures. Copper also increased the PC contents of *R. serpentina* cultures but is much less active than cadmium (1). Previous studies on copper-treated intact plants have shown that they contain 'MT-like porteins' which bind copper but which have never been purified to homogeneity (2). The fact that some of these isolates contain high levels of glutamic acid/glutamine, cysteine and glycine, the three amino acids found in PC's, suggests that when intact plants are treated with copper they respond in the same way as plant cell cultures on treatment with cadmium and increase their PC content. We have investigated this possibility by purifying further the copper complex present in roots of copper tolerant *Mimulus guttatus* when grown in the presence of copper (3).

MATERIALS AND METHODS

Growth of Plants

Stocks of metal-tolerant and non-tolerant isogenic lines of *M. guttatus* were grown as previously described (3). Roots were obtained by growing shoots, with basal leaves removed, in 10 litres of Long Ashton Solution (diluted 1:10) for 14 days. Plants were then transferred to fresh 1/10 Long Ashton Solution containing 10 µM copper for a further 4 days, with a change of culture solution every second day.

Extraction

Roots were harvested, washed in distilled water, blotted and frozen in liquid nitrogen and immediately reduced to a powder which was added to a previously hydrated (overnight) slurry of polyvinylpyrrolidone (1.5g g^{-1} root) in 50mM Hepes, pH 6.5, 10mM sodium metabisulphite (7cm^3 g^{-1} PVP). The slurry was homogenised for 1 minute, strained through muslin and the filtrate centrifuged at 10,000g for 30 minutes at 4°C, the supernatant shell-frozen, freeze dried and analysed by gel filtration and anion exchange chromatography.

Chromatography

Gel filtration. Samples(10cm^3) were applied to a Sephadex Superfine G-50 column (2.6 x 200cm) in 50mM ammonium bicarbonate, pH 7.8, 2mM dithiothreitol (DTT) and eluted with the same buffer at 30cm^3h^{-1} collecting 10cm^3 fractions. Copper was monitored by atomic absorption spectrophotometry and protein by measurement of absorbances at 280mμ. Fractions containing copper-inducible complexes were pooled and exchange into the appropriate buffer ready for anion exchange chromatography.

HPLC anion exchange. An LKB 5PW HPLC column (0.75cm x 7.5cm) was used with the following buffer system. Buffer A, 10mM Bis-tris, pH6.0, 2mM DTT: Buffer B, 10mM Bis-tris, pH 6.0, 2mM DTT & 1M KCl. The column flow rate was 1 cm^3 min^{-1}; eluates were monitored at 280mμ and 2cm^3 fractions collected.

Molecular weight estimation. A TSK SW-2000 HPLC gel-permeation column was used equilibrated with 10mM DTT, 10mM KH_2PO_4 and 6M guanidine pH 6.5 and calibrated with standards (1,400-45,000 daltons). The copper-containing peptide was incubated in the elution buffer for 90 minutes at 37°C before application to the column.

RESULTS AND DISCUSSION

Figure 1 shows the results of an analysis by gel filtration of an extract of roots of copper-tolerant *M. guttatus* grown for 10 days in normal levels (0.1μM) of copper. Before analysis copper sulphate was added to the extract to determine the nature of any *in vitro* binding of the metal which could have taken place during homogenisation. Two copper-containing components (peaks 1 and 4) eluted off the column. The results of a similar analysis of an extract of roots grown in the presence of 10μM copper for 4 days also contained these two components, and two additional ones (peaks 2 and 3, Fig. 2). The material present in peak 2 was

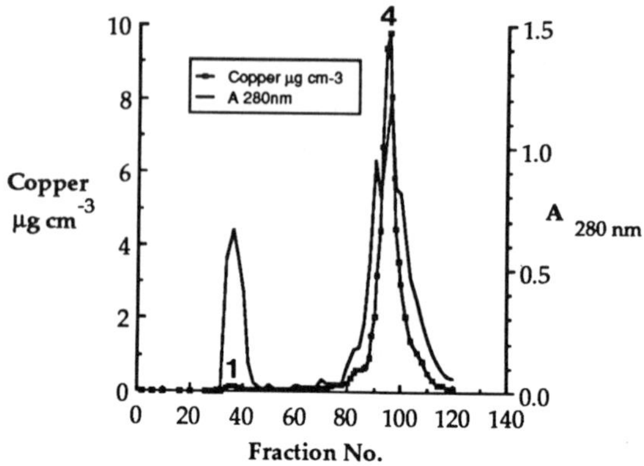

FIGURE 1. Gel filtration, G-50 Sephadex-Superfine column (2.6 x 200cm) of an extract of roots grown in 0.1µM copper in Long Ashton Solution for 14 days.

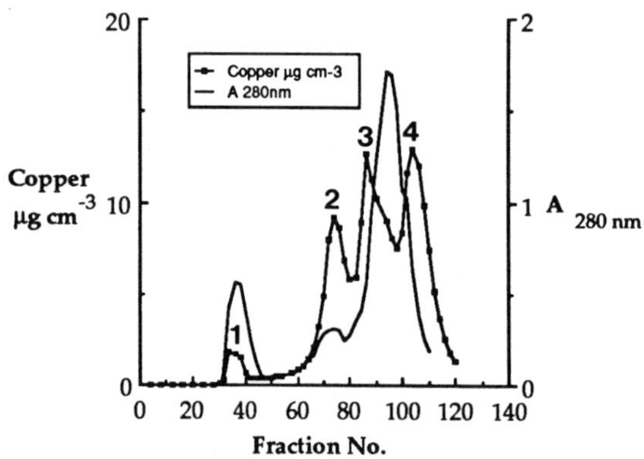

FIGURE 2. Gel filtration, G-50 Sephadex-Superfine column (2.6 x 200cm) of an extract of roots grown in 0.1µM copper in Long Ashton Solution for 10 days and then transferred to 10µM for a further 4 days.

purified further by anion exchange chromatography at pH 6.0, the major copper-containing component on this chromatogram eluted at approximately 300mM KCl (peak A, Fig. 3). The two fractions containing the highest concentrations of copper from this peak were pooled, desalted and either used for the determination of amino acid compositions (columns 1 and 2, Table 1) or for a second anion exchange chromatography at pH 7.0 using a linear gradient of increasing molarity of KCl; the results of this second ion exchange step are shown in Fig. 4. One major copper-containing component was detected (peak B) which eluted at the same molarity of KCl as the major copper-containing component shown on Fig. 3. The two fractions containing the highest concentration of copper from peak B were pooled, desalted and the amino acid composition of the isolate determined (Column 3, Table 1).

The amino acid composition of the copper-containing component isolated by gel filtration and only one ion exchange step suggests that it could be a copper phytochelatin, though amino acids other than those found in phytochelatin were present. However a second ion exchange step at pH 7.0 reduced these amino acids to negligible levels such that 93% of the amino acid residues present could be accounted for by the three amino acids found in phytochelatins. The ratio of glutamic acid/glutamine: cysteine:glycine was 3:3:1. Molecular weight determinations on this isolate under denaturing conditions indicated an apparent value of 1600 suggesting that the material isolated was a dimer.

The material present in peak 3 on gel filtration chromatograms of copper-grown roots (Fig. 2) when analysed by anion exchange chromatography at pH 6.0 behaved in a similar way to the copper-peptide purified as described above, only one major copper-containing peak was detected which eluted at 300mM KCl and its amino acid composition suggested that this isolate may also be a phytochelatin with glutamic acid/glutamine:cysteine:glycine ratio of 2:2:1, in this case 76% of the amino acid residues were accounted for by the three amino acids found in phytochelatins.

Roots of copper-tolerant *M. guttatus* when treated with copper, compared with untreated ones, contain increased amounts of a copper-containing peptide. One of these has been purified to homogeneity, the amino acid composition of this peptide suggests that it is copper-containing phytochelatin. The structure of this peptide is being studied further and its role in copper tolerance evaluated.

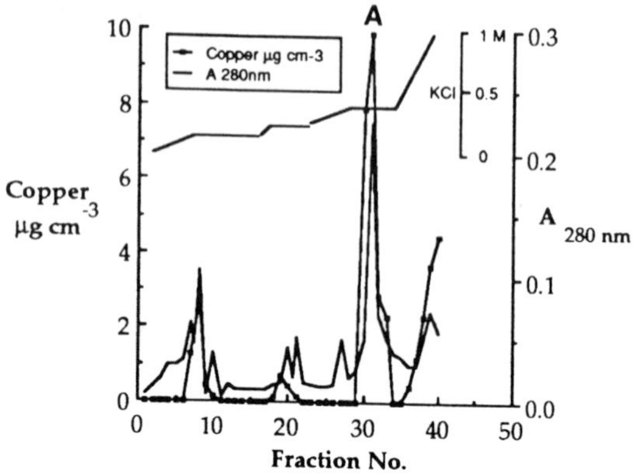

FIGURE 3. Anion exchange chromatography of peak 2 (Fig. 2) on a LKB 5PW HPLC column (0.75cm x 7.5cm). Buffer A, 10mM Bis-tris, pH 6.0, 2mM DTT: buffer B, 10mM Bis-tris, pH 6.0, 2mM DTT and 1M KCl. Column flow rate $1 cm^3 min^{-1}$, fraction size $2cm^3$.

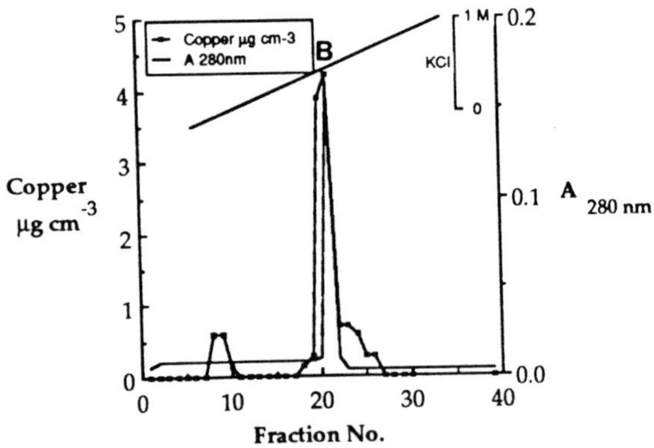

FIGURE 4. Anion exchange chromatography of peak A (Fig. 3). Column conditions as above except a linear gradient of KCl was used at pH 7.0.

TABLE 1

AMINO ACID COMPOSITION[a] OF COPPER-CONTAINING PEPTIDES ISOLATED FROM COPPER-GROWN ROOTS OF COPPER TOLERANT *M. GUTTATUS*.

	% Total residues		
	(1)	(2)	(3)
CYS	36.2	35.6	39.1
ASX	4.7	5	1.7
THR	1.5	1.4	0.7
SER	3.5	3.3	1.5
GLX	30.8	32.9	39.1
PRO	0.0	0.0	0.0
GLY	13.4	15.0	13.4
ALA	2.5	2.5	1.8
VAL	1.4	-	0.0
MET	0.0	0.0	0.0
ILE	1.0	0.9	0.4
LEU	1.2	1.4	0.8
TYR	0.3	0.0	0.0
PHE	0.8	0.0	0.5
HIS	0.0	0.0	0.0
TRP	0.0	0.0	0.0
LYS	1.2	1	0.8
ARG	0.8	0.3	0.0

1.2. Sephadex peak 3 (Fig.1) purified further by anion exchange chromatography pH 6.0. 1,2, different samples.

3. As in (1) above but purified further by anion exchange chromatography, pH 7.0.

[a]Peptides were hydrolysed in evacuated sealed tubes in constant boiling HCl at 110°C for 24 hours. Oxygen was removed by bubbling oxygen-free nitrogen through the sample. Cysteine was oxidised to cysteic acid. Amino acids were separated by ion exchange chromatography and estimated by ninhydrin using an automatic amino acid analyser.

REFERENCES

1. Grill E, Winnacker E-L, Zenk MH (1987). Phytochelatins a class of heavy metal-binding peptides from plants, are functionally analogous to metallothioneins. Proc Natl Acad Sci 84:439.
2. Robinson NJ, Jackson PJ (1986). "Metallothionein-like metal complexes in Angiosperms : Their structure and function". Physiol Plant 67:499.
3. Robinson NJ, Thurman DA (1986). Isolation of a copper complex and its rate of appearance in roots of *Mimulus guttatus*. Planta 169:192.

Metal Ion Homeostasis: Molecular Biology and Chemistry, pages 375–384

METAL RESISTANCE IN ASPERGILLUS NIDULANS

A.B. Tomsett, K.E. Hodges, R.N. Cooley,[1] and D.A. Thurman

Departments of Genetics and Botany, University of Liverpool, Liverpool L69 3BX, U.K.

ABSTRACT Eight cobalt-resistant mutants of Aspergillus nidulans have been isolated which can grow on higher levels of cadmium than can wild-type strains. In each case, resistance results from a single gene mutation: these identify two new loci cobA and cobB, on chromosome VI. In addition, a mutant isolated as cadmium-sensitive has been partially characterised. This mutant strain is also sensitive to a series of other metals. In an attempt to unravel the molecular mechanisms determining metal resistance/sensitivity, we are currently examining sulphur metabolism. Three of the four metal resistant loci identified in this organism thusfar appear to have altered responses for sulphur metabolism.

INTRODUCTION

We are undertaking a study of the range of metal-resistant mechanisms available to a single organism, namely the filamentous fungus Aspergillus nidulans, and to correlate these with cellular metabolism. Several strategies for metal tolerance have been described (1) by collating the data from a variety of organisms including both prokaryotes and eukaryotes. For many of the previously used organisms, analysis is limited. We have chosen A. nidulans because it is amenable to genetical, biochemical and molecular biological techniques, all of which will be required to unravel

---

[1]Present address: Department of Genetics, University of Birmingham, Birmingham B15 2TT, U.K.

the complex interrelationship of metal tolerance and metal homeostasis. Our approach is to artificially induce mutations in laboratory strains of this organism to identify the total number of genes in which alleles can confer metal tolerance, and subsequently to attempt to define both the tolerance mechanism and their normal physiological function.

We have previously reported (2) the isolation and genetic characterisation of cadmium-resistant mutants of *A. nidulans*. Thirteen cadmium-resistant mutants were isolated which could grow on higher levels of cadmium than the wild-type strain. These defined two new genes, *cadA* on chromosome IV and *cadB* on chromosome VI. Each of these genes must confer resistance by separate mechanisms because the phenotype of *cadA cadB* double mutants is additive rather than epistatic (in which case the phenotype of the double mutant would be identical to one of the two single mutants).

Metal-resistant mutants have been isolated previously from *A. nidulans* but the strains have been lost (3). These results revealed, however, that zinc, cobalt, nickel, iron, manganese and tin-resistant mutants can be isolated from this organism. Here, we report on preliminary experiments with a group of cobalt-resistant mutants of *A. nidulans*, and on a cadmium-sensitive mutant strain.

MATERIALS AND METHODS

Strains and genetic techniques. *Aspergillus nidulans* strains used in this work carried markers in general use which have been described previously (4). The genetic techniques used were modified after Pontecorvo *et al*. (5). Haploidization of diploid strains for assigning markers to linkage groups has been described and was induced by 750 µg/litre benlate in complete medium (see 2).

Media and supplements. Standard media were as described previously (2). For tests of cobalt toxicity, low phosphate minimal media were used which contain one quarter of the normal phosphate levels. This media did not drastically alter the growth of *A. nidulans* in absence of metal and enabled metal toxicity tests to be undertaken in the absence of excessive metal precipitation. Cadmium toxicity tests were carried out on normal minimal medium.

Assessment of metal tolerance. Two indices of metal tolerance have been used. The ability to germinate in the presence of metal is measurable by counting the number of colonies growing on metal-containing media relative to a control plate. The relative growth rate (hyphal extension) of the germinated spores is assessed by calculating the area of the colony in the presence of metal relative to a control. Details of these techniques have been described previously (2).

Isolation of cobalt-resistant mutant strains. A conidial suspension at 10^8 conidia/ml was exposed to 750 joules/m^2 ultra-violet light. The conidia were then added to molten cobalt-containing overlay agar at 49^oC as 10^5 conidia per plate. Resistant colonies were selected from plates containing 6 mM cobalt chloride on normal phosphate containing minimal media.

Isolation of the cadmium-sensitive mutant. Mutagenesis was achieved by exposing a conidial suspension, at 10^8 conidia per ml, to 250 joules/m^2 ultra-violet light. In order to identify cadmium-sensitive mutants, mutagenised conidia were first overlayed in the absence of cadmium at 10^2 conidia per plate. Colonies were velvet replicated onto plates containing 150 µM cadmium. Mutants growing less strongly than the wild-type parental strain were retested on master plates. Colonies consistently showing weak growth on 150 µM cadmium were classified as cadmium-sensitive mutants.

Growth tests. Phenotypic tests were carried out in petri dishes with solid media. For selected strains, the phenotype of the particular allele was scored by replica-plating progeny from a cross in which the relevant allele segregated, so that any modification of the phenotype on differently supplemented test media could be detected.

RESULTS AND DISCUSSION

Analysis of Cobalt-Resistant Mutants.

Phenotypic tests of wild-type and cobalt-resistant strains. After UV mutagenesis and selection on 6 mM cobalt chloride on normal minimal medium, ten cobalt resistant mutants were obtained. The toxicity of cobalt ions to these mutants and the wild-type strain from which they were derived has been measured in two distinct assays. Figure 1 presents data for the inhibition of conidial germination in the presence of the metal on low phosphate media. The wild-type (C5, *bi*A1) strain showed 50% inhibition at 700 µM cobalt and complete inhibition by 2 mM. Two representative mutant strains, HM44 (*bi*A1 *cob*A^{R}10) and HM35 (*bi*A1 *cob*B^{R}1), showed 50% inhibition at approximately 2.8 mM and complete inhibition at 4 mM cobalt. All ten mutant strains show an increase in germination frequency in the presence of cobalt compared to the wild-type.

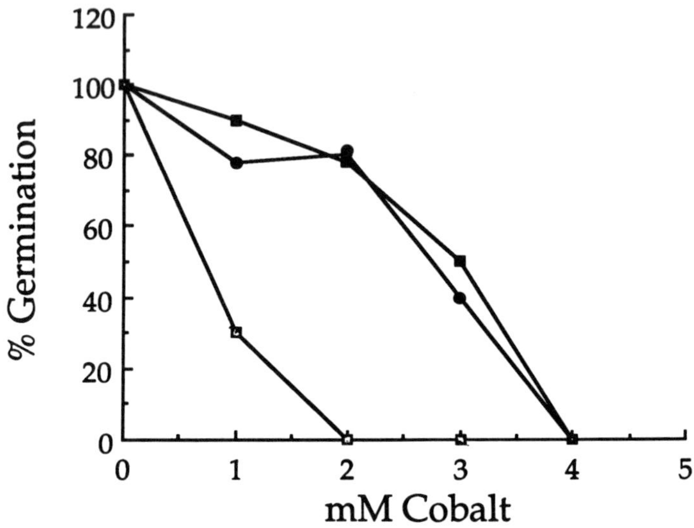

FIGURE 1. Inhibition of conidial germination in the presence of cobalt: □ wild-type (C5, *bi*A1); ● HM35 (*bi*A1 *cobB*^{R}1); ■ HM44 (*bi*A1 *cobA*^{R}10).

Figure 2 presents data for the inhibition of hyphal growth in the presence of metal on low phosphate media. The concentration of cobalt yielding 50% inhibition of hyphal extension is 1 mM for wild-type (C5), 2.3 mM for HM56 (*paba*A1 *w*A3 *cobB*^{R}5), 3.3 mM for HM57 (*paba*A1 *w*A3 *cobA*^{R}7), and 4 mM for HM58 (*paba*A1 *w*A3 *cobA*^{R}7 *cobB*^{R}5). All ten mutants show some increase in hyphal growth in the presence of cobalt compared to the wild-type.

Genetic analysis of cobalt-resistant mutants. All ten cobalt-resistant mutants were backcrossed to the wild-type strain L20 (*paba*A1 *w*A3). Eight of the ten crosses gave 1:1 segregations characteristic of a single gene mutation. *cob*^{R}3 and *cob*^{R}8 gave unusual segregations and have not been examined further. Complementation tests have not been possible because cobalt-resistance is a semi-dominant phenotype.

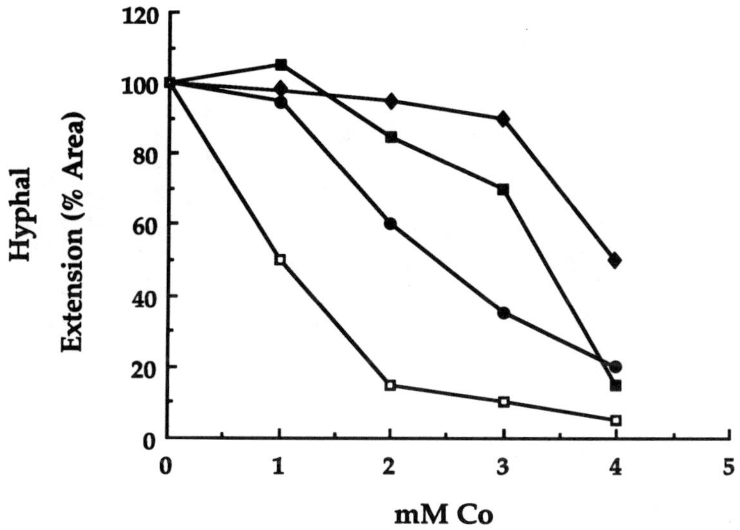

FIGURE 2. Inhibition of hyphal growth in the presence of cobalt: □ wild-type (C5, biA1); ● HM56 (pabaA1 wA3 cobB^{R}5); ■ HM57 (pabaA1 wA3 cobA^{R}7), ◆ HM58 (pabaA1 wA3 cobA^{R}7 cobB^{R}5).

Instead, the mutants have been assigned to loci by inter-crossing the mutant strains. Crosses yielding no recombinant progeny (100 tested) of either wild-type (cob-sensitive) or double resistant mutant phenotype indicated the two mutants reside at the same locus (though not necessarily in the same gene). Crosses between mutants at different loci resulted in a 3:1 segregation of cob^R:wild-type(cob^S) progeny. On this basis, two loci were defined: cobA having six representative mutant alleles, and cobB having two representative mutant alleles. From such crosses, cobA^R cobB^R double mutants were isolated and these showed an additive phenotype (Figs. 2 and 3). The additive phenotype indicates that cobA^R and cobB^R confer tolerance by separate mechanisms. cobA^R cobB^R double mutants when backcrossed to wild-type yield the expected 1:2:1 ratio of double mutants:single mutants:wild-type progeny.

By haploidization, both cobA and cobB appear to map to linkage group VI. At present, their exact location is unknown but they are not tightly linked to each other and they are both distant from the cadB locus, which confers cadmium

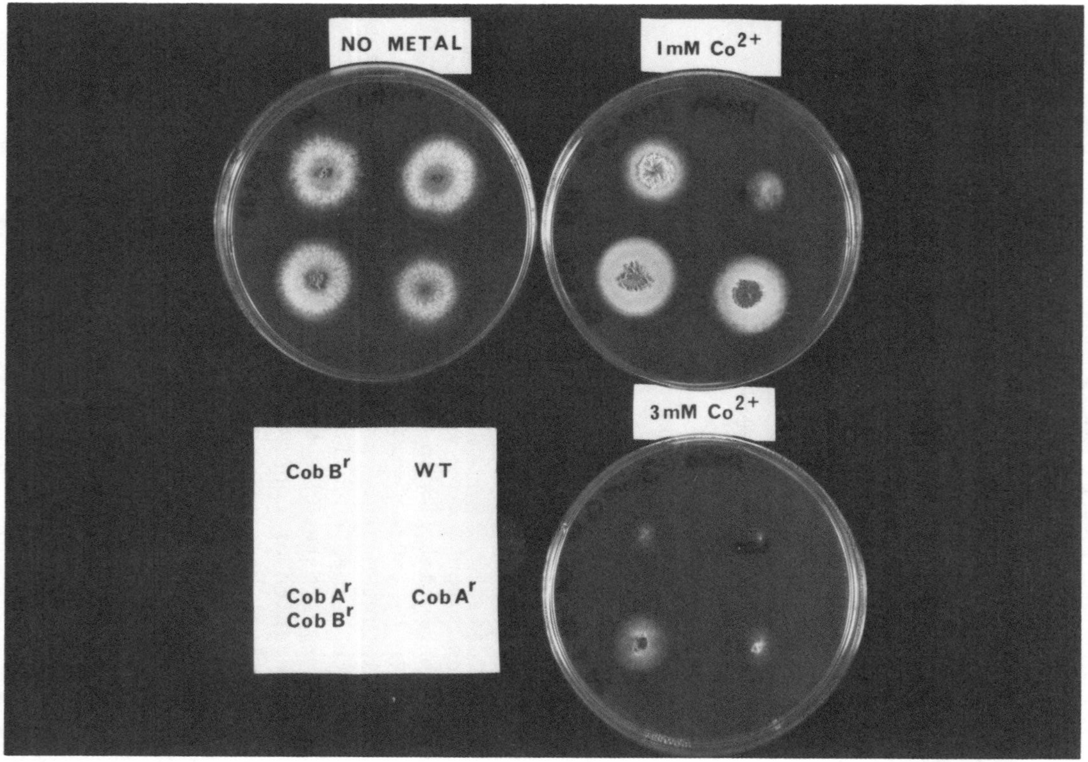

FIGURE 3. Growth test of wild-type and cobalt-resistant mutants.

tolerance on *A. nidulans*.

Analysis of the Cadmium-Sensitive Mutant.

Isolation and phenotypic tests of the cadmium-sensitive mutant. In order to isolate cadmium-sensitive mutants, wild-type (C5) conidia were UV irradiated and plated in the absence of cadmium. Master plates were constructed from the colonies produced and these were replicated onto plates containing 150 µM cadmium. In an attempt to avoid selecting morphological or leaky auxotrophic mutants only those colonies which produced a wild-type morphology in the absence of cadmium were chosen. Only one cadmium-sensitive mutant, *cad*C^{S}1, was isolated in this way.

*cad*C^{S}1 was tested for inhibition of hyphal extension in the presence of 150 µM cadmium, 150 µM copper, 350 µM nickel, 3 mM cobalt, 3 mM zinc, 6 mM lead and also on plates lacking metal. The results of these growth tests are summarised as growth rates in Table 1. *cad*C^{S}1 grew at a slower rate than the wild-type in the presence of cadmium and copper and failed to grow in the presence of the other metals tested.

TABLE 1
GROWTH RATE OF WILD-TYPE AND *cad*C^{S}1 MUTANT STRAINS IN THE PRESENCE OF VARIOUS METALS

Metal	Growth rate (cm day^{-1}) Wild-type (C5)	*cad*C^{S}1
0	1.40	1.22
150 µM Cd^{2+}	0.29	0.08
150 µM Cu^{2+}	0.77	0.13
350 µM Ni^{2+}	1.08	0.00
3.0 mM Co^{2+}	1.25	0.00
3.0 mM Zn^{2+}	1.25	0.00
6.0 mM Pb^{2+}	1.28	0.00

The cadmium-sensitive mutant was assayed for inhibition of germination (Table 2) and was found to be more sensitive than the wild-type, 50% inhibition being achieved by approximately 200 µM cadmium. Table 3 shows that this mutant was also more sensitive than the wild-type in the presence of cadmium when assayed for inhibition of hyphal extension.

TABLE 2
INHIBITION OF GERMINATION OF WILD-TYPE AND *cadC*^{S}1 STRAINS IN THE PRESENCE OF CADMIUM

Cadmium concentration (µM)	Percent germination Wild-type (C5)	*cadC*^{S}1
0	100	100
100	111	88
200	89	49
300	55	7
400	27	0
500	5	0

TABLE 3
INHIBITION OF HYPHAL GROWTH (AREA) OF WILD-TYPE AND *cadC*^{S}1 STRAINS IN THE PRESENCE OF CADMIUM

Cadmium concentration (µM)	Percent hyphal growth Wild-type (C5)	*cadC*^{S}1
0	100	100
50	51	41
100	20	5
150	12	1
200	6	1
250	4	1
300	1	1
350	1	0

Genetic analysis of the cadmium-sensitive mutant. Wild-type strains were routinely distinguished from cadmium-sensitive strains by growth in the presence of 50 and 100 µM cadmium after point inoculation from master plates. When cadC^{S}1 was backcrossed to wild-type strains the sensitive phenotype was recovered as half the total number of progeny scored suggesting that it arose by mutation at a single gene locus.

The cadC^{S}1 mutation was assigned to its linkage group by the use of standard techniques for the haploidization of diploids. Analysis of haploid sectors segregating from a diploid constructed between a cadC^{S}1 strain (carrying markers fwA1, pabaA1, uaY9) and strain R153 (wA3, pyroA4) allowed cadC^{S}1 to be assigned to chromosome VIII. We have not been able to determine the exact location of this mutation on chromosome VIII because it has not shown linkage to any marker tested thusfar. It is clear however that cadC^{S}1 is not an allele of cadA or cadB which lie on linkage groups IV and VI respectively.

Metal Resistance and Sulphur Metabolism.

We have started to investigate the relationship between metal resistance and alterations to sulphur metabolism. We have previously described the very tight linkage between cadB mutants and the methB gene (2). The methB gene is involved in the synthesis of methionine from cysteine, and the possibility that cadB is a gene integral to sulphur metabolism is attractive. As yet we have not been able to discover any clues to its normal cellular function.

We have crossed cobA^{R} and cobB^{R} mutants to a series of strains carrying mutations defective in various steps of sulphur metabolism. In crosses to methA and methB (both of which are unable to convert cysteine to cystathionine), no meth^{-} cob^{R} progeny were obtained. In crosses to mecA and mecB, which are defective in the conversion of homocysteine to cystathionine and cystathionine to cysteine respectively, cobA^{R} mutants were found to have a 'mec'-like phenotype, i.e. selenate hypersensitivity in the presence of methionine (6). Selenate toxicity presumably results from the incorporation of selenium into selenocysteine which consequently kills the organism. The wild-type can grow on methionine plus selenate because they make cysteine from methionine. mecA and mecB are selenate hypersensitive because they cannot make cysteine from methionine. We know that cobA^{R} mutants are not alleles of mecA or mecB because these two genes are on chromosome I and crosses show the expected high recombination frequencies. We are currently testing the possibility that it is an allele of a third gene *mecC*.

At present, we have no conclusions as to the normal physiological role of *cobA* and *cobB*. From the preliminary data presented here, it appears that both *cobA* and *cobB* play some role in sulphur metabolism and that one class of mutant alleles at these loci result in cobalt resistance.

ACKNOWLEDGMENTS

This work was supported by a Natural Environment Research Council studentship to K.E.H. and a Science and Engineering Research Council studentship to R.N.C.

REFERENCES

1. Wood JW (1984). Microbiological strategies in resistance to metal ion toxicity. In Sigel H (ed): "Metal Ions in Biological Systems", vol 18, New York: Marcel-Dekker, p 233.
2. Cooley RN, Haslock HR, Tomsett AB (1986). Isolation and characterization of cadmium-resistant mutants of *Aspergillus nidulans*. Current Microbiol 13:265.
3. Elorza MV (1969). Toxicidad de los iones metalicos para *Aspergillus nidulans*. Microbiol Espan 22:131.
4. Clutterbuck AJ (1974). *Aspergillus nidulans*. In King RC (ed): "Handbook of Genetics", vol I, New York: Plenum, p 447.
5. Pontecorvo G, Roper JA, Hemmons LM, Macdonald KD, Bufton AWJ (1953). The genetics of *Aspergillus nidulans*. Adv Genet 5:141.
6. Pieniazek NJ, Kowalska IM, Stepien PP (1973). Deficiency in methionine adenosyltransferase resulting in limited repressibility of methionine biosynthetic enzymes in *Aspergillus nidulans*. Molec gen Genet 126:367.

Metal Ion Homeostasis: Molecular Biology
and Chemistry, pages 385–394

CHARACTERIZATION OF A ZN-BINDING PEPTIDE ASSOCIATED WITH A DECLINE DISORDER OF CITRUS[1]

K.C. Taylor[2], L.G Albrigo[3], C.D. Chase[2]

[2]Department of Vegetable Crops, University of Florida, Gainesville FL 32611.
[3]Citrus Research and Education Center, University of Florida, Lake Alfred, FL 33850

ABSTRACT Citrus blight is a decline disorder of unknown etiology. Zn redistribution within the tree is the earliest known symptom. Leaves are often Zn deficient while Zn accumulates in the phloem. Phloem Zn was associated with a highly anionic Zn binding fraction isolated (DEAE Sephadex chromatography) from phloem tissue extracts of healthy and blight-affected citrus. The partially purified Zn binding fractions from healthy and blight extracts had an apparent MW of 4.5 kDa (gel filtration) and similar amino acid compositions, which had high percentages of asx and gly. This suggests they are may not be $(\gamma\text{-glutamylcysteinyl})_n$ glycine peptides. A higher level of Zn binding peptide was associated with the occurrence of blight and the peptide isolated from phloem tissue of blight-affected citrus had a higher specific Zn content than that from healthy.

INTRODUCTION

Citrus blight, a complex decline disorder, is a major cause of mature tree loss in the worldwide citrus industry (1). No causal organism has been determined. However, transmission of the disorder from blight to healthy trees has been affected through root grafts (2). The earliest known symptom of the citrus blight syndrome is an accumulation of Zn in the phloem 40-50 cm above the bud

---

[1]Supported by USDA Competitive Res. Grant #86-CRCR-1-2081

union (3). This may be accompanied by Zn deficiency in the leaves (4). Zn redistribution occurs 1-3 years prior to xylem plugging which reduces water conductivity and causes severe and progressive decline of the canopy (3). Of significance is the finding that ATP, carbonic anhydrase (5) and IAA oxidase (6) activities are decreased in declining trees. Carbonic anhydrase and IAA oxidase are Zn-requiring enzymes (7). These decline stage metabolic changes appear secondary to the earliest known change associated with the citrus blight syndrome, i.e., trunk phloem Zn accumulation (3). Therefore, Zn accumulation appears closer in symptom ontogeny to the primary cause of blight. This report characterizes a Zn sequestering agent which has been associated with the occurrence of citrus blight.

Previous work (8) indicated that blight-affected grapefruit and orange phloem extracts bound at least 5 x more added Zn than healthy phloem extract and approximately 3 x more Pb and more that 25 x more Cd. In addition, the binding by phloem extract was temperature stable and reversibly pH 2 labile. These binding characteristics were similar to those in the (γ-glutamylcysteinyl)$_n$ glycine peptides [(γEC)$_n$G] which are reported responsible for Cd binding in plants (9-13). This study was undertaken based on the hypothesis that the Zn sequestering molecule associated with citrus blight might be (γEC)$_n$G.

PARTIAL PURIFICATION

Phloem tissue samples were collected (8) from 'Valencia' sweet orange [*Citrus sinensis* (L.) Osbeck] on rough lemon [*C. jambhiri* (L.)], the most blight susceptible rootstock commonly used in Florida (1, 14). Tissue was extracted in a Tris buffer as described earlier (8). Extracts from healthy and blight-affected trees were loaded onto a DEAE Sephadex column (2.4 x 20 cm) on an equivalent protein basis (30 mg total protein, each). Fractions (6.5 ml) were eluted with a 0.25-2.0 M NaCl gradient in 0.05 M Tris-HCl, pH 7.8 at 4°C at 50 ml/hr.

One Zn peak was resolved from both healthy and blight material. The blight peak contained almost 2 x as much total Zn as the healthy (determined by atomic absorption spectrophotometry) (Fig. 1). All Zn loaded onto the columns was accounted for in the elution profiles. For the blight curve, 86% of the total Zn loaded was contained in the Zn peak and for healthy, 71% was contained in the Zn peak.

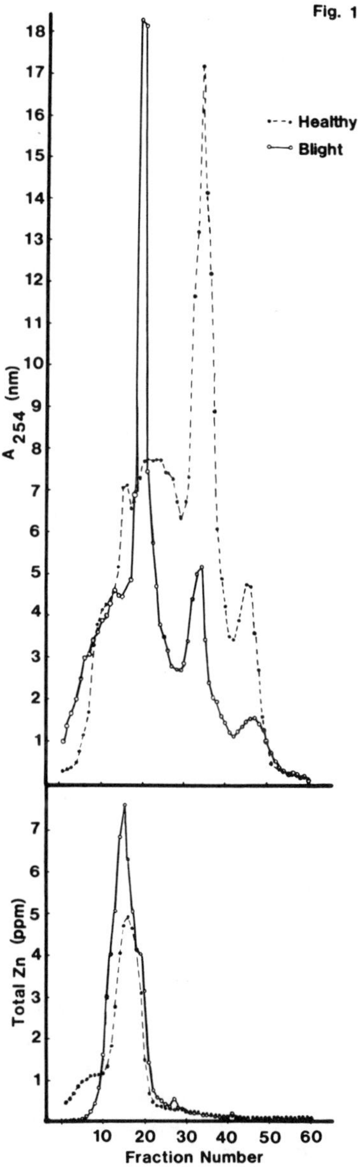

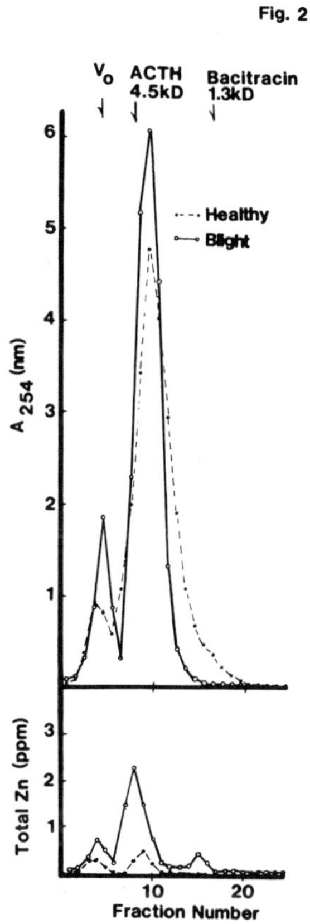

FIGURE 1. Ion exchange chromatography (IEC) of extracts from phloem tissue of healthy and blight-affected citrus.

FIGURE 2. Gel filtration chromatography of the Zn containing peak isolated in previous IEC.

These Zn containing peaks were coincident with peaks of apparent mercaptide as determined by absorption at 254 nm (Fig. 1). The blight mercaptide peak was substantially more attenuated than the healthy peak. The Zn-mercaptide was highly anionic, requiring 0.4 M NaCl for elution. Additional peaks, detected at 254 nm, were eluted in both healthy and blight material but were not associated with Zn (Fig 1). It is of interest that these non-Zn containing peaks are more evident in healthy than blight. The material in the Zn containing peak associated with blight may be a modified form of that found in the non-Zn containing peaks that are characteristic of the healthy condition.

Appropriate fractions from the Zn containing peak from the ion exchange column were pooled, PEG-concentrated (15), reconstituted with additional reductant (2-mercaptoethanol) and placed on a Bio-Rad P-6 column (2.4 x 20 cm) for gel filtration. Again the columns were loaded on an equivalent total protein basis (6.4 mg). Elution of 4.5 ml fractions was with 0.05 M Tris-HCl, pH 7.8 at 4°C at 10 ml/hr. Approximately 5 x as much Zn was found in the blight peak compared to healthy and was associated with an apparent mercaptide (Fig. 2). The Zn sequestering molecule had an apparent molecular weight of 4.5 kDa.

AMINO ACID ANALYSIS

Peak fractions were pooled, lyophilized, and hydrolyzed for amino acid analyses. DMSO hydroysis was used to determine cysteine as cysteic acid (16) and a standard HCl hydrolysis (17) was used for the other amino acids. The results of these 2 hydrolysis were combined and reported in Table 1.

The peaks for the gel filtration profile contained similar peptides that were richest in aspartate or asparagine and glycine. Comparing these compositions with those for similarly purified Cd-binding $(\gamma EC)_n G$ peptides isolated from suspension cultures of higher plant cells (15,18), it seems unlikely that the Zn sequestering peptide isolated from citrus phloem is a $(\gamma EC)_n G$ peptide. The ratio of cysteine, glutamate, and glycine for $(\gamma EC)_n G$ peptides is generally n:n:1 n= 2-8 (18,19,20), and the ratio of these residues from our amino acid composition is approximately 1:1:2.

TABLE 1
AMINO ACID COMPOSITION

	(nmole %)	
	HEALTHY	BLIGHT
CYS	3.96	4.24
ASX	11.34	12.01
THR	5.63	5.55
SER	7.00	6.57
GLX	5.92	4.56
PRO	7.10	7.11
GLY	10.16	10.72
ALA	8.45	7.22
VAL	5.87	6.21
MET	1.09	0.64
ILE	5.12	5.16
LEU	7.36	7.25
TYR	4.41	3.03
PHE	4.99	5.44
HIS	1.42	1.44
LYS	5.51	4.84
ARG	3.41	6.69
TOTAL	98.74	98.68

PROTEIN AND ZINC ACCOUNT

Protein and Zn contents were accounted for throughout purification (Table 2). Remembering that equivalent amounts of protein were applied to the surface of the chromatography columns, it is clear that the partially purified blight fraction contained much more peptide than healthy. There was 2.5x more total protein and 4.5x more mg protein/g fresh weight in the blight peak. In addition each mole of the blight peptide was associated with approximately 6.5x more Zn than the healthy peptide.

TABLE 2
Protein[a] and Zinc Account

	HEALTHY			BLIGHT		
	Total Protein (mg)	ug Zn/ug Protein	mg Prot/g Fresh Weight	Total Protein (mg)	ug Zn/ug Protein	mg Prot/g Fresh Weight
Crude Extract	30.0	0.011	3.00	30.0	0.013	5.84
IEC Peak	6.41	0.035	0.71	7.66	0.048	1.44
Gel Filtration Peak	0.02	0.136	0.002	0.05	0.854	0.009

[a]Protein determined by Bio-Rad coomassie reagent.

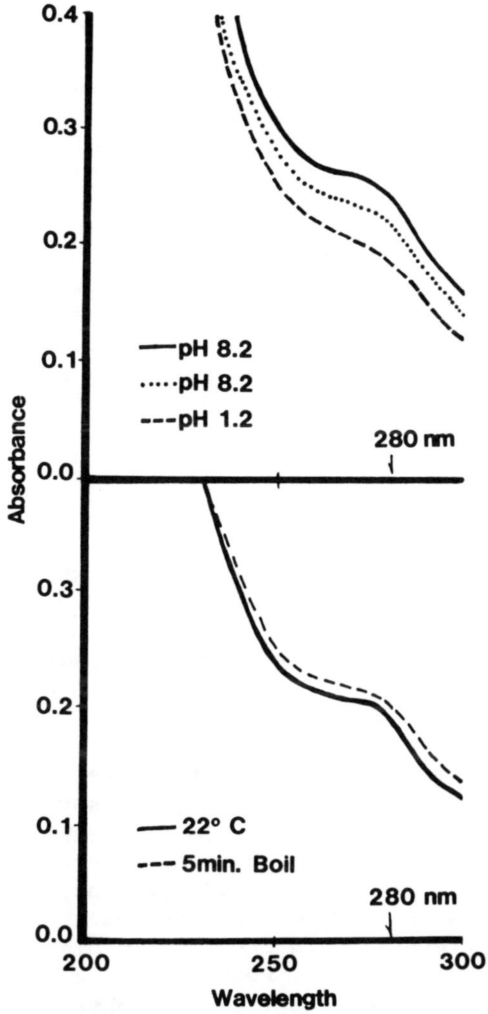

FIGURE 3. Determination of temperature and pH stability of Zn-binding by partially purified peptide.

CHARACTERIZATION

A preliminary experiment was performed to assess the temperature and pH stability of the Zn binding of the partially purified peptide. When the partially purified peptide was given a 5 min. boiling treatment the binding was unaffected (Fig. 3). When the pH was changed from 8.2 to 1.2, the binding decreased. With increase of pH back to 8.2 from 1.2 the Zn binding was partially restored. The dilution due to additions of acid and base was not sufficient to account for this partial restoration of Zn binding.

As in the crude phloem extracts (8), the partially purified peptide is temperature stable and somewhat reversibly pH 1.2 labile. These data are comparable to those reported by Rauser (11) for partially purified Cd binding peptide which was ultimately shown to be $(\gamma EC)_nG$ peptide in its purified form (21).

CONCLUSION

A Zn binding peptide (apparent MW= 4.5 kDa) has been partially purified from both healthy and blight-affected tissue and shown to be associated with the occurrence of citrus blight. The peptide was correlated with citrus blight by the increased level of the peptide and the increased specific Zn content of the peptide in blight-affected trees. Because this peptide was found in healthy trees it may have a normal role in the tree's metabolism, i.e. one not associated with the blight disorder. Like the $(\gamma EC)_nG$ peptides, it was highly anionic, temperature stable and reversibly pH 1.2 labile. However, the amino acid composition did not have a ratio of cysteine, glutamate, and glycine suggestive of $(EC)_nG$. Rauser, et al (21) have reported an aspartate/ asparagine rich Cd binding protein. This peptide may be similar. Possibly a new group of metal binding peptides has been observed in higher plants.

At this time we have achieved an approximate 65-fold purification, based on specific Zn content of the partially purified blight peptide. Following final purification and determination of an amino acid sequence, an oligonucleotide probe will be synthsized and used to ask whether the Zn sequestering peptide associated with citrus blight is a primary gene product or if, like the $(\gamma EC)_nG$ peptides it is a product of a biosynthetic process.

Literature Cited

1. Smith, PF (1974). History of citrus blight in Florida. Citrus Industry 55(9):13-14, 16, 18-19: (10):9-10, 13, 14 (11):12-13.

2. Tucker, DPH, RF Lee, LW Timmer, LG Albrigo, and RH Brlansky (1984). Experimental transmission of citrus blight. Plant Disease Reporter 68:979-980.

3. Albrigo, LG, JP Syvertsen, and RH Young (1986). Stress symptoms of citrus trees in successive stages of decline due to blight. J Amer Soc Hort Sci 111(3):465.

4. Rhoads, AS (1936). Blight: A non-parasitic disease of citrus. Florida Agri Exp Sta Bull 296.

5. Bausher, MG. (1979). Changes in ATP levels and carbonic anhydrase in the presence of citrus blight. Plant Physiol 63 (suppl.):115.

6. Bausher, MG (1982). Changes in IAA-oxidase activity and the severity of citrus blight. Hortscience 17 (suppl.):497.

7. Randall, PJ and D Bouma (1973). Zinc deficiency, carbonic anhydrase, and photosynthesis in leaves of spinach. Plant Physiol 52:229-232.

8. Taylor, KC, LG Albrigo, CD Chase (1988). Zinc complexation in the phloem of blight-affected citrus. J Amer Soc of Hort Sci 113:407.

9. Murasugi, A, C Wada, Y Hayashi (1981). Purification and unique properties in UV and CD spectra of Cd-binding peptide 1 from _Schizosaccharomyces pombe_. Biochem Biophys Res Commun 103:1021.

10. Grill, E, EL Winnacker, and MH Zenk (1985). Phytochelatins; The principal heavy-metal complexing peptides of higher plants. Science 230:674-676.

11. Jackson, PJ, CJ Unkefer, JA Doolen, K Watt, NJ Robinson (1987). Poly(gamma-glutamyl-cysteinyl)glycine: Its role in cadmium resistance in plant cells. PNAS 84:6619.

12. Rauser, WE (1984). Isolation and partial purification of cadmium-binding protein from roots of the grass *Agrostis gigantea*. Plant Physiol 74:1025.

13. Wagner, GJ and MM Trotter (1982). Inducible cadmium binding complexes of cabbage and tobacco. Plant Physiol 76:797-805.

14. Young, RH, LG Albrigo, DPH Tucker and GL Williams (1980). Incidence of citrus blight on carrizzo and citrange and some other rootstocks. Proc Fla State Hort Soc 93:14.

15. Robinson, NJ, DA Thurman (1986). Isolation of a copper complex and its rate of appearance in roots of *Mimulus guttatus*. Planta 169:192.

16. Spencer, RL, F Wold (1969). A new convenient method for estamation of total cystine-cysteine in proteins. Anal Biochem 32:185.

17. Hirs, CHW (1967). Determination of cysteine as cysteic acid. Methods Enzymol 11:59-62.

18. Huang B, E Hatch and PB Goldsbrough (1987). Selection and characterization of cadmium tolerant cells in tomato. Plant Science 52:211.

19. Steffens JC, DF Hunt and BG Williams (1986). Accumulation of non-protein metal binding polypeptides (gamma-glutamyl-cysteinyl)$_n$-glycine in selected cadmium-resistant tomato cells. J Biol Chem 261:13879.

20. Grill, E, E-L Winnacker, MH Zenk (1986). Synthesis of seven different homologous phytochelatins in metal-exposed *Schizosaccharomyces pombe* cells. FEBS 197:115.

21. Rauser, WE, PE Hunziker and JHR Kagi (1986). Reverse-phase high-performance liquid chromatography of Cd-binding proteins from the grass *Agrostis gigantea*. Plant Science 45:105.

IV. MECHANISMS OF METAL DETOXIFICATION

Metal Ion Homeostasis: Molecular Biology and Chemistry, pages 397–405

SEPARATION OF THE MERCURY-BINDING AND DNA-BINDING ACTIVITIES OF THE MER R REGULATORY PROTEIN[1]

Lisa M. Shewchuk,[2] John D. Helmann,[2] Wilma Ross,[3] Soon-Jung Park,[3] Anne O. Summers,[3] and Christopher T. Walsh[2]

[2]Department of Biological Chemistry and Molecular Pharmacology, Harvard Medical School, Boston, Massachusetts, 02115, and [3]Department of Microbiology, University of Georgia, Athens, Georgia, 30602.

ABSTRACT Genetic and biochemical experiments support a structural model for the bacterial MerR regulatory protein in which metal-binding and DNA-binding are separable functions. Mercury titration experiments establish that MerR protein can specifically bind a maximum of 1 mercuric ion per dimer. Mutations which remove any of three separate cysteine residues greatly decrease the affinity of purified MerR protein for mercuric ion yet have little or no effect on affinity for the DNA operator site.

INTRODUCTION

The essential role of many metal ions in biological systems, coupled with their potential toxicity, has led to the evolution of sophisticated systems for the detection and regulation of intracellular metal ion levels. These "metal-ion

[1] This work supported by grants from the National Institutes of Health to C.T.Walsh (no. GM20011) and A.O. Summers (no. GM28211).

homeostasis" systems often involve metal-specific changes in gene expression mediated by complex transciptional and translational control mechanisms.

Although there is no known physiological role for mercury, the prevalence of ionic mercury in the environment has given rise to complex sensing and detoxification mechanisms in many bacterial species. We have focused on the role of mercuric ion in the regulation of gene expression in the bacterial transposons, Tn21 and Tn501.

The primary sensor of intracellular mercuric ion in these bacterial systems is the MerR regulatory protein. This protein acts to stimulate transcription of the mercury detoxification genes in a mercuric ion-dependent manner and also autoregulates its own synthesis.

Overexpression and purification of the MerR regulatory proteins from both Tn501 and Tn21 has demonstrated that these proteins bind to a specific DNA operator site (1,2). This operator overlaps the promoter for the transport and detoxification genes (P_T) and lies immediately downstream of the divergently oriented promoter for the merR structural gene (P_R)(Figure 1). The 144 amino acid MerR protein exists as a 32 kD dimer in solution, and binds to the symmetrical dyad in the operator site, presumably as a dimer (1,2). In vivo experiments demonstrate that in the absence of mercuric ion, this protein represses transcription from both P_R and P_T. In the presence of mercuric ion, however, transcription from P_T is specifically stimulated approximately 100-fold.

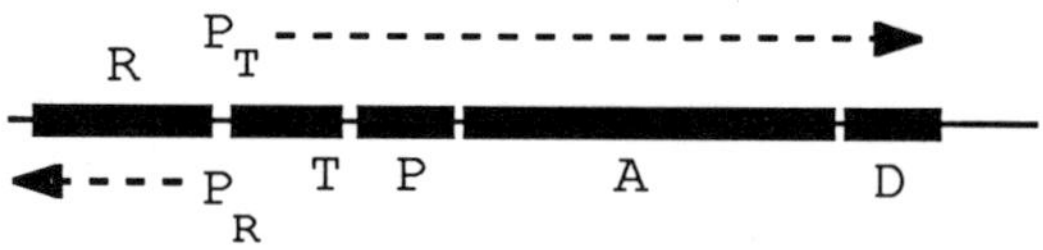

Figure 1. Diagram of the mercury resistance determinant of Tn501 illustrating the divergently oriented transcription units controlled from a central promoter/operator region.

To determine the mechanism of mercuric ion-dependent activation of transcription by MerR, we sought to isolate mutant proteins which were altered in their regulatory properties. The sequences of these mutations suggest that at least three of the four cysteine residues in MerR are important for transcriptional activation, but apparently not critical for repression. Further, a mutant in a proposed helix-turn-helix region (residues 9 through 29) can neither repress nor activate transcription (W. Ross, S.J. Park, and A. O. Summers, unpublished observations).

To test the basis of these phenotypes we have overproduced and purified a number of mutant MerR proteins and investigated the ability of the purified proteins to bind mercuric ion and the DNA operator site. Since the MerR proteins encoded by Tn501 and Tn21 are 95% identical at the amino acid level, we have used both proteins in our in vitro studies.

METHODS

The production and in vivo characterization of the mutant Tn21 MerR proteins will be described elsewhere (W. Ross, S.J. Park and A. O. Summers, in preparation). Plasmids containing the mutationally altered MerR structural genes were subcloned into a derivative of pTO90-16 (1) and the proteins overexpressed and purified by modifications of the previously described procedure (1).

DNA-binding was assayed using a gel shift assay in which a dilution series of purified MerR protein was incubated with an operator containing DNA fragment (from plasmid pJOE114, 3) and electrophoresis performed at 4° C. Autoradiography was used to estimate the concentration of MerR protein sufficient to shift one-half of the labeled DNA fragment (K_d apparent). In some cases, DNAseI footprinting was also used to demonstrate that the region bound by the protein was the operator region.

Mercuric ion binding was assayed by incubating purified MerR proteins with a 10 fold molar excess of $^{203}Hg^{II}$. Samples were sequentially diluted to the final indicated protein concentrations in a buffer containing 50 mM Tris

HCl, pH 7.5, 250 mM NaCl, 5% (w/v) glycerol, and 1 mM 2-mercaptoethanol. After incubation for at least 30 minutes on ice, 100 microliter aliquots were injected on a BioSil TSK-125 HPLC sizing column equilibrated in the same buffer but lacking thiols. The free and bound mercuric ion were then quantitated by liquid scintillation counting and protein determined by the BioRad protein assay as standardized against the peptide bond absorbance (A^{205}) of purified MerR protein.

Sizing chromatography of several purified MerR mutant proteins (as described above) revealed two protein peaks which eluted with mobilities corresponding to ca. 16 kD and ca. 32 kD. This was interpreted as a monomer:dimer equilibrium and was observed to be concentration dependent.

RESULTS

Using the previously described overexpression system (1), a series of MerR proteins have been overexpressed and purified. The sequences of the wild-type Tn21 MerR protein and some of the relevant mutations are shown in Figure 2.

```
1                                                50
MENNLENLTIGVFAKAAGVNVETIRFYQRKGLLREPDKPYGSIRRYGEAD

51                                               100
VVRVKFVKSAQRLGFSLDEIAELLRLDDGTHCEEASSLAEHKLKDVREKM
      V                             Y
101                                          144
ADLARMETVLSELVCACHARKGNVSCPLIASLQGEAGLARSAMP
                Y        Y
```

Mutation	Phenotype	
A 60 V	$I^{-/+}$	R^{+}
C 82 Y	I^{-}	R^{+}
C 117 Y	I^{-}	R^{+}
C 126 Y	I^{-}	R^{+}

Figure 2. The sequence of MerR from Tn21 and the location and phenotype of selected mutations are indicated. The two potential helix-turn-helix regions are underlined and cysteine residues are

in bold. Phenotypes refer to the ability of the mutationally altered proteins to induce (I) or repress (R) transcription in vivo.

The results of in vivo analyses indicate that mutations in three of the four cysteine residues prevent the mercuric ion-dependent induction of transcription, yet retain the ability of MerR to repress transcription in the absence of mercuric ion. This result suggests that these mutant proteins may still retain DNA-binding activity yet be deficient in their ability to bind mercuric ion. To test this idea directly, the affinity of each of these mutant proteins for both mercuric ion and the DNA operator site was determined (Table 1). All three cysteine to tyrosine mutations greatly reduce the affinity of the protein for mercuric ion, but do not greatly alter the affinity for the DNA operator site.

In contrast, an alanine to valine change at position 60 yields protein with wild-type affinity for mercuric ion, yet with somewhat reduced affinity for the DNA operator (ca. 10 fold). This mutation greatly reduces the ability of MerR to stimulate transcription, but is less severe in its effect than alteration of the cysteines 82, 117, or 126 (Figure 2). Since this protein still represses transcription efficiently in the absence of mercuric ion and can still bind mercuric ion in vitro, it may be specifically deficient in its ability to stimulate transcription by RNA polymerase.

Mutation	Mercuric Ion Binding	Dimerization	DNA-Binding (~ K_d)
wild-type	++	++	~1 x 10^{-10}
A 60 V	++	+	~1 x 10^{-9}
C 82 Y	+	+	~7 x 10^{-10}
C 117 Y	+	+	~1 x 10^{-10}
C 126 Y	+/-	++	~1 x 10^{-10}

Table 1 In vitro properties of purified MerR proteins from Tn501 (wild type) and Tn21 (mutants).

To quantitate the ability of wild-type and mutationally altered MerR polypeptides to bind mercuric ion, an HPLC gel filtration assay has been developed. In this assay, MerR protein and radiolabeled mercuric ion are preincubated at various concentrations and free and bound mercuric ion are quantitated. With the wild type MerR protein from Tn501, addition of saturating amounts of mercuric ion never leads to more than one mercuric ion per dimer. This suggests that mercuric ion may be binding at the dimer interface to stabilize MerR protein in a conformation competent for transcriptional activation.

The affinity of all three cysteine to tyrosine mutant proteins for mercury is greatly reduced as judged by a quantitative mercuric ion binding assay (e.g. compare binding at 10^{-6} M MerR in Figure 3). However, in vitro it is possible to saturate at least two of these three cysteine to tyrosine mutant proteins with high concentrations of mercuric ion (Figure 3).

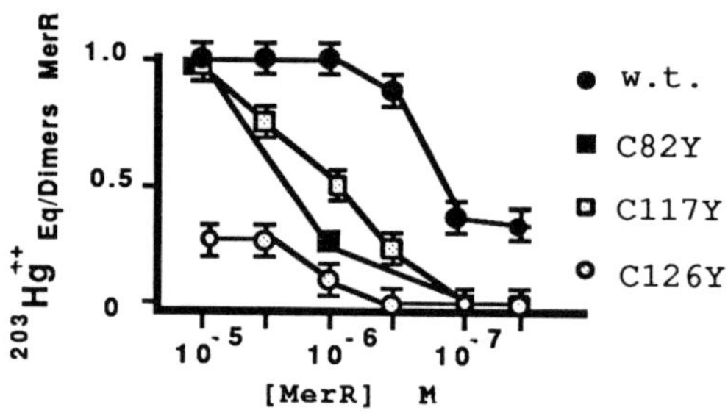

Figure 3 Binding of mercuric ion to MerR proteins as assayed by HPLC gel filtration.

These results strongly support the contention that the primary reason the cysteine to tyrosine mutants are unable to activate transcription in vivo is due to a lowered affinity for mercuric ion. In addition, this establishes that the stoichiometry of mercuric ion binding is one ion per dimer. The precise relationship between these 6 cysteine residues (3 per monomer) and the single bound mercuric ion is currently under investigation. However, these preliminary results suggest a critical role for cysteine 126 since Y126 is the least able to bind mercuric ion. This is corroborated by studies of a proteolytic fragment of MerR containing residues 1 through 120 (or 121) (data not shown).This fragment retains identical DNA binding properties as judged by DNAse I footprinting, yet is unable to bind mercuric ion at a protein concentration of 10^{-6} M in the HPLC size exclusion chromatography assay.

A subset of the mutational changes we have investigated affect the ability of MerR protein to form stable dimers in solution at low protein concentrations. Mutants with severe defects in dimerization would be predicted to be deficient in DNA-binding and should be unable to either induce or repress transcription. Representatives of this class of mutation (I^-R^-) have been isolated and are under study.

Current studies are directed at combining the in vivo analysis of MerR structural mutants with in vitro characterization of the purified mutant proteins. In this way, the structural features important for protein:protein, protein:metal, protein:DNA, and protein:RNA polymerase interactions can be determined.

DISCUSSION

Biochemical characterization of purified MerR proteins from Tn501 and Tn21 has established that these regulatory proteins bind very tightly to their DNA operator site both in the presence and absence of bound mercuric ion (1,2, unpublished results). However, the presence of bound mercuric ion is necessary for in vivo activation of transcription and may act to stabilize an active conformer of the protein. To dissect the functional domains of the MerR protein a combination of genetic and biochemical assays have been developed to monitor mercuric ion binding, DNA binding, dimerization, and transcriptional activation and repression.

The results presented here demonstrate that the purified MerR protein binds a maximum of one mercuric ion per dimer. Mercuric ion binding is separable from DNA binding by both genetic and biochemical criteria. Mutations of three separate cysteine residues differentially impair the affinity of pure MerR for mercuric ion yet have little or no effect on DNA operator binding. In vivo these cysteine mutants fail to activate transcription but are capable of transcriptional repression. In addition, limited proteolysis experiments support a model in which an amino terminal region contains the specific operator recognition elements while the carboxyl terminal cysteines are involved in mercuric ion binding. In contrast, other altered proteins have been characterized that retain mercuric ion binding, yet are deficient in operator recognition.

The ability of MerR to function as a metal-dependent transcriptional activator is reminiscent of other metalloregulatory proteins. The transcription of eukaryotic metallothionein genes, for example, is regulated by metal-dependent DNA binding proteins (4,5). In contrast to the MerR protein, these factors appear to require metal for DNA-binding (4,5). Analogously, a very large class of eukaryotic transcriptional regulatory proteins have sequences suggestive of zinc-binding domains (6) and some of these proteins have also been shown to require metal for DNA-binding in vitro (7,8,9). Similarly, the E. coli iron regulatory protein, fur, has a metal-dependent DNA-binding activity (10), although this protein functions only to repress, rather than to induce or repress, transcription.

The bifunctional regulation mediated by the MerR protein provides a unique system in which DNA-binding is physically uncoupled from transcriptional stimulation. Therefore, this is a particularly attractive system for investigation of the unique metal-induced conformational changes required to activate transcription. The combination of genetic and biochemical approaches presented in outline here promise to define the precise molecular nature of the MerR regulatory switch.

REFERENCES

1. O'Halloran T, Walsh CT (1987). Metalloregulatory DNA-Binding Protein Encoded by the merR Gene: Isolation and Characterization. Science 235:211.
2. Heltzel A, Gambill D, Jackson WJ, Totis PA, Summers AO (1987). Overexpression and DNA-Binding Properties of the mer-Encoded Regulatory Protein from Plasmid NR1 (Tn21). J Bact 169:3379.
3. Brown NL, Ford SJ, Pridmore RD, Fritzinger DC (1983). Nucleotide Sequence of a gene from the Pseudomonas Transposon Tn501 Encoding Mercuric Reductase. Biochemistry 22:4089.
4. Sequin C, Hamer DH (1987). Regulation in Vitro of Metallothionein Gene Binding Factors. Science 235:1383.
5. Andersen RD, Taplitz SJ, Wong S, Bristol G, Larkin B, Herschman HR (1987). Metal-Dependent Binding of a Factor In Vivo to the Metal-Responsive Elements of the Metallothionein 1 Gene Promoter. Mol Cell Biol 7:3574.
6. Evans RM, Hollenber SM (1988). Zinc Fingers: Gilt by Association. Cell 52:1.
7. Hanas JS, Hazuda DJ, Bogenhagen DF, Wu FYH, Wu CW (1983). Xenopus Transcription Factor A Requires Zinc for Binding to the 5 S RNA Gene. J Biol Chem 258:14120.
8. Kadonaga JT, Carner KR, Masiarz FR, Tjian R (1987). Isolation of cDNA Encoding Transcription Factor Sp1 and Functional Analysis of the DNA Binding Domain. Cell 51:1079.
9. Sabbah M, Reduilh G, Secco C, Baulieu EE (1987). The Binding Activity of Estrogen Receptor to DNA and Heat Shock Protein (M_r 90,000) Is Dependent on Receptor-bound Metal. J Biol Chem 262:8631.
10. Bagg A, Neilands JB (1987). Ferric Uptake Regulation Protein Acts as a Repressor, Employing Iron(II) as a Cofactor To Bind the Operator of an Iron Transport Operon in Escherichia coli. Biochemistry 26:5471.

Metal Ion Homeostasis: Molecular Biology and Chemistry, pages 407–416

IN VITRO ANALYSIS OF THE MerR METALLOREGULATORY PROTEIN: A HEAVY METAL-RESPONSIVE REGULATOR OF GENE EXPRESSION[1]

Diana M. Ralston, Betsy Frantz, Myung Shin, Jeffrey G. Wright, and Thomas V. O'Halloran

Department of Chemistry and
Department of Biochemistry, Molecular and Cellular Biology,
Northwestern University, Evanston, IL., 60208.

ABSTRACT Although the mechanisms by which gene expression is regulated by heavy metals are unknown, it is likely that metalloregulatory proteins mediate cellular responses. Metalloregulatory proteins can be defined as multifunctional heavy metal signal receptors where one of the functions is to directly regulate gene expression in response to changes in the concentrations of heavy metal ions. The bacterial mercury resistance (mer) operon is the best understood heavy metal detoxification system and opportune for the investigation of metal-sensing and response. The Tn501 mer operon encodes an inducible, plasmid-mediated resistance to inorganic mercury. Genetic analysis has determined that the promoter of the structural genes, which are directly responsible for mercuric ion detoxification, as well as the promoter for the merR gene are under the control of MerR metalloregulatory protein.[1 2 3 4] In the absence of mercuric ion, the overlapping, divergently-oriented promoters are repressed by MerR, but in the presence

---

[1]This work was supported by research grants from the National Institutes of Health (GM-38784), the National Science Foundation Presidential Young Investigator Award to T.V.O., the General Electric Corporation, and the Searle Scholars Program of the Chicago Communittee Trust. We acknowledge an ARCS Foundation Inc. predoctoral scholarship for D.M.R..

of mercuric ion, the transcription of structural genes isselectively activated. Using the *in vitro* techniques ofphysical biochemistry and molecular biology, we are investigating the mechanism by which submicromolar concentrations of mercuric ion are detected by the bacterial cell and signal the activation of *mer* expression. We propose a model for *mer* operon regulation by MerR.

INTRODUCTION

Cellular apparati for detection of changes in the cellular levels of essential and toxic metals are central components in homeostasis and heavy metal resistance systems. These components may also coordinate an appropriate cellular response by altering the expression of relevant genes. Proteins which carry out these activities have been isolated and characterized from two metal-responsive systems. They are the prokaryotic MerR and Fur proteins, which regulate mercury detoxification and iron uptake, respectively. Both proteins have been purified to homogeneity and studied *in vitro*.[5 6] Although MerR and Fur share the ability to bind DNA and to act as direct receptors for the appropriate metal ions, their effects on gene expression are quite different. The Fur protein is a transcriptional repressor in the presence of Fe(II) but does not apparently regulate gene expression in the absence of iron.[6] MerR, on the other hand, is a transcriptional activator in the presence of Hg(II), and also functions as a repressor in the absence of Hg(II).[1 2]

The MerR and Fur proteins may be prototypes of an entire class of metalloregulatory proteins which operate by a variety of mechanisms to effect the expression of appropriate genes. Other heavy metal homeostasis and detoxification systems are likely to employ similar proteins. The metallothionein genes in eukaryotes which are strongly induced by cadmium appear to be regulated at the transcriptional level by a metalloregulatory protein.[7] Iron-responsive regulation appears to be involved in the expression of ferritin and the transferrin receptor in eukaryotic cells.[8] In this instance, the putative metalloregulatory protein binds in the absence of Fe(II) to a stem-loop structure in the 5′ sequence of ferritin mRNA preventing translation, and the same or similar protein may bind to identical structures in the 3' sequence of transferrin receptor mRNA providing message stabilization. It has been suggested that iron-binding to the protein results in the dissociation of the protein/RNA complex.[7] Finally, the transcription of the ferritin light subunit gene has been shown to be stimulated in the presence of iron.[9] This result suggests a role for a metalloregulatory protein controlling transcriptional initiation.

Several other metal-responsive sytems which may utilize metalloregulatory proteins are listed in Table 1. and have recently been reviewed.[10] Our investigations with the mer system are aimed at elucidating the molecular mechanism by which the MerR protein is converted from a repressor to an activator of transcription, as a means of understanding how an inorganic signal is transduced into a biological response.

TABLE 1
SYSTEMS EXHIBITNG METAL-REPONSIVE GENE EXPRESSION

System	Refs.
Prokaryotes	
Arsenic: Resistance operon	11 12
Cadmium: Heavy metal, heat shock, oxidative stress regulon	13 14 15 16 17 18 19
Copper: Resistance operon	20 21 22 23
Iron: Super oxide dismutase	24 25
Iron: Uptake regulon	6 26 27 28 29
Molybdenum: Nitrogenase	30 31 32
Molybdenum: Nitrate reductase	33 34
Nickel: Hydrogenase	35 36
Tellurium: Resistance operon	37 38 39
Eukaryotes	
Iron	
1. ferritin	8 9 40 41 42 43
2. transferrin receptor	8 44 45
3. respiratory proteins	46 47 48 49
Copper	
1. metallothionein	50
2. plastocyanin and cyt c552	51 52 53
Cadmium	
1. metallothionein	7 50
2. heat shock	

THE *mer* OPERON

Several bacterial mercury resistance operons have been isolated.[54 55 56] Broad spectrum operons provide resistance to both organomercurials and inorganic mercury. Narrow spectrum resistance, however, confers the ability to detoxify inorganic mercuric ion salts, only. In both systems, proteins are induced in the presence of mercury to scavenge mercury from the periplasmic space (*merP* gene product) and pass it through the cell membrane (*merT* gene product) to the flavoenzyme mercuric ion reductase (*merA* gene product) which reduces Hg(II) to volatile Hg(0). Genetic analysis by transcriptional and translational fusions has determined that none of the structural genes play a role in regulation but that a divergently-oriented open reading frame, encoding a 15.9 KD protein, is necessary for both repression and activation of the *mer* genes.[1 2 56 57] Furthermore, this regulatory protein, MerR, negatively regulates its own expression (the *merR* gene) in a Hg-independent manner.[2] *In vivo* experiments could not determine whether or not other cellular factors were involved in *mer* operon activity.

MerR Purification and Physical Characterization

Further study of *mer* regulation was facilitated by MerR purification and *in vitro* analysis. Purification was possible once sequences involved in negative autoregulation were removed and the *merR* gene was inserted into an overproducing vector. Extracts of cells with this vector determined that MerR was produced as 20% of total protein.[5] Chromatography results in purification of crude extracts to a single band on silver-stained SDS-PAGE gels (Wright and O'Halloran, unpublished results). Gel filtration of the protein indicates that it exists as 32KD dimer in solution. When solutions of the MerR protein are dialyzed against thiol-containing buffer, a complex can be formed containing 0.5 Hg(II)/monomer which has been found to be biologically active in the transcription assay described below. Analysis of these complexes by EXAFS studies indicate two short Hg-S bonds at 2.37(2)Å and two longer Hg-S bonds at 2.51(2)Å. These results suggest a distorted four-coordinate geometry, however the degree of heterogeneity in these samples has not yet been determined.

Transcriptional Analysis

Purified MerR protein has been analyzed for activity *in vitro* by a run-off assay in which a DNA fragment containing the overlapping promoter sequences of the divergently-oriented *merR* and *merTPAD* genes of the Tn501 *mer* operon is incubated with E. coli RNA polymerase and ribonucleotides, including ^{32}P-labeled UTP. Transcription products may be visualized after separation by gel electrophoresis and autoradiography. If the MerR protein is not included in the incubation, a transcript of the appropriate size from the *merR* promoter is detected. This signal is greatly reduced when MerR is included in the incubation. No signal corresponding to the *merTPAD* promoter is visible in the absence of Hg(II). However, if both MerR and mercuric chloride are incubated with DNA template and RNA polymerase, a transcript is produced corresponding in size to the expected *merT* message. The transcriptional behavior of the promoter fragments under these conditions directly reproduces the *in vivo* response and demonstrates that no other cellular factors are involved in regulation at the *mer* operon. Activation of the *merTPAD* promoter is possible with nanomolar concentrations of mercuric ion in the presence of a large excess of dithiothreitol and is very specific for mercury. Of ten heavy metals of similar electron configuration tested, only cadmium at levels ten-fold higher than Hg(II) may activate MerR/promoter complexes for transcription by RNA polymerase at *merTPAD*. The 0.5 Hg(II)/monomer MerR complex described above produces similar amounts of *merT* transcript whether or not additional Hg is added to the incubation.

Gel Retardation of Protein-DNA Complexes

Using the gel retardation method described by Fried and Crothers in which the mobility of radioactively end-labeled promoter DNA can be retarded by bound proteins, the apparent equilibrium dissociation constant of MerR has been determined to be 1.4×10^{-10}M.[59] In the presence of mercuric ion, this value is reduced three fold. This technique has also been utilized to determine that the fraction of active protein in binding promoter DNA sequences is 80%.

Footprinting Analysis

Extensive analysis by a variety of footprinting techniques has provided information about the specific molecular interactions between mer promoter sequences, MerR, and RNA polymerase in the presence and absence of Hg(II). The methods used include protection from DNAse I, protection from and interference by methylation by dimethyl sulfate, and high resolution protection from hydroxyl radical. The results indicate that while acting both as a repressor and an activator, MerR contacts its operator sequence primarily along a single face of the DNA. This sequence is an interupted, perfect palindrome corresponding to the region between -35 and -10 of the merTPAD promoter and also to -6 to +22 relative to the merR gene. Suprisingly, no difference in MerR-DNA contacts are apparent in the presence of Hg(II) except for a reduction in affinity of MerR for its operator sequence, as discussed above.

The footprinting reactions described above were carried out on DNA complexes formed in the presence of MerR and/or RNA polymerase with or without Hg(II) and separated by the gel shift method. In this manner, individual complexes can be resolved and analyzed. Moreover, the abundance of each complex may be observed. A small amount of DNA/MerR/RNA polymerase complex is formed in the absence of Hg(II).In the presence of Hg(II), a dramatic increase in the amount of this complex is apparent. Further analysis demonstrates that the mercury-induced complex is trancriptionally competent, while that formed in the absence of Hg(II) is not. High resolution footprinting of these complexes suggests that MerR contacts on the back face of the DNA helix are lost in the presence of Hg(II) when RNA polymerase is also present.

MerR is not an Alternate Sigma Factor

The operator sequences of other activator proteins, such as CAP and Lambda CII, lie upstream of the -35 RNA polymerase recognition sequence in their target promoters. The unusual location of the MerR operator sequence, spanning the -35 to -10 sequences of the RNA polymerase binding site for the merTPAD promoter, suggested that MerR may function as an alternate sigma factor to confer specific promoter recognition to core RNA polymerase. Furthermore, systems which use alternate sigma factors, such as heat shock and sporulation, are stress-induced, as is the mer operon. However, analysis by in vitro transcription in which core RNA polymerase, MerR, and sigma 70, the major template recognition subunit of RNA polymerase were incubated individually and in combination with template DNA determined that transcription of the mer operon is dependent on the presence of sigma 70 in the transcriptional apparatus. This result was confirmed by the analysis of the protein composition from a gel slice of the transcriptionally competent Hg/MerR/DNA/RNA polymerase complex. Both sigma 70 and MerR were found in addition to the $\alpha_2\beta\beta'$ subunits of RNA polymerase.

MerR is a Genetic Switch

Based on our experimental results, the following model is proposed for mer operon regulation (Fig. 1). Footprinting analysis indicates that in the absence of mercuric ion, the MerR protein is bound at its operator site in the mer operon and gel shift analysis shows that RNA polymerase may also associate at low frequency (Fig. 1B). However, the presence of MerR prevents transcription of either merR or merTPAD. In the absence of MerR transcription of the merR gene occurs (Fig. 1A). When the bacterium comes into contact with mercuric ion in the environment and as little as a nanomolar concentration of the metal enters the cell, the mer operon is activated (Fig. 1C). Mercuric ion is bound with high affinity by MerR, resulting in a conformational change that may either alter the structure of promoter DNA or expose a favorable surface for protein-protein interaction with RNA polymerase in such a way as to result in transcriptional activation as determined by in vitro transcription assays. This process appears to require a cooperative interaction between MerR, mercuric ion, RNA polymerase, and DNA, since all components are necessary for a strong formation of this complex in gel shift assay. A system of this nature would permit a rapid response to the presence of mercuric ion, and the scavenging and reduction to Hg(0) carried out by the merTPAD gene products would quickly clear the environment of the toxic substance.

ACKNOWLEDGMENT

We thank Dr. R.R. Burgess (U. of W.I.) for his gift of purified core and holo RNA polymerase, and sigma 70.

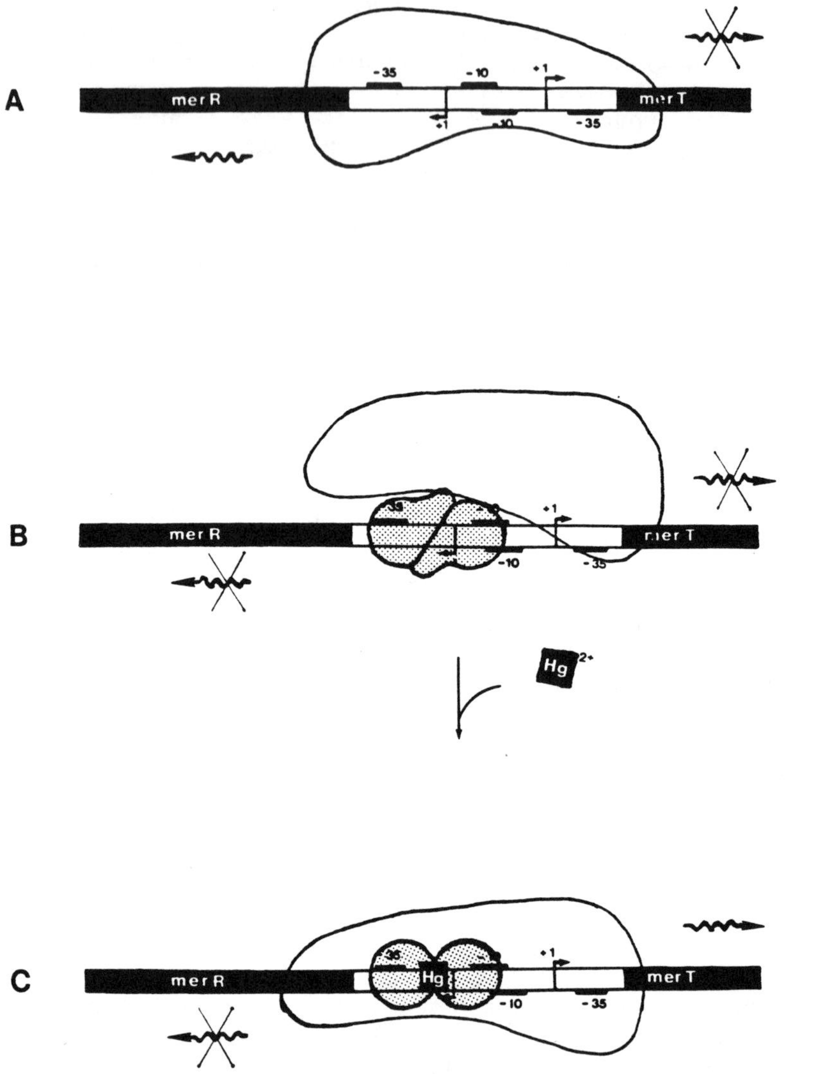

FIGURE 1. Schematic model of <u>mer</u> operon regulation by MerR. Stippled shapes represent the MerR protein, shaded squares represent Hg(II), and the unshaded shape represents E. coli RNA polymerase. Wavy arrows represent transcription except where "x's" have been drawn through them. In the latter cases, little or no transcription is observable.

REFERENCES

1. TJ Foster, M Nakahara, AA Weiss, and S Silver (1979), J.Bacteriol., 140:167.
2. NN Ni'Bhriain, S Silver, TJ Foster (1983), J. Bacteriol., 155:690.
3. TJ Foster, and NL Brown(1985),J. Bacteriol., 163:1153.
4. PJ Barino and AO Summers (1983), Gene, 25:209.
5. T O'Halloran and C Walsh (1987), Science, 235:211.
6. A. Bagg, JB Neilands (1987), Microbiol.Reviews, 51:509.
7. C Seguin and DH Hamer (1987), Science, 235:1383.
8. JL Casey, MW Hentze, DM Koeller, SW Caughman, TA Rouault, RD Klausner, JB Harford(1988),Science,240:924.
9. K White and HN Munro (1988), J. Biol. Chem.,263:8938.
10. TV O'Halloran, in "Metal Ions in Biological Systems", H Sigel, ed., (Marcel Dekker, NY) vol 25, in press.
11. BP Rosen, U Weigel, C Karkaraia, and P Gangola (1988), J.Biol. Chem., 263:3067.
12. S Silver, K Budd, KM Leahy, WB Shaw, D Hammond, RP Novick, GR Willsky, NH Malamy and H Rosenberg (1981) J. Bacteriol., 146:983.
13. S Silver (1985), in "Environmental Inorganic Chemistry", KJ Irgolic and AE Martel, eds., (VCH Publishers, Deerfield Beach, FL) 513.
14. RD Novick and C Roth (1968), J.Bacteriol., 95:1335.
15. H Nakahara, T Ishikawa, Y Sari, I Kondo, H Kozukoe, and S Silver (1977), Appl. Environ. Microbiol., 975.
16. RA van Bogelen, PM Kelley, FC Neidhart (1987), J. Bacteriol., 167:26.
17. S Lindquist (1986), Annu. Rev. Biochem., 55:1151.
18. FC Neidhart, RA van Bogelen and V Vaughn (1985), Annu. Rev. Genet., 295.
19. MF Christman, RW Morgan, FS Jacobson and BN Ames (1985), Cell, 41:753.
20. TJ Tetaz and RKJ Luke (1983), J Bacteriol., 154:1263.
21. D Rouch, J Camakaris, BTO Lee, and RKJ Luke (1985), J. Gen. Microbiol., 131:939.
22. D Rouch, J Camakaris, and BTO Lee (1988), J. Cell. Biochem. Suppl., 120:328 (abstracted).
23. CL Bender, and DA Cooksey (1987), J. Bacteriol.,169:470.
24. CS Moody and HM Hassan(1984),J. Biol.Chem.,259:12,821.
25. HM Hassan and CS Moody (1987),J. Biol. Chem.,262:17,173.
26. V Braun (1985), Trends in Biochem Science, 75.
27. A Bagg and JB Neilands (1987), Biochem., 26:5471.
28. A Bindereif, JB Neilands (1985),J.Bacteriol., 162:1039.
29. SB Calderwood, JJ Mekalanos(1987),JBacteriol.,169;4759.

30. RE Premakumar, M Lemos, and PE Bishop (1984), Biochim. Biophys. Acta, 797:64.
31. RD Joerger, R Premakumar, and PE Bishop (1986), J. Bacteriol, 168:673.
32. MR Jacobsen, R Premakumar, and PE Bishop (1986), J. Bacteriol., 167:480.
33. S Iuchi and ECC Lin (1987), J. Bacteriol., 169.
34. S Iuchi, EEC Lin (1987),Proc.Natl.Acad.Sci.USA,84:3901.
35. LW Stults, WA Sray, and RJ Maier (1986), Arch. Microbiol., 146:280.
36. P Schonheit, J Moll, and RK Thauer (1979), Arch. Microbiol., 123:105.
37. MG Jobling, DA Ritchie (1987),Mol. Gen. Genet., 208:228.
38. AO Summers and GA Jacoby (1977), J. Bacteriol., 129:276.
39. D Taylor and AO Summers (1979), J. Bacteriol., 137:1430.
40. EC Theil (1987), Annu. Rev. Biochem., 56:289.
41. LF Dickey, Y Wang, GE Schol, IA Wortman III, and EC Beal (1988), J. Biol Chem., 263:3071.
42. MW Hentze, TA Rouault, SW Coughman, A Dancis, JB Harford, and RD Klausner (1987). Proc. Natl. Acad. Sci. USA, 84:6730.
43. MW Hentze, SW Coughman, TA Rouault, JG Barriocanal, A Dancis, JB Harford, RD Klausner (1987),Science,238:1570.
44. JL Casey, B Dijeso, K Rao, RD Klausner, and JB Harford (1988), Proc. Natl. Acad. Sci. USA, 85:1787.
45. DO Owen and LC Kuhn (1987), EMBO J., 6:1287.
46. L Guarente, B Lalonde, P Gifford, E Alani (1984), Cell, 36:503.
47. L Guarente and TL Mason (1983), Cell, 32:1279.
48. VJ Dwarki, VNK Francis, GJ Bhat, and G Padmanaban (1987), J. Biol. Chem., 262:16,958.
49. AM Meyers, MD Crivellone, TJ Koerner, and A Tzagoloff (1987), J. Biol. Chem., 262:16,822.
50. DH Hamer (1986), Annu. Rev. Biochem., 55:913.
51. S Merchant and L Bogorad (1987), EMBO J., 6.
52. S Merchant and L Bogorad (1987),J.Biol.Chem., 262:9062.
53. S Merchant and L Bogorad (1986), Mol.Cell.Biol., 6:462.
54. AO Summers, S Silver (1978),Annu.Rev.Microbiol.,32:637.
55. JB Robinson, OH Tuovinen(1978),Microbiol. Reviews,48:95.
56. S Silver and TK Misra, in "Genetic Control of Environmental Pollutants", GS Owens and A Hollander, eds., (Plenum Press, NY) 23.
57. PA Lund, SJ Ford, and NL Brown (1986), J. Gen. Microbiol., 132:465.
58. PA Lund and NL Brown (1987), Gene, 49:579.
59. M Fried and DM Crothers (1981), Nuc. Acids Res., 9:6505.

Metal Ion Homeostasis: Molecular Biology and Chemistry, pages 417–426

PLASMID-DETERMINED HEAVY METAL RESISTANCES: ARSENIC, CADMIUM AND MERCURY[1]

Tapan K. Misra, Giuseppina Nucifora, Lien Chu and Simon Silver

Department of Microbiology and Immunology
University of Illinois, Chicago, IL 60680

ABSTRACT Bacterial resistances to heavy metals are often governed by genes on plasmids and transposons. The arsenic and mercury resistance operons are inducible; the cadA system for Cd^{2+} resistance is constitutive. Bacteria accumulate arsenate via the chromosomally-determined phosphate transport systems and resistant cells then efflux arsenate and arsenite out by an ATPase arsenic pump. The structural genes, arsA, arsB, arsC, for the ars operon of *Escherichia coli* plasmid R773 were identified from DNA sequencing analysis. The ArsA protein has two putative adenylate binding sites homologous to those in the β subunit of the mitochondrial and bacterial oxidative phosphorylation ATPase. The ArsB protein is an integral membrane protein. ArsA and ArsB proteins form a complex with arsenite pumping activity. ArsC protein confers additional arsenate resistance and efflux. The DNA sequence of the cadA cadmium-resistance determinant from *Staphylococcus aureus* plasmid pI258 contains one open reading frame encoding a polypeptide of 727 amino acids with significant amino acid homologies to the sequences of K^+ ATPases of bacteria (and other E1E2 cation-translocating ATPases of microbial and animal cells, such as the yeast H^+ ATPase, the Na^+/K^+ ATPase and the muscle Ca^{2+} ATPase). The mer operon (for mercurial resistance) contains 5 to 7 genes. The first mer structural genes encode proteins with Hg^{2+} transport function. In absence of this function, the cells are Hg^{2+} sensitive. The

[1]This work was supported by grants from the National Science Foundation (DMB86-04781) and National Institutes of Health (AI 24795).

enzyme mercuric reductase (governed by the merA gene) is a flavoprotein that reduces Hg^{2+} to Hg^0. Hg^0 is less toxic and volatile. The mercuric reductase has significant sequence homology with a variety of oxido-reductases of bacterial and eukaryotic sources, including human glutathione reductase. The enzyme organomercurial lyase (governed by the merB gene) breaks the C-Hg bond in organomercurials such as phenylmercury. The merR regulatory genes are significantly different between gram positive and gram negative bacteria.

INTRODUCTION

Metal ions predate life. Cations and anions usually enter into microbial cells through chromosomally encoded specific transport systems. Many toxic ions sneak into the cell using transport systems for normally required nontoxic ions. Examples are arsenate, which enters the cell using phosphate transport machinery; Cd^{2+}, which enters the cell by competing with Mn^{2+} for transport and chromate which enters via the sulfate transport system. Resistance genes for metal ions and organometallic compounds and antibiotics are often encoded in plasmids or transposons. Transport genes for Hg^{2+} are located in the same operon that contains genes involved in the detoxification of Hg^{2+} by reduction to Hg^0. Resistances for Hg^{2+}, and AsO_2^- and AsO_4^{3-} are coded for by inducible operon-like systems in both gram positive and gram negative bacteria. Cells become resistant to toxic ions using three basic mechanisms: by i) restricting the entry of the specific ions, ii) specifically and efficiently deporting unwanted and otherwise harmful ions as soon as they enter into the cell, and iii) modifying chemically or binding to cellular factor(s) resulting in a form that is no longer toxic to the cell. AsO_2^- and AsO_4^{3-} and Cd^{2+} are effluxed out of the cell by specific ATPase driven pumps. Hg^{2+} and organomercurials are chemically modified by enzymes (mercuric reductase and organomercurial lyase) synthesized in resistant cells. These plasmid mediated heavy metal resistance systems have been periodically reviewed by others and ourselves (1,2).

ARSENIC RESISTANCE

Arsenate and arsenite resistances result from reduced

net accumulation of these ions by induced resistant *Staphylococcus aureus* and *Escherichia coli* cells containing the plasmids pI258 and R773 respectively (3). The biochemical basis for the reduced accumulation of arsenate and arsenite ions was determined to be an ATPase efflux system (4,5): An *E. coli* strain lacking the ability to synthesize ATP from respiratory substrates could efflux out arsenate when in media allowing glycolytic ATP synthesis (containing glucose) but not with succinate as the sole carbon source. The efflux of arsenate was blocked in *S. aureus* in presence of protein ionophores at neutral or acidic pH but not at alkaline pH. From these studies it has been suggested that ATP is the driving force for extrusion of arsenicals (4-6).

Genetic, cloning and DNA sequencing studies showed that a 4.3 kilobase pairs (kb) *Hin*dIII fragment contained the *ars* structural genes from plasmid R773 (7). The DNA sequence of the 4.3 kb fragment containing the *ars* operon revealed three open reading frames for genes *arsA*, *arsB* and *arsC* (8). The amino acid sequence of the ArsA protein deduced from the DNA sequence has two regions homologous to the adenylate binding sites of ATP-binding proteins, including the β subunit of the oxidative phosphorylation ATPase of mitochondria and bacteria (8, and Fig. 1). Another feature of the *arsA* gene is internal two-fold symmetry with 46% nucleotide identities and 23% amino acid identities when the

PROTEIN	SEQUENCE					
	REGION I			REGION II		
ArsA	13	FTGKGGVGKTSISCATAIR	31	121	DEFTGLLTDTS	131
ArsA	332	LMGKGGVGKTTMAAAIAVR	350	455	TGHTLLLLDAT	465
Mdr (human)	424	LVGNSGCGKSTTVQL	438	551	RNPKILLLDE	560
Mdr (human)	1066	LVGSSGCGKSTVVQLL	1082	1195	RQPHILLLDE	1204
F1-ATPase β (*E. coli*)	131	LFGGAGVGKTVNMMELIRN	149	218	GRDVLLFVDN	227
Mitochondrial ATPase β (bovine heart)	154	LFGGAGVGKTVFIMELINN	172	248	GQDVLLFIDN	257

FIGURE 1. Sequence homology among the ATP binding regions of some transport ATPases (compiled from refs. 8-12).

sequences are optimally aligned (8 and Fig. 2). The ArsA ATPase polypeptide has been purified and identified as an

```
ATGCAATTCTTACAGAATAT CCCCCCTTATCTGTTTTTTA CGGGTAAAGGAGGCGTGGGT AAAACCTCTATTTCCTGCGC   80
    :  :    :  :::    :         :  :  : :  ::::::::: ::::::::  ::::::  : ::  :    ::
GATGATATTGCCCGTAATGA ACATGGCCTGATTATGCTGA TGGGTAAAGGTGGCGTGGGG AAAACCACGATGGCTGCTGC 1037

CACGGCGATCCGTCTGGCAG AACAGGGGAAACGGGTGCTG CTGGTCAGTACCGATCCGGC CTCAAACGTAGGCCAGGTGT  160
::  ::  :: : ::::: : :   :::       :: :   :::   :   : ::::: ::   :  :  :  ::  :
CATTGCTGTCAGGCTGGCCG ACATGGGATTTGATGTCCAT CTGACAACATCTGATCCTGC GGCGCATCTCAGCATGACCC 1117

TTAGCCAGACGATTGGCAAT ACCATTCAGGCAATAGCCTC TGTTCCTGGATTGTCGGCTC TTGAGATTGATCCTCAGGCT  240
: : :   :   ::  :::              :                                  ::: ::::::::: :
TGAACGGTAGCCTTAACAAC CTGCAGGTCAGCAG------ -------------------- ----GATCGATCCTCACGAG 1167

GCTGCACAACAGTACCGGGC CAGAATCGTTGACCCTATTA AAGGCGTCCTGCCTGATGAC GTTGTTTCCAGCATCAACGA  320
:   :  :::  :: ::         :  ::::  : :     :   :  :::   :: :   :       :    :  : ::
GAAACGGAACGCTATCGTCA GCATGTTCTTGAAACAAAGG GAAAAGAACTGGACGAAGCG GGAAAACGCCTGCTGGAAGA 1247

ACAACTGTCAGGTGCATGCA CAACAGAGATTGCGGCTTTT GATGAATTTACCGGATTACT GACAGATACCTCTCTGCTGA  400
  :  :        : :::: :    :: ::::: : :::   ::   ::: :  :  :  :     ::  :   :   :
GGACTTACGCTCACCTTGCA CTGAGGAAATTGCTGTTTTC CAGGCCTTTTCACGGGTGAT TCGTGAAGCGGGTAAACGTT 1327

CGCGGTTTGACCATATCATT TTTGATACCGCGCCGACGGG TCACACCATTCGCCTTCTGC AGTTGCCGGGCGCCTGGAGT  480
    : :                 ::::: :: ::::: ::  ::::: :    :: :::   :     : ::::: :
TCGTGGTGATG--------- ---GATACGGCTCCGACCGG ACACACGCTGTTGCTGCTGG ATGCTACAGGCGCGTACCAC 1395

AGCTTTATTGACAGCAATCC CGAGGGCGCGTCCTGTCTCG GCCCAATGGCGGGGCTGGAA AAGCAGCGTGAACAGTATGC  560
 ::   ::::  :  ::     :  :       :  : ::   : ::  :  :  ::::     ::
CGCGAAATTGCGAAGAAAAT GGGAGAAAAAGGCCATTTCA CCACACCGATGATGCTGCTT CAG----------------- 1458

CTATGCTGTTGAAGCGTTGT CTGATCCAAAGCGTACCCGA CTGGTATTAGTTGCTCGTCT GCAAAAATCCACGTTGCAGG  640
                       :::::  :::: ::   :  :: :: : ::  : :  :  : :::   :  :  :::  :
-------------------- --GATCCGGAGCGCACTAAA GTGTTACTGGTCACGCTGCC GGAAACCACGCCCGTGCTTG 1516

AAGTCGCCCGGACTCATCTG GAACTTGCCGCCATCGGCCT TAAAAATCAGTACCTGGTTA TTAATGGCGTTCTGCCCAAA  720
:::  ::       ::     :: ::::        ::: : : :      :  :  : ::: : :::  :   ::  :::
AAGCGGCAAATTTACAGGCC GACCTTGAGCGTGCAGGCAT TCACCCCTGGGGCTGGATTA TCAATAACAGCCTTTCCATT 1596

ACTGAAGCGGCAAACGATAC ACTGGCTGCTGCAATATGGG AGCGTGAACAGGAGGCGCTG GCCAATCTTCCCGCTGATCT  800
 :             :::::  :   :  :     :  :   ::: : :::  :::  ::   :
GCG------------GATAC CCGTTCGCCACTGCTTCGTA TGCGGGCACAACAGGAACT TCC----------------- 1646

TGCAGGTTTGCCAACTGACA CATTATTTCTCCAGCCGGTC AATATGGTCGGTGTGTCTG CGCTCAGCAGGCTTCTCTCCA  880
 :::: :        : : : :   :   : :::::::                   :::     :::: :     :
-GCAGATAGAGAGCGTTAAA CGCCAGCACGCCAGCCGTGT CGCTCTTGTTCCCGTACTG GCGTCAGAACCAACCGGCATC 1725

CTCAGCCTGTAGCGTCGCCA TCCTCGGATGAATACCTGCA ACAGCGGCCTGATATTCCG TCACTTTCTGCGTTGGTT 957
  ::  ::  : :  :     :
GACAAACTCAAACAACTTGC TGGGTAA 1752
```

FIGURE 2. Alignment of the first 957 bp of the *arsA* ATPase gene with the remaining 795 bp (modified from ref. 8).

arsenite stimulated ATPase (13). An analogous situation was found with the bacterial ribose transport ATP-binding protein (14) and the mammalian multidrug resistance (Mdr) protein associated with some tumor cells which has become refractory to chemotherapy by pumping out various drugs (9,10,15). Here also, we see internal two-fold duplications of adenylate binding sequences.

The arsB gene is required for both arsenate and arsenite resistances. The arsC gene, however, is required for arsenate resistance but not for arsenite resistance, as confirmed by deletion mapping using nested clones generated for DNA sequencing (8). The ArsA plus ArsB protein complex is the first example of an anion transporting ATPase; has no significant homology with cation transporting ATPases, such as E1E2 and F0F1 ATPases, other than the recognizable ATP binding sequence (Fig. 1).

CADMIUM RESISTANCE

Plasmid-determined Cd^{2+} resistance has been found both in gram-positive bacteria and in gram-negative bacteria. Unlike arsenic resistance (above) or mercury resistance (below) for which there appear to be single plasmid mechanisms common to all bacteria, there are several different systems for cadmium resistance (1). There are two Cd^{2+} resistance determinants in the S. aureus plasmid pI258. A high level of Cd^{2+} resistance is expressed by the cadA locus and a low level resistance is mediated by a genetically distinct locus cadB (16). The basis for Cd^{2+} resistance by the cadA locus is again efflux and this was initially proposed to be chemiosmotic electroneutral Cd^{2+}-for-$2H^+$ exchange system (17). A 3.5 kb DNA fragment containing the cadA locus was cloned and sequenced. B. subtilis cells containing the hybrid plasmid conferred resistance to and reduced uptake of Cd^{2+}. The nucleotide sequence of the cadA DNA fragment contains two open reading frames (ORF). Further subcloning experiments showed that only one of the ORFs, which is 2184 nucleotide pairs long (including the stop codon), is necessary and sufficient to confer resistance to Cd^{2+}. The deduced amino acid sequence of this ORF is highly homologous to those of various E1E2 ATPases of bacterial, lower eukaryotic, and animal origin. The closest homology is found with Streptococcus faecalis K^+ ATPase (Fig. 3) (18). The putative 727 amino acid cadA ATPase has the conserved aspartate residue at position 415. Phosphorylation occurs at this position in other well-studied E1E2 ATPases (19). The seven amino acid residues following the aspartyl residue are highly conserved in all E1E2 ATPases. This class of ATPases has very short membrane spanning regions (three, four, or five pairs of potential hydrophobic alpha-helical structures). The putative Cd^{2+} ATPase could have three pairs of membrane spanning sequences as predicted from computer analysis.

The proline residue (at position 371, see Fig. 3) is highly conserved and in the case of putative Cd^{2+} ATPase it is present in the second pair of membrane-spanning region (not shown). Experiments are in progress to demonstrate directly that the cadA gene product belongs to the family of E1E2 ATPases.

```
        MSEQKVKLMEEEMNVYRVQGFTCANCAGKFEKNVKKIPGVQDAKVNFGASKIDVYGNASV  60

Cd2+    EELEKAGAFENLKVSPEKLANQTIQRVKDDTKAHKEEKTPFYKKHSTLLFATLLIAFGYL 120

Cd2+    SHFVNGEDNLVTSMLFVGSIVIGGYSLFKVGFQNLIRFDFDMKTLMTVAVIGATIIGKWA 180
                                                                  :.
K+                   MELKQKSPAMMTLIAMGITVAYVYSVYSFIANLISPHTHVMDFFWE  47

Cd2+    EASIVVILFAISEALERFSMDRSRQSIRSLMDIAPKEA-LVRRNGQEIIIHVDDIAVGDI 239
         :...::.. ... .:. ..........: .. :...  ....: : .. ....  ::
K+      LATLIVIML-LGHWIEMNAVSNASDALQKLAELLPESVKRLKKDGTEETVSLKEVHEGDR 106

Cd2+    MIVKPGEKIAMDGIIVNGLSAVNQAAITGESVPVSKAVDDEVFAGTLNEEGLIEVKITKY 299
        .::..:.:.. ::.: .: . :...:.:::  :.:.:.:.:..:..:..: ::...:
K+      LIVRAGDKMPTDGTIDKGHTIVDESAVTGESKGVKKQVGDSVIGGSINGDGTIEITVTG- 165

Cd2+    VEDTTITKIIHLVEEAQGERAPAQAFVDKFAKYYTPIIMVIAALVAVVPPLFFGGSWDTW 359
        ... .. :....: .:::::... . . :: ::.   . .:. ...: .. ::...  :.
K+      TGEMVTCKVMEMVRKAQGEQSQLEFLSDKVAKWLFYVALVV-GIIAFIAWLFLANLPDA- 223
                    *                                         PO4
Cd2+    VYQGLAVLVVGCPCALVISTPISIVSAIGNAAKKGVLVKGGVYLEKLGAIKTVAFDKTGT 419
        . . ..:....:: :: ...:. ...... :::.:.:.:.   .:. ...... .:::::
K+      LERMVTVFIIACPHALGLAIPLVVARSTSIAAKNGLLLKNRNAMEQANDLDVIMLDKTGT 283

Cd2+    LTKGVPVVTDFEVLNDQVEEKELFSIITALEYRSQHPLASAIMKKAEQDNIPYSNVQVEE 479
        ::.:  .::..:.:... .:.:... :.::: ...:::: .::.  ....:  .  :..:
K+      LTQGKFTVTGIEILDEAYQEEEILKYIGALEAHANHPLAIGIMNYLKEKKI--TPYQAQE 341

Cd2+    FTSITGRGIKGIVNGTTYYIGSPKLFKELNVSDFSLGFE-NNVKILQNQGKTAMIIGTEK 538
         . ..: :....:....        :...: ... ..: .. ...:  ..::.:. .. ...
K+      QKMLAGVGLEATVEDKD-----VKIINEKEAKRLGLKIDPERLKNYEAQGNTVSFLVVSD 396

Cd2+    TILGVIAVADEVRETSKNVIQKLHQLGIKQTIMLTGDNQGTANAIGTHVGVSDIQSELMP 598
        ...:::..: .. ..:. :: ... .:  ..::::::. .:.:.....:...   ..:.:
K+      KLVAVIALGDVIKPEAKEFIQAIKEKNI-IPVMLTGDNPQAAQAVAEYLGINEYYGGLLP 455

Cd2+    QDKLDYIKKMQSEYDNVAMIGDGVNDAPALAASTVGIAMGGAGTDTAIETADIALMGDDL 658
        .:: . ...   .. ..: :.:::.::::.:: .:.:.:. :::::.::..:::..: ..:
K+      DDKEAIVQRYLDQGKKVIMVGDGINDAPSLARATIGMAI-GAGTDIAIDSADVVLTNSDP 514

Cd2+    SKLPFAVRLSRKTLNIIKANITFAIGIKIIALLLVIPGWLTLWIAILSDMGATILVALNS 718
        ...    . :...: . . .:. .. : .:::. :. .: :.    :::   ..:..:..
K+      KDILHFLELAKETRRKMIQNLWWGAGYNIIAIPLA-AGILAPIGLILSPAVGAVLMSLST 573

Cd2+    LRLMRVKDK 727
        .
K+      VVVALNALTLK 583
```

Figure 3. Comparison of cadA ATPase with the K^+ ATPase from S. faecalis. (:) identities, (•) conservative replacements. PO_4, presumed site of phosphorylation, conserved proline (from Nucifora et al., in preparation) (*).

MERCURY RESISTANCE

The genetics and molecular biology of mercurial resistances have been studied in both gram-positive and -negative bacteria (for reviews see refs 1,2,20,21). The best studied mercurial resistance systems in gram-negative bacteria are those with the resistance factor R100 and the transposon Tn501. The complete nucleotide sequence of the mer operon of R100 and Tn501 are known. These two operons are functionally very similar, so also are their DNA sequences. These operons have a regulatory gene, merR, followed by an operator/promoter region and genes (merT and merP) that code for proteins that transport Hg^{2+} into the cells. The next common gene is the merA, for mercuric reductase that reduces Hg^{2+} to less toxic and volatile Hg^{0}. The merR gene product is a trans-acting inducer-repressor. In absence of Hg^{2+} it works as a repressor but in the presence of Hg^{2+} it acts as an activator. The regulation of mer operon has been a topic of recent interest among various groups (1,2,20-22).

R100	MENNLENLTIGVFAKAAGVNVETIRFYQRKGLLREPDKPYGSIRRYGEAD	50
Tn501	*********************************L****************	
pDU1358	**K******************************P****************	
R100	VVRVKFVKSAQRLGFSLDEIAELLRLDDGTHCEEASSLAEHKLKDVREKM	100
Tn501	*T**R*********************E***********************	
pDU1358	*T**R**************************************Q******	
R100	ADLARMETVLSELVCACHARKGNVSCPLIASLQGEAGLARSAMP	144
Tn501	********A***********R*************G*S**G****	
pDU1358	T*************F*****Q**************KEPRGADAV	

FIGURE 4. Alignment of amino acid sequences of MerR from plasmids of gram negative bacteria. *, residues identical to those of R100 (ref 23 and Nucifora et al., in preparation).

The mer operon of the Serratia plasmid pDU1358 has two mer determinants (narrow spectrum and broad spectrum) (22). The merB in the broad spectrum mer determinant codes for the organomercurial lyase enzyme that breaks the C-Hg bond producing Hg^{2+} (22). The broad spectrum mer determinant thus confers resistance to a wider range of organomercurials including phenylmercuric acetate and the operon is also inducible with phenylmercury. Recently, we have determined the nucleotide sequence of the merR regulatory gene of pDU1358. A comparison of the deduced amino acid sequence of the broad spectrum MerR with those of the narrow spectrum MerR of R100 and Tn501 (Fig. 4) shows the striking difference in the amino acids at the carboxy terminus. Cysteine at position 115 is missing in pDU1358 MerR. Other three cysteines at positions 82, 117 and 126 are conserved in the three MerR sequences. These cysteines might be the target amino acids for Hg^{2+} binding. Deletion of 54 nucleotides from the 3′ end of the broad spectrum merR gene made the system inducible with Hg^{2+} but not with phenylmercuric acetate. The conclusion that the C-terminal region of the pDU1358 MerR protein is responsible for responses to phenylmercury is drawn from deletion, subcloning and complementation studies (a detailed report will be published elsewhere; Nucifora et al., in preparation).

REFERENCES

1. Silver S, Misra TK (1988). Plasmid-mediated heavy metal resistances. Annu Rev Microbiol 42:717.
2. Walsh CT, DiStefano MD, Moore MJ, Shewchuk LM, Verdine GL (1988). Molecular basis of bacterial resistance to organomercurial and inorganic mercuric salts. FASEB J 2:124.
3. Silver S, Budd K, Leahy KM, Shaw WV, Hammond D, Novick RP, Willsky GR, Malamy MH, Rosenberg H (1981). Inducible plasmid-determined resistance to arsenate, arsenite and antimony(III) in Escherichia coli and Staphylococcus aureus. J Bacteriol 146:983.
4. Mobley HLT, Rosen BP (1982). Energetics of plasmid-mediated arsenate resistance in Escherichia coli. Proc Natl Acad Sci USA 79:6119.
5. Silver S, Keach D (1982). Energy-dependent arsenate efflux: the mechanism of plasmid-mediated resistance. Proc Natl Acad Sci USA, 79:6114.
6. Rosen BP, Borbolla MG (1984). A plasmid-encoded arsenite pump produces arsenite resistance in Escherichia coli. Biochem Biophys Res Commun 124:760.

7. Mobley HLT, Chen CM, Silver S, Rosen, BP (1983). Cloning and expression of R-factor mediated arsenate resistance in Escherichia coli. Molec Gen Genet 191: 421.

8. Chen CM, Misra TK, Silver S, Rosen BP (1986). Nucleotide sequence of the structural genes for an anion pump. The plasmid encoded arsenical resistance operon. J Biol Chem 261:15030.

9. Gros P, Croop J, Housman, D (1986). Mammalian multidrug resistance gene: complete cDNA sequence indicates strong homology to bacterial transport proteins. Cell 47:371.

10. Gerlach JH, Endicott JA, Juranka PF, Henderson G, Sarangi F, Deuchars KL, Ling V (1986). Homology between P-glycoprotein and a bacterial haemolysin transport protein suggests a model for multidrug resistance. Nature 324:485.

11. Kanazawa H, Kayano T, Mabuchi K, Futai M (1981). Nucleotide sequence of the genes coding for α, β and subunits of the proton-translocating ATPase of Escherichia coli. Biochem Biophys Res Commun 103:604.

12. Runswick MJ, Walker JE (1983). The amino acid sequence of the β subunit of ATP synthase from bovine heart mitochondria. J Biol Chem 258:3081.

13. Rosen BP, Weigel U, Karkaria C, Gangola P (1988). Molecular characterization of an anion pump. The arsA gene product is an arsenite (antimonate)-stimulated ATPase. J Biol Chem 263:3067.

14. Buckel SD, Bell AW, Rao JKM, Hermodson MA (1986). An analysis of the structure of the product of the rbsA gene of Escherichia coli K12. J Biol Chem 261:7659.

15. Choi K, Chen CJ, Kriegler M, Roninson IB (1988). An altered pattern of cross-resistance in multi-drug-resistant human cells results from spontaneous mutations in the mdr1 (P-glycoprotein) gene. Cell 53:519.

16. Smith K, Novick RP (1972). Genetic studies on plasmid-linked cadmium resistance in Staphylococcus aureus. J Bacteriol 112:761.

17. Tynecka Z, Gos Z, Zajac J (1981). Energy-dependent efflux of cadmium coded by a plasmid resistance determinant in Staphlylococcus aureus. J Bacteriol 147:313.

18. Solioz M, Mathews S, Furst P (1987). Cloning of the K^+-ATPase of Streptococcus faecalis. J Biol Chem 262:7358.

19. Walderhaug MO, Dosch DC, Epstein W (1987). Potassium transport in bacteria. In Rosen BP, Silver S (eds): "Ion Transport in Bacteria", San Diego: Academic Press, p 84.
20. Foster TJ (1987). Genetics and biochemistry of mercury resistance. CRC Crit Rev Microbiol 15:117.
21. Summers AO (1986). Organization, expression and evolution of genes for mercury resistance. Annu Rev Microbiol 40:607.
22. Griffin H, Foster TJ, Silver S, Misra TK (1987). Cloning and DNA sequence analysis of the mercuric and organomercurial resistance determinants of plasmid pDU1358. Proc Natl Acad Sci USA 84:3112.
23. Misra TK, Brown NL, Fritzinger DC, Pridmore RD, Barnes WM, Haberstroh L, Silver S (1984). Mercuric ion-resistance operon of plasmid R100 and transposon Tn501: The beginning of the operon including the regulatory region and the first two structural genes. Proc Natl Acad Sci USA 81:5975.

Metal Ion Homeostasis: Molecular Biology and Chemistry, pages 427–438

IN VIVO FOOTPRINTING OF MerR BINDING TO THE *mer* REGULATORY REGION[1]

Andreas Heltzel, Paul Totis, and Anne O. Summers

Department of Microbiology, University of Georgia, Athens, Georgia 30602

ABSTRACT The *mer* regulatory region includes the *in vitro* binding site (a symmetrical sequence with 7-bp inverted repeats separated by 4-bp) governing two divergently oriented promoters. Using *in vivo* DNA methylation of uninduced and mercury-induced cells, we have shown that in the presence of MerR and regardless of mercury being present, five out of eight guanine residues in the symmetrical dyad are protected. Additional mercury-independent *in vivo* interactions at the promoter for the structural genes which include two adenine residues hypersensitive to dimethyl sulfate attack and one protected guanine in the -35 region on the template strand suggest RNA polymerase contacts. All interactions are absent in a *merR* deletion mutant; thus, MerR binds to the same general region as a repressor or as an activator for the expression of the structural genes.

INTRODUCTION

In R100 (Tn*21*), the expression of genes involved in mercury resistance (Fig. 1) is regulated by the *trans*-acting function of the *merR* gene, a DNA-binding protein (1,2) which acts as a repressor in

---

[1]This work was supported by GM28211 from NIH.

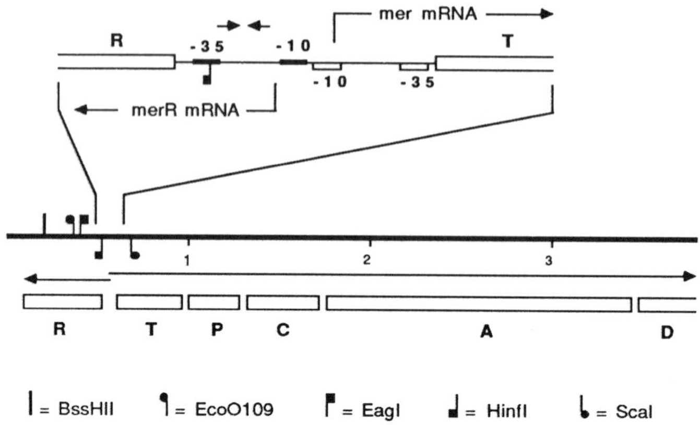

FIGURE 1. Map of the *mer* operon from Tn*21*.

the absence of and as an activator in the presence of Hg(II) (Fig. 2) (3,4,5). MerR also negatively autoregulates its own synthesis (Fig. 2) (6, A. Heltzel, unpublished results) in an Hg(II)-independent manner.

The *mer* regulatory region consists of two divergently oriented promoters (7), P_{TPCAD} for the expression of the structural genes and P_R for *merR* expression (Fig. 1). Using an *in vitro* MerR-*merO/P* gel electrophoresis binding assay, we previously established that MerR binds specifically to the *mer* regulatory region (Fig. 3) (1). Other laboratories (2, G. Verdine *et al.*, pers. comm.; T. O'Halloran *et al.*, this volume) using DNase, hydroxyl radical and *in vitro* DMS protection experiments have shown that, in the absence of Hg(II), purified MerR protein binds to the region of dyad symmetry located between the -10 and -35 sites of the structural gene promoter. However, in the presence of Hg(II), *in vitro* protection experiments (2, T. O'Halloran *et al.*, this volume) as well as gel electrophoresis binding assays (Fig. 3) (1) did not indicate a shift in relative occupancy or dissociation of MerR from the *merO/P* region. In

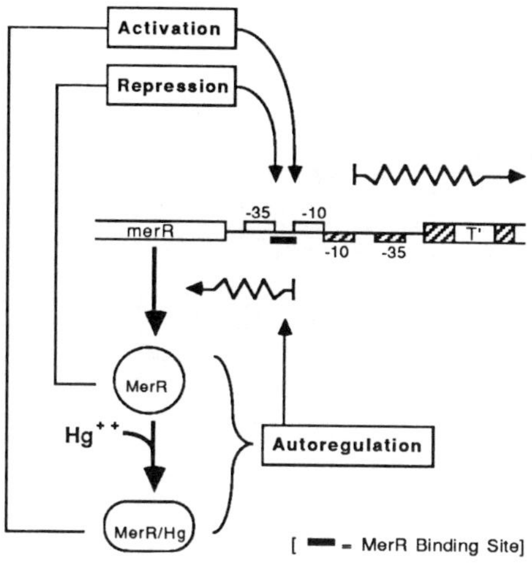

FIGURE 2. MerR regulation of the *mer* operon.

addition, *in vitro* methylation protection at the structural gene promoter using purified MerR and RNA polymerase indicated that RNA polymerase contacts at the promoter region are Hg(II)-dependent (O'Halloran *et al.*, this volume).

Our work extends the *in vitro* results by using *in vivo* footprinting at biologically relevant mercury concentrations. Methylation of *mer* promoter DNA inside the cell reveals MerR and possible RNA polymerase contacts in both the absence and the presence of Hg(II). Absence of all contacts in a *merR* deletion mutant confirms that all *in vivo* interactions at the promoter are MerR-dependent.

METHODS

Strains and the construction of plasmids used in this paper have been described in detail elsewhere (1). A *merR* deletion mutant was constructed by deleting *merR* leftward from *Eag*I (Fig. 1). MerR-DNA binding assays using crude lysates have been described (1).

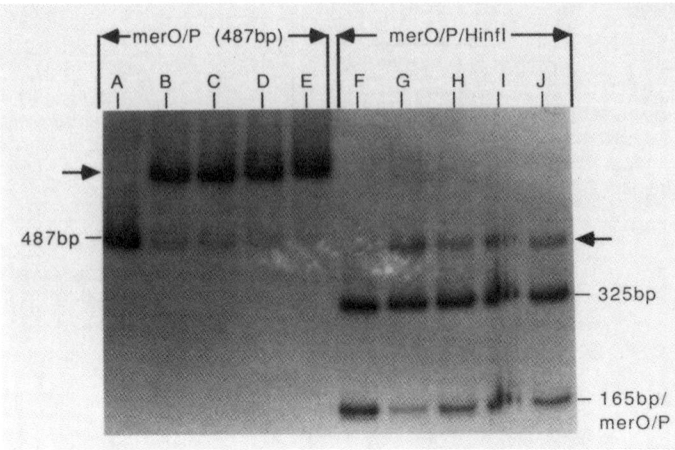

FIGURE 3. Gel retardation assay: A/F = DNA controls with 10uM $HgCl_2$; E/G = MerR-lysate controls without $HgCl_2$; lanes B and H, C and I, and D and J contain 10, 1, and 0.1uM $HgCl_2$, respectively.

In vivo footprinting using partial methylation by dimethyl sulfate (DMS) was done on cells containing *merR* and *merO/P* on a multicopy plasmid and *merTPCAD* in *trans* on a monocopy plasmid. Cells were grown in rich or minimal glucose media to mid-logarithmic phase, concentrated 100-fold, and methylated for 30 seconds at 37°C using 0.5% DMS.

For Hg(II)-induced cells, $HgCl_2$ was added for specified time intervals (0, 45, 90, and 180 seconds with 5uM $HgCl_2$) or at indicated concentrations (0.001uM to 10uM $HgCl_2$ for 90 seconds) before methylation. After cell lysis (lysozyme/ Triton X-100), plasmid DNA was 5' or 3' end-labeled at the *Eco*O109 site, cleaved at methylated bases using the piperidine reaction, and analyzed on sequencing gels (9).

RESULTS

Methylation protection and enhancement at

purine positions can reflect interactions of DNA-binding proteins with target DNA. Reaction of DMS with duplex DNA results in preferential methylation at the N-7 position of guanine in the major groove and at a lower rate at the N-3 position of adenine in the minor groove (10).

In both the absence or presence of Hg(II) we observe the same methylation patterns, e.g. MerR protects at least five G residues (-17, -18, -29, -30, -31) in the symmetrical region between the -10 and -35 hexamers of P_{TPCAD} (Fig. 4). Strong G protections are observed at -18 and -29 whereas weaker protection patterns are found for G's at positions -17, -30, and -31. The five *in vivo* contacts are symmetrically oriented on a planar helix map (Fig. 6) indicating that binding to the region of inverted repeats occurs in a similar manner with contact sites in adjacent major grooves. In the -35 region of P_{TPCAD}, an additional protected G (-33) is observed which is offset slightly from the helix face defined by G contacts in the region of dyad symmetry. A slight loss of protection of these G's occurs in the presence of biologically relevant Hg(II) concentrations (Fig. 4). Two hyperreactive A residues are observed in the -35 region of the template strand which remain reactive with or without Hg(II) (Fig. 4). One enhanced A is located with the inverted repeats on the same helical face (Fig. 6).

In order to determine whether *in vivo* promoter contacts were MerR-specific, a *merR* deletion mutant was constructed containing the *merO/P* region on an *Eag*I-*Sca*I fragment (Fig. 1). *In vivo* methylation analyses of this mutant confirms that differences in reactivities of purine residues are MerR-dependent (Fig. 5, bottom strand only). G protections and enhancement of A's are lost if *merR* is not present in *cis;* however, the original *in*

FIGURE 4. *In vivo* dimethyl sulfate methylation patterns at *merO/P* (DMS lanes = uninduced): (**A**) induction using 5uM $HgCl_2$ for specified time intervals (45, 90, and 180 seconds); (**B**) 90 seconds induction times using different $HgCl_2$ concentrations (0.001 to 10uM $HgCl_2$). (See page 432.)

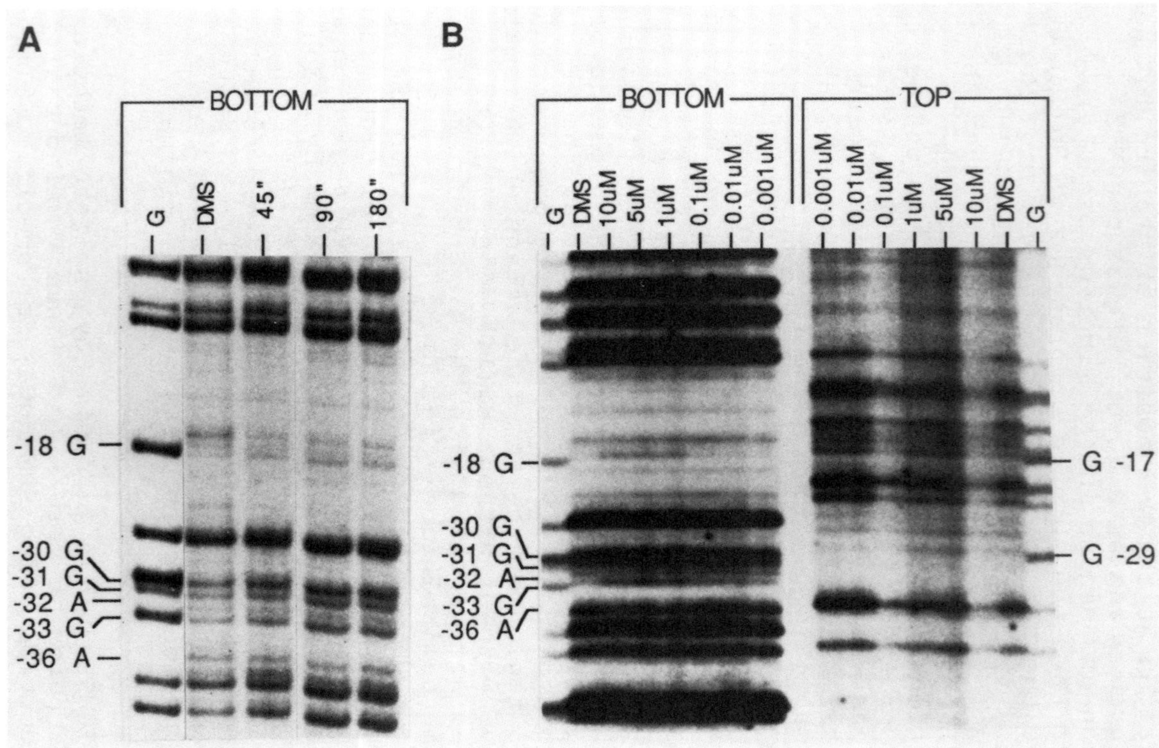

FIGURE 4. (See page 431 for legend.)

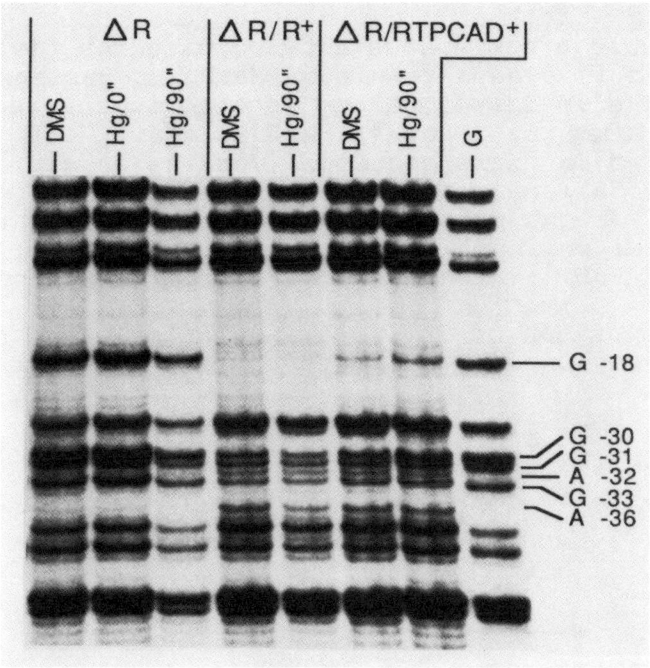

FIGURE 5. *In vivo* dimethyl sulfate methylation pattern using a *merR* deletion mutant and 5uM $HgCl_2$ (bottom strand only).

vivo methylation pattern (Fig. 4: *merR* present in *cis*) is partly restored if *merR* is provided in *trans* on a multicopy plasmid carrying *merR* alone or the entire *mer* operon. Comparing the *in vivo* promoter contacts with *merR* in *cis* (Fig. 4) or *merR* in *trans* to the *merR* deletion mutant (Fig. 5: R^-/R^+; $R^-/RTPCAD^+$), the hyperreactivities of A's are identical. However, the intensity of G reactivities differ slightly. These differences may arise from lower MerR occupancy of the target *merO/P* region (on the same replicon as the *merR* deletion mutant) due to competition from additional *merO/P* copies in *trans*.

DISCUSSION

Figure 6 summarizes altered base reactivities to DMS *in vivo* as a result of MerR interaction at P_{TPCAD}. The *in vivo* contacts located in the region of dyad symmetry (Fig. 4) include nearly all documented *in vitro* contacts (Verdine *et al.*, pers. comm.; O'Halloran *et al.*, this volume), regardless of Hg(II) being present. However, *in vivo,* three additional purines in the -35 region (Fig. 4: G -33; A's -36, -32) are reactive with or without Hg(II).

Comparison of *in vivo* methylation patterns at the -35 region of P_{TPCAD} with *in vivo* interactions

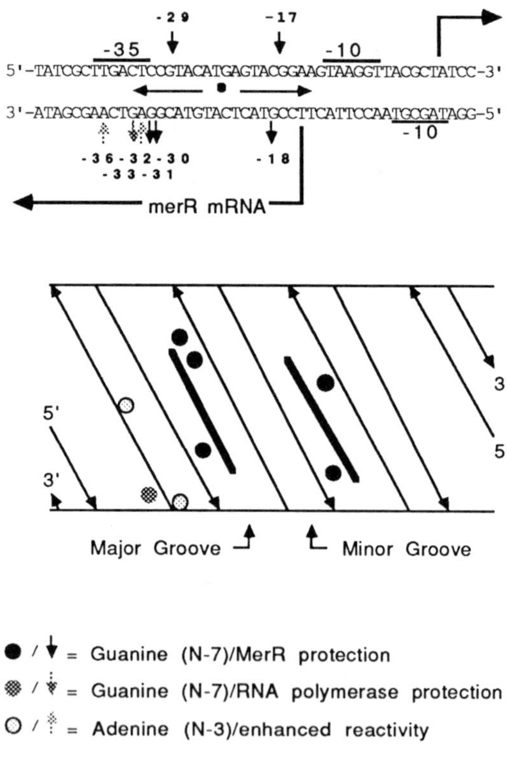

FIGURE 6. Summary of *in vivo* interactions at the *mer* regulatory region.

of RNA polymerase at the *lac* UV5 promoter (11) suggests that at least two purines in P_{TPCAD} (G -33 and A -36) are possible sites of *in vivo* RNA polymerase interaction with the *mer* promoter. Since DNA supercoiling has been proposed to strengthen promoter recognition by RNA polymerase (11,12), the *in vivo* reactivity of purines in the -35 region of P_{TPCAD} arises from bound RNA polymerase. Thus, *in vivo*, RNA polymerase binding is dependent on MerR but independent of Hg(II). Interestingly, O'Halloran *et al.* (this volume) do not detect protection or reactivity in these three purines *in vitro* unless Hg(II) is provided along with RNA polymerase and MerR. This discrepancy which may arise from technical differences (e.g. linear DNA *in vitro* vs. supercoiled DNA *in vivo*) will be resolved by genetic studies in progress.

In *E. coli* promoters that use sigma-70, the -35 and -10 regions and the length of the spacer separating them are important for efficient transcription initiation (8). The -35 and -10 regions of P_{TPCAD} have complete sequence homology at positions identical with the most highly conserved bases of the *E. coli* consensus promoter (7,8); however, the spacing is 19-bp rather than the ordinary spacing of 17+/-1 base pairs (7,8). Studies with the *lac* p^s promoter indicate that shortening the spacer region to a length of 17-bp increases the rate of open complex formation by RNA polymerase *in vitro* (13). Lund *et al.* (pers. comm.) shortened the spacing between the -10 and -35 sites of P_{TPCAD} by a single base pair deletion of one of the two Cs at positions -30 or -31. The promoter strength of this mutant is approximately 100-fold higher than the wild-type P_{TPCAD}. We hypothesize that due to the length of the spacer region, in the absence of MerR, RNA polymerase cannot efficiently contact both -10 and -35 regions of P_{TPCAD} simultaneously and thus cannot efficiently form an open complex (13).

Our current model for regulation of the structural genes (Fig. 7) involves MerR binding to the same operator site in the absence or presence of Hg(II). Without Hg(II), MerR binding appears to increase the affinity of RNA polymerase for the -35 site of P_{TPCAD}. RNA polymerase binding at this

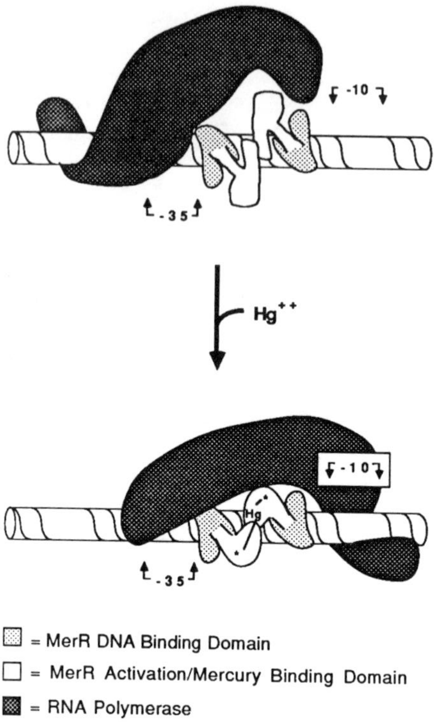

FIGURE 7. Proposed model for transcriptional repression and activation at the structural gene promoter.

region may be mediated by MerR-dependent changes in the DNA helix which would allow favorable RNA polymerase-DNA contacts; however, open complex formation would be prevented by the repressor conformation of MerR. In order to stimulate transcription, an Hg(II)-induced conformational change in the activation domain of MerR could position bound RNA polymerase to allow favorable contacts with the -35 and -10 regions for efficient open complex formation. We are currently investigating *cis*- as well as *trans*-acting mutants which are defective in induction/activation functions to elucidate the apparent cooperative interaction of MerR and RNA polymerase at the *mer* regulatory region.

REFERENCES

1. Heltzel A, Gambill D, Jackson WJ, Totis PA, Summers AO (1987). Overexpression and DNA-Binding Properties of the *mer*-Encoded Regulatory Protein from Plasmid NR1(Tn*21*). J Bacteriol 169:3379.
2. O'Halloran TV, Walsh C (1987). Metalloregulatory DNA-Binding Protein Encoded by the *merR* Gene: Isolation and Characterization. Science 235:293.
3. Foster TJ, Nakahara M, Weiss AA, Silver S (1979). Transposon A-Generated Mutations in the Mercuric Resistance Genes of Plasmid R100-1. J Bacteriol 140:167.
4. Misra TK, Brown NL, Fritzinger DC, Pridmore RD, Barner WM, Haberstroh L, Silver S (1984). Mercuric Ion Resistance Operons of Plasmid R100 and Transposon Tn*501*: the Beginning of the Operon Including the Regulatory Region and the First Two Structural Genes. Proc Natl Acad Sci USA 81:5975.
5. Barrineau PJ, Summers AO (1983). A Second Positive Regulatory Function in the *mer* (Mercury Resistance) Operon. Gene 25:209.
6. Foster TJ, Brown NL,(1985). Identification of the *merR* Gene of R100 by Using *mer-lac* gene and Operon Fusions. J Bacteriol 163:1153.
7. Summers AO (1986). Organization, Expression, and Evolution of Genes for Mercury Resistance. Annu Rev Microbiol 40:607.
8. McClure WR (1985). Mechanism and Control of Transcription Initiation in Prokaryotes. Annu Rev Biochem 54:171.
9. Maxam AM, Gilbert W. (1980). Sequencing End-Labeled DNA with Base-Specific Chemical Cleavages. In Grossman L, Moldave K (eds): "Methods in Enzymology: Nucleic Acids Part 1," New York: Academic Press, p 499.
10. Lawley PD, Brookes P (1963). Further Studies on the Alkylation of Nucleic Acids and their Constituent Nucleotides. Biochem J 89:127.

11. Borowiec JA, Gralla JD (1986). High Resolution of *Lac* Transcription Complexes inside Cells. Biochemistry 25:505.
12. Borowiec JA, Gralla JD (1987). All Three Elements of the *lac* p^s Promoter Mediate its Transcriptional Response to DNA Supercoiling. J Mol Biol 195:89.
13. Stefano JE, Gralla JD (1982). Spacer Mutations in the *lac* p^s Promoter. Proc Natl Acad Sci USA 79:1069.

Metal Ion Homeostasis: Molecular Biology and Chemistry, pages 439–446

GENETIC AND MOLECULAR BASIS OF COPPER RESISTANCE IN *ESCHERICHIA COLI*[1]

D. Rouch, B.T.O. Lee, and J. Camakaris

Department of Genetics, University of Melbourne, Parkville, Victoria 3052 Australia

ABSTRACT The study of copper resistance mechanisms has been facilitated by the availability of the plasmid, pRJ1004, which confers an inducible copper resistance in *E. coli*. Two determinants contribute to the resisance (1) *cdr*, which encodes DNA repair functions and (2) *pco*, which codes for an inducible mechanism which results in decreased intracellular Cu accumulation. The *pco* determinant has been cloned and contains at least four genes (*pco*A, B, C and R). The *pco*C gene product is an inducible 26 Kd Cu-binding protein. The *pco*R gene has a *trans*-acting regulatory role. The roles of *pco*A and *pco*B are not yet known. Expression of *pco* genes is controlled largely at the transcription level. At least two chromosomal gene products are essential for the functions of *pco* encoded Cu resistance. Kinetic analysis has shown that the reduced Cu accumulation is due to an energy dependent enhanced efflux of Cu. We propose that the mechanism of Cu resistance conferred by *pco* involves intracellular Cu binding, exclusion by enhanced efflux and chemical modification such that the excluded Cu is not recognised by chromosome encoded Cu uptake systems.

INTRODUCTION

Copper is an essential trace element which is also toxic in its free ionic form. Detoxifying mechanisms should be inducible in order to avoid interference with normal copper

---

[1]This work was supported by grants from the Australian Research Grants Scheme and the Murdoch Institute.

homeostasis. Thus copper as an essential trace element poses different problems to the purely toxic heavy metals such as cadmium and mercury (1).

Escherichia coli K-12 exhibits decreased survival at copper concentrations greater than 4 mM. The presence of the plasmid pRJ1004 extends the range of copper concentration cells can tolerate to 22 mM, and the copper resistance is inducible (2).

The present work describes the genetic and molecular characterisation of copper resistance encoded by pRJ1004.

RESULTS

Genetic Analysis of pRJ1004 and Cloning of the Major Copper-resistance Determinant

Tn 10 transposon insertion mutagenesis (3) of pRJ1004 defined two classes of mutants (i) where Cu resistance was markedly reduced, (ii) where Cu resistance was only slightly reduced. This provided evidence for two Cu-resistance determinants. The major Cu-resistance determinant was designated *pco* and the minor Cu-resistance determinant *cdr*. The latter encodes functions associated with repair of copper damage to DNA.

pco was cloned into pBR322 (ppA87), and into a low copy number vector pMU530 (ppA173). See Fig. 1. An inverse relationship was found between the copy number of the plasmid

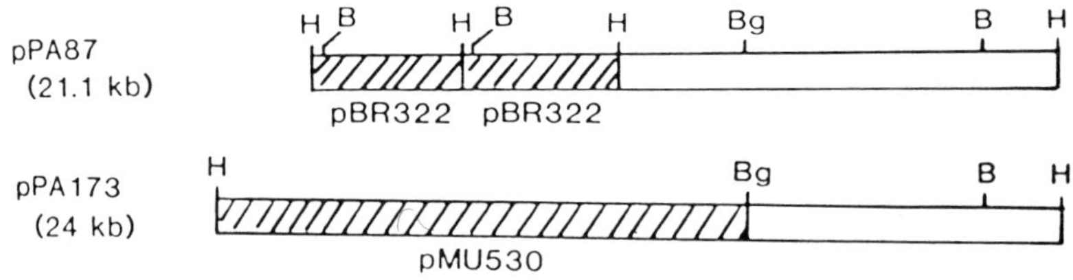

FIGURE 1. Structures of *pco* clones. Vector DNA's are shaded, insert DNA's are unshaded. Some restriction sites are shown (B, BamH1; Bg,Bgl II; H, Hind III).

and the level of Cu-resistance produced by the pco inserts. In order to estimate the size of pco, vectors containing pco inserts were subjected to transposon insertion mutagenesis using Tn 1000 (4). Tn 1000 insertions over a 5.8 kb region inactivated Cu-resistance, thus defining the pco region.

Complementation analysis was performed between Tn 1000 derivatives of ppA87 and ppA173 which have lost the Cu-resistance phenotype (Fig. 2). These studies defined four genes, specified pcoA, pcoB, pcoC and pcoR. All four genes are required intact in order to achieve expression of the Cu-resistance phenotype.

Physiological Studies on pco

The presence of either pRJ1004 or ppA173 (contains pco insert) results in a three-fold reduction in cellular accumulation of ^{64}Cu (Fig. 3). Therefore pco is responsible for the reduced ^{64}Cu accumulation, and reduced Cu accumulation is an important component of the resistance mechanism.

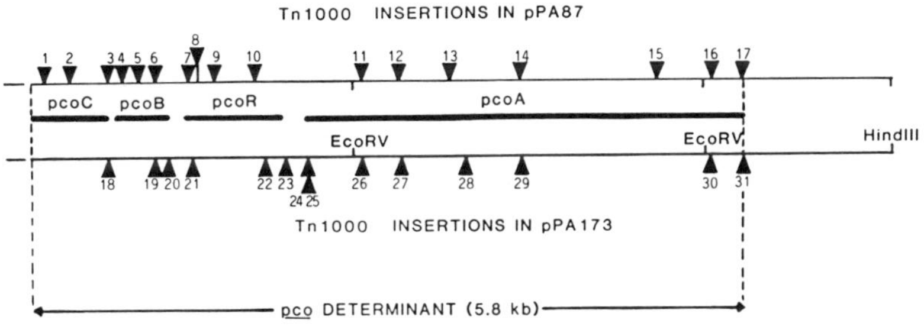

FIGURE 2. Tn 1000 transposon insertions used in complementation analysis of the pco determinant. Insertions are indicated by shaded triangle symbols.

Reduced ^{64}Cu accumulation may result from reduced uptake or enhanced efflux. Studies on initial (10 second) rates of uptake of ^{64}Cu into parental E. coli cells revealed the presence of a saturable Cu uptake system with an apparent

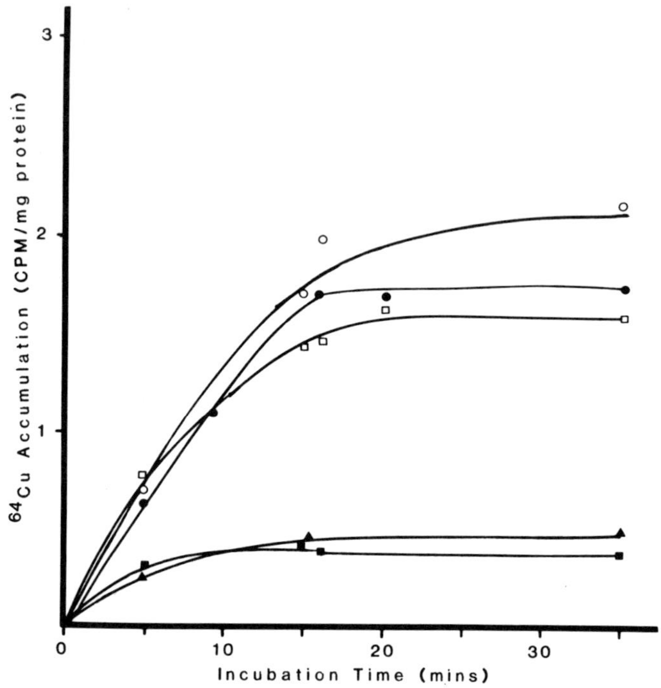

FIGURE 3. Effect of pRJ1004 and ppA173 on ^{64}Cu accumulation. O, parental strain; ●, +pRJ1004; □, parental strain (induced); ■, +pRJ1004 induced; ▲, +ppA173 (induced).

Km of 7 μM and an apparent Vmax of 2.5 μ moles/minute/gram dry weight. These kinetic parameters were not altered by the presence of pRJ1004 or plasmids containing pco inserts (± preinduction). Therefore reduced Cu accumulation due to pco is not a result of decreased rate of uptake, and the most likely explanation is enhanced efflux of Cu. In the presence of arsenate or the uncA mutation (5) there was an increase in ^{64}Cu accumulation by resistant strains. This shows that the resistance mechanism is dependent on ATP or a related high energy compound. Furthermore the resistance was not expressed in two strains bearing mutations in chromosomal Cu transport genes (6), indicating that the products of these genes are necessary for expression of *pco* encoded resistance.

Nature of *pco* Gene Products

^{35}S methionine labelled proteins encoded by *pco* plasmids were analysed using mini-cells (7). Only one inducible ^{35}S labelled protein was detected. This protein had a molecular weight of 26 Kd and was not detectable in uninduced cells or in cells containing vector without the *pco* insert.

Expression of the 26 Kd protein was examined in Tn 1000 insertion mutants of each of *pcoA*, *pcoB*, *pcoC* and *pcoR*. This protein was not produced in *pcoC* mutants but was produced in the other mutants (Fig. 4). (Low expression in the *pcoC6* mutant suggests insertion in the promoter region of the gene.) This suggests that it is the product of the *pcoC* gene.

Copper-binding proteins were analysed by incubating whole cells in the presence of ^{64}Cu. 34,000 g. supernatants were fractionated by gel filtration on Sephadex G-75. Copper-induced cells bearing pRJ1004 had two resolvable Cu-binding peaks - one eluting at 11 Kd and the other at 26 Kd. However the 11 Kd peak was also present in Cu-induced parental (plasmidless) cells. No resolvable (from void and total volume peaks) peaks were present in supernatants from uninduced cells. These data suggest that the *pcoC* gene product is an inducible 26 Kd Cu binding protein, whilst the inducible 11 Kd protein is encoded by a chromosomal gene.

Studies on Regulation of *pco* Expression

Both the transcription inhibitor rifampicin and the translation inhibitor chloramphenicol inhibited synthesis of copper-inducible ^{35}S methionine labelled proteins in mini-cells containing *pco*. This suggests that control of *pco* expression is primarily at the transcription level. This was confirmed by mRNA studies where, following induction of *pco* containing strains, there was a significant increase in a number of mRNA's hybridisable to a nick-translated *pco* insert. These mRNA's were not present in parental strains (not containing *pco*).

Since expression of *pco* is subject to transcriptional control, promoter-*lac* fusions were constructed in order to

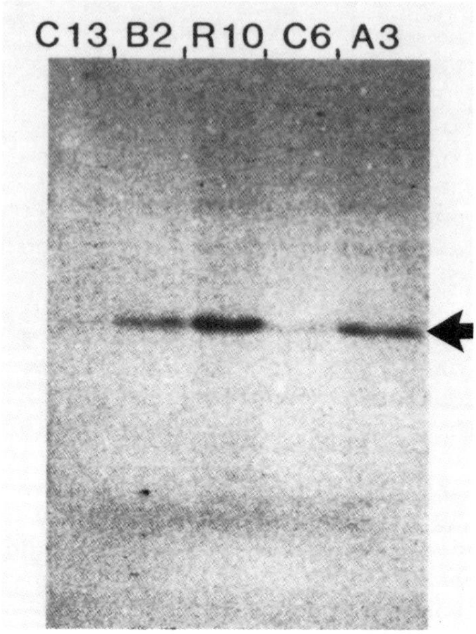

FIGURE 4. Expression of the 26 Kd protein by _pco_ mutants. Mini-cells containing derivatives of the _pco_ clone ppA173 with insertions in each of the four _pco_ genes were induced with 0.8 mM Cu and labelled with ^{35}S methionine. Mini-cell extracts were then electrophoresed through a 17.5% SDS-polyacrylamide gel and autoradiographed. Arrowhead indicates position of the 26 Kd protein. C13, _pcoC_ insertion; B2, _pcoB_ insertion; R10, _pcoR_ insertion; C6, _pcoC_ insertion; A3, _pcoA_ insertion.

investigate regulatory elements. Fragments prepared from restriction digests of ppA87 were inserted adjacent to _lac_ structural genes in a vector with no endogenous _lac_ promoter. Ligated DNA's were transformed into a _recA_ strain containing ppA87 in order to provide control by the putative _trans_-acting regulatory element on _pco_. Copper-inducible _lac_ gene expression was detected on _X-gal_ plates and quantified using β-galactosidase assays. These studies in combination with transposon insertional inactivation of regulatory genes revealed the presence of two trans-acting regulatory genes which repress expression from the promoter-lac fusions - pcoR (approximately 0.7 kb in a complementation group previously defined) and cutR which maps to 97 minutes on the E. coli chromosome.

DISCUSSION

Using complementation analysis four genes have been identified which are present on the major Cu-resistance determinant, pco. These genes pcoA, pcoB, pcoC and pcoR are all required for expression of Cu-resistance. The pcoR gene codes for a trans-acting repressor. It is not required for Cu induction as the pcoC gene product is expressed in mini-cells which contain an insertionally inactivated pcoR gene. The requirement for pcoR in expression of the Cu-resistance phenotype suggests that it may code for a structural component involved in the resistance mechanism. The cutR regulatory gene provides interaction with normal copper transport.

Our current model for the pco encoded resistance mechanism is that excess Cu is sequestered by the inducible 26 Kd protein (pcoC gene product). This protein then presents Cu to an energised efflux/modification system. Evidence for modification comes from the fact that Cu is observed to be deposited as an insoluble complex surrounding Cu-resistant colonies. The efflux modification system is probably encoded by pcoB, pcoA and at least two chromosomal genes (which code for components in normal Cu transport).

Cooksey (8) has recently characterised a plasmid which confers Cu resistance in Pseudomonas syringae; but that system exhibits a number of differences to the pco system.

Further work is directed towards sequencing pco, characterising pco gene products and mechanisms of resistance, and elucidating the metal-responsive regulatory elements which "link" the inducible plasmid mediated copper resistance system with normal copper transport in E. coli.

REFERENCES

1. Brown NL (1985). Bacterial resistance to mercury-reductio ad absurdum? Trends in Biochemical Sciences 10:400.
2. Rouch D, Camakaris J, Lee BTO, Luke RJK (1985).

Inducible plasmid-mediated copper resistance in Escherichia coli. J Gen Microbiol 131:939.

3. Kleckner N, Barker DF, Ross DG, Botstein D (1978). Properties of the translocatable tetracycline resistance element Tn 10 in Escherichia coli and bacteriophage lambda. Genetics 90:427.
4. Guyer MS (1978). The $\gamma\delta$ sequence of F is an insertion sequence. J Mol Biol 126:347.
5. Mobley HLT, Rosen BP (1982). Energetics of plasmid-mediated arsenate resistance in Escherichia coli. Proc Natl Acad Sci (USA) 79:6119.
6. Rouch D, Camakaris J, Lee BTO (1988). Copper transport in Escherichia coli. In Winge D, Hamer D (eds): "Metal Ion Homeostasis: Molecular Biology and Chemistry," New York: Alan R. Liss, p __.
7. Jackson WJ, Summers AO (1982). Biochemical characterisation of HgCl2 - inducible polypeptides encoded by the mer operon of plasmid R100. J Bacteriol 151:962.
8. Mellano MA, Cooksey DA (1988). Nucleotide sequence and organisation of copper resistance genes from Pseudomonas syringae pv. tomato. J Bacteriol (in press).

V. PHYSIOLOGY AND DISEASES OF UNBALANCED METAL ION HOMEOSTASIS

Metal Ion Homeostasis: Molecular Biology and Chemistry, pages 449–458

A POSSIBLE ROLE FOR INTRACELLULAR CU-BINDING PROTEINS IN CU-UPTAKE

F. Palida, G. Waldrop and M. Ettinger

Department of Biochemistry, SUNY/Buffalo, Buffalo, New York 14214

ABSTRACT Copper uptake by hepatocytes and fibroblasts was compared to determine the biochemical basis of rapid, preferential Cu-uptake by the liver. Initial rates of Cu-uptake from His_2Cu are significantly higher with hepatocytes. Net Cu-accumulation at equilibrium is also higher with hepatocytes when albumin is in the extracellular medium. Fibroblasts from the brindled mouse mutant model of Menkes disease accumulate excess copper. Normal fibroblasts with comparably elevated metallothionein (MT) levels take up copper normally. Thus, the data with both normal and mutant cells are consistent with a role for intracellular proteins other than MT in determining net Cu-uptake.

A fundamental question of metal biochemistry is: How is a specific metal ultimately incorporated into a specific metalloprotein within a specific cell type? Implicit in this question are several related questions. How is the metal transported into the cell and within the cell? What factors are required for metal incorporation, e.g. is a specific enzyme or metal-transfer factor involved? Also, how does a cell regulate the amount of metal that it contains? Since the transport systems which mediate metal-uptake may also mediate metal-efflux, cells must have mechanisms for trapping a metal. Related to a trapping function may be a buffer function which protects a cell against potentially toxic effects of a metal. Specific cell types may also have mechanisms for regulating a balance between uptake, storage and secretion or excretion. A key question about intracellular metal metabolism is: How is the distribution of the metal among various possible end-products regulated? Another regulation aspect is whether or not a metal regulates the synthesis of a metalloprotein, and little is known about co-translational versus post-translational metal incorporation.

The above questions suggest that several specific factors may be involved in the cellular biochemistry of a specific metal. For copper, the fact that there are at least three, distinctly different inherited diseases indicates that there are at least three essential proteins required for normal Cu-metabolism. Wilson's disease is characterized by impaired incorporation of copper into ceruloplasmin and defective secretion of copper into bile (1,2). Copper accumulates in the liver and eventually spills into blood. The clinical manifestations of the disease are hepatic and neurologic Cu-toxicity. Menkes disease is a fatal, X-linked, in-born error of Cu-metabolism which is characterized by a variety of connective tissue disturbances and impaired development of the brain (3,4). The connective tissue abnormalities have been attributed to decreased activity of the Cu-enzyme, lysyl oxidase; the biochemical basis of the effect on the brain remains unknown. Another X-linked disease of Cu-metabolism affecting connective tissue is Ehlers-Danlos type lX (5). Finally, Rauch has described an inherited disease of Cu-metabolism unique to mice (6,7). The toxic-milk syndrome is fatal to all off-spring of affected mothers due to lack of copper in the milk. The important point about these diseases is that they indicate that there are at least three, intracellular proteins that are essential for cellular Cu-metabolism since none of the defects seem to involve defects in plasma membrane transport systems. The proteins associated with these diseases are expected to have one or more of the functions discussed above.

Our work has focused on Cu-transport, the proteins that copper binds to when it first enters cells, and the biochemical basis of Menkes disease. Rather unexpectedly, all three subjects seem to converge on the possible role of intracellular proteins in determining the net amount of copper that a cell takes up and retains.

A striking fact about Cu-transport is that the liver has a unique ability to rapidly and preferentially sequester a large fraction of the copper that is absorbed from the gastrointestinal tract. This is easily demonstrated by the kinetics of the disappearance of copper from plasma following an intravenous injection, and the concurrent uptake in the liver. Copper leaves the plasma with a half-life of ~10 min and approximately 85% of the administered dose is taken up by the liver (8). While this fact has been known for more than thirty years, the biochemical basis of rapid, preferential uptake by the liver is unknown. In principle, both kinetic and thermodynamic contributions are possible. Hepatocytes may have an especially efficient Cu-transport system which transports copper more rapidly than the Cu-transport systems of other cells, and/or hepatocytes may contain one or more high affinity Cu-ligands which are especially affective at trapping copper within hepatocytes.

We began our Cu-transport studies by characterizing the kinetics of initial rates of ^{64}Cu-uptake and ^{64}Cu-efflux by isolated rat hepatocytes from serum-free media (9,10,11). The data were consistent with a single transport system which facilitated both Cu-uptake and Cu-efflux. There was no evidence for uptake by simple diffusion. The apparent Km for Cu-uptake was 11 μM. Subcellular fractionation studies indicated that net, intracellular uptake was being measured as opposed to surface binding, and atomic absorption analyses confirmed net uptake rather than isotopic exchange. Energetically, Cu-uptake and Cu-efflux are both passive (9,10); and independent of transmembrane potential (9). Experiments in which Na^+ in the incubation media was replaced by choline demonstrated that Cu-transport was also Na^+-independent (12). Similar Km values for Cu-uptake have been reported for mouse hepatocytes (13), lymphoblasts (14), and the hypothalamus (15). Cu-transport is strictly passive in each of these systems.

The Cu-binding ligands of plasma must be considered because how the liver mobilizes copper from specific Cu-ligands may be fundamental to the basis of rapid, preferential Cu-uptake. While the exact distribution of copper in plasma has not been unambiguously determined, the basic picture seems clear. Most (65-90%) of plasma copper is tightly bound to ceruloplasmin. While a Cu-transport function has been proposed for ceruloplasmin, it is clearly not the species which transports copper from the gastrointestinal tract to the liver; copper is incorporated into ceruloplasmin within the liver. It has been known for years that albumin is the principal species which transports copper to the liver, and the half-life of albumin-Cu is the same as the half-life of the non-ceruloplasmin pool of plasma (16,17). Recently, a high molecular weight protein ("transcuprein") was reported which bound about 26 % of a radioactive dose at a time when albumin had 69% of the dose (18). The possible role of this protein in Cu-transport remains to be determined. Recently, Gordon and Cousins suggested that it may represent an artifact based on experiments in which it was not detected (19). About 10% of plasma copper is coordinated to amino acids (20). Histidine as the His_2Cu complex has a Kd $\sim 10^{18}$ (21) which is comparable to the Cu-affinity of albumin (KD $\sim 10^{17}$ (22)). Lau and Sarkar also showed that a ternary, albumin-Cu-His complex (KD $\sim 10^{23}$ (22)) was also possible. Most importantly, Sass-Kortsak and co-workers demonstrated that several amino acids, especially histidine facilitated Cu-uptake by several cell types when the incubation media contained serum or albumin (23,24).

The kinetic effect of albumin on Cu-transport by hepatocytes is to markedly inhibit initial rates of uptake (12). The data are consistent with a mechanism in which albumin complexes extracellular copper thereby decreasing the effective concentration for uptake. Histidine as the His_2Cu complex inhibits Cu-transport slightly relative to the transport of free Cu(II) (12). However, in the presence of albumin, histidine facilitates Cu-uptake, i.e. histidine (as the His_2Cu complex which would form in plasma under the conditions of the experiments) increased the rate of albumin-inhibited (12) Cu-uptake. The effect was histidine specific (12), and the facilitating effect of the entire low molecular weight fraction of rat plasma could be accounted for by its histidine content (12).

Based on the above results, a model was proposed for how the liver rapidly sequesters copper from plasma (12). According to this model, neither albumin-Cu nor albumin-His-Cu is directly available for Cu-transport. As suggested by Lau and Sarkar, histidine may react with the albumin-Cu-His complex to form the His_2Cu complex (25), or this complex may be in equilibrium with the albumin-Cu species. It was proposed that the His_2Cu complex interacts directly with the putative, plasma membrane Cu-transport protein on hepatocytes (12). Experiments in which the doubly-labeled $^{3}His_2{}^{64}Cu$ was used showed that copper was transferred as the free ion rather than as the intact His_2Cu complex (12). This suggests that a rapid ligand displacement reaction occurs with the Cu-transport protein. The question remained: Does this model have any bearing on rapid, preferential uptake by the liver?

Fibroblasts were chosen recently as an example of an extra hepatic cell type because of their bearing on Cu-metabolism and Menkes disease as discussed below. The kinetics of free Cu(II)-transport (initial rates) by fibroblasts were virtually identical to the kinetics of free Cu(II)-transport by hepatocytes as indicated by the kinetic parameters: Km 6.9 + 0.2M; Vmax 2.6 + 0.1 nmol/min/mg protein. Moreover, Cu-uptake is strictly passive as with hepatocytes. The marked inhibitory effect of albumin on the initial rates of Cu-uptake by fibroblasts was also the same as was observed with hepatocytes.

One striking difference that is observed between Cu-transport by hepatocytes and fibroblasts is in the kinetics of Cu-uptake from the His_2Cu complex. As noted above, in the absence of albumin, histidine has only moderate effects on Cu-transport by hepatocytes. However, histidine markedly inhibits Cu-transport by fibroblasts; namely Vmax is inhibited 5-fold by His_2 Cu relative to the Vmax for free Cu(II). The significance of this property is clearly demonstrated by the effect of histidine on albumin-inhibited Cu-uptake. Under conditions in which histidine doubles Cu-uptake by hepatocytes, histidine further inhibits Cu-uptake by fibroblasts by a factor of two. This can be attributed to the kinetic difference in the rate of displacing copper from the His_2Cu complex as reflected by the effect on Vmax. Since the kinetic parameters for free Cu(II) transport are similar, fibroblasts and hepatocytes may contain iso-transport proteins which differ specifically in the rate of the requisite ligand-displacement reaction.

The question remains: are there intracellular Cu-ligands unique to the liver (or in high concentration in the liver) which help the liver sequester Cu-preferentially? Another striking difference between Cu-uptake by fibroblasts and hepatocytes is the effect that extracellular albumin has on net Cu-accumulation as steady-state or equilibrium is approached. Although albumin markedly inhibited the initial rate of Cu-uptake by hepatocytes (5 μM ^{64}Cu (II) + 5 μM albumin), by 4h, net ^{64}Cu-accumulation was comparable with or without albumin. In marked contrast, albumin markedly limited net Cu-accumulation by fibroblasts under the same experimental conditions, i.e. the net copper taken up at 4h in the presence of 5 μM albumin was about 10% that of net Cu-uptake without albumin. Similar results were also recently reported by McArdle, et al.

The differential effects of albumin on net

Cu-accumulation by hepatocytes and fibroblasts are consistent with an equilibrium involving extracellular and intracellular copper with extracellular albumin and intracellular Cu-binding proteins. The fact that Cu-transport is mediated in both directions, passively allows for such an equilibrium. According to this interpretation, hepatocytes are predicted to contain one or more Cu-binding proteins which are either liver-specific or are in higher concentrations in hepatocytes than in other cell types as exemplified by fibroblasts. Sephadex G150 and Superose-12 profiles of the distibution of ^{64}Cu among cytosolic proteins from hepatocytes, fibroblasts or whole kidney reveal that hepatocytes contain a 38 kd Cu-binding protein fraction which is low or absent in fibroblasts or kidney. This Cu-binding component is not Cu/Zn-superoxide dismutase which elutes after the Cu-binding fraction. Finally, implicit in the above interpretation is the concept that intracellular proteins can determine how cells equilibrate with extracellular copper.

To examine the possibility that intracellular proteins may influence net Cu-uptake, it is useful to consider Menkes disease. The brindled mouse is an excellent animal model of Menkes disease (4,26). In Menkes patients and brindled mice, some cell types are high in copper (e.g. fibroblasts, kidney) while other cell types are low in copper (e.g. liver and brain). The simplest assumption is that both cell types express the primary defect; differences in Cu-levels would reflect differences in Cu-metabolism by the cell types involved (13). It should be noted that the kinetics of Cu-transport, per se, both uptake and efflux, are normal in both hepatocytes (13) and lymphoblasts (14).

The excess copper in Menkes cell types with elevated copper is bound to metallothionein (MT) which is also elevated. However, the primary defect does not seem to be related to MT because no MT genes are found on the X-chromosome (27), and the MT levels are consistent with the elevated Cu-levels (28). Fibroblasts from Menkes patients or the brindled mice take up excess copper when they are incubated with copper for 24h in serum-containing media, i.e. under conditions in which Cu-uptake by normal fibroblasts is limited by albumin in the extracellular medium (29). The question we posed was: Is this excess Cu-accumulation due to the elevated MT-levels found in these cells, i.e. is MT an example of an intracellular Cu-binding protein which, at least in part, determines how much copper is accumulated by fibroblasts? To address this question, MT was induced in normal cells by incubating confluent cultures with 400 μM Cu(II) or 100 μM Zn(II) for 48h (the maximal concentrations which were not toxic) or with dexamethasone. Normal, brindled, and normal fibroblasts with elevated MT were incubated for 24h under conditions in which the Menkes phenotype of excess Cu-accumulation is observed. Radioimmune assays (Dr. J. Garvey, Syracuse University) confirmed that the MT-levels in each cell type remained constant throughout the incubation period. The Menkes phenotype was NOT exhibited by the normal fibroblasts which contained comparable MT-levels to the Menkes fibroblasts (3-5 fold elevated), i.e. net Cu-accumulation by these cells was similar to net Cu-accumulation by normal fibroblasts with normal levels of MT. These results indicate that excess Cu-accumulation by Menkes or brindled fibroblasts is associated with the primary defect rather than the secondary increase in MT-levels. The results imply that one function of the protein associated with the primary defect is to determine how much copper is taken up by fibroblasts, and by inference, the other cell types with elevated Cu-levels in Menkes disease.

In an attempt to isolate and identify the protein associated with the primary defect of Menkes disease and the liver-specific Cu-binding protein, hepatocyte and kidney cytosols were applied to FPLC columns. Briefly, starting-buffer equilibrated cytosols were applied to a Mono-Q column (Pharmacia), and eluted with a linear salt gradient. At least 15-protein fractions were clearly detected by A_{280} absorbance. The pooled protein fractions were either re-chromatographed on the anion-exchange column or applied directly to a Superose-12 column. Individual fractions were then applied to SDS-PAGE-gels (reducing) which were Ag-stained. Several differences were detected between the samples from normal and brindled mice, three of which were reproducible. An ~80 kd band was reproducibly elevated in the brindled mouse liver and kidney samples. As suggested by Leone and Hamer, this may represent heat-shock protein, HSP-84 (30). A ~55 kd component was consistently higher in normal mice than the brindled mice liver cytosol (~5 times higher). This component was not detected in the kidney. Thus, this difference apparently represents a liver-specific, secondary response. The best candidate for a protein associated with the primary defect or a protein which is secondarily affected in both those cells which are high, i.e. kidney and those which are low in Cu, i.e. liver was a 48 kd component which was consistently higher (about 10X) in the control than brindled samples. This component eluted at about 0.25 M salt from the Mono-Q column and at about 64 Kd on the Superose column. Work is in progress to purify this component to homogeneity.

Any working hypothesis of the biochemical basis of Menkes disease must be consistent with the following facts: (1) Some cell types are high in copper while others are low in copper, (2) the primary defect is probably expressed in both cell types as further evidenced by the similarities in the major differences detected in specific components with liver and kidney samples (i.e. the 48 and 80 kd components), (3) one function of the protein associated with the primary defect is to determine how much copper is taken up by fibroblasts and other cell types, (4) the protein associated with the primary defect also seems to be required to deliver copper to sites of Cu-incorporation into Cu-proteins as evidenced by the low lysyl oxidase activity from fibroblasts (31) and low plasma ceruloplasmin activity from hepatocytes (3) and, (5) the liver has unique activities in Cu-metabolism (1). A working hypothesis which is consistent with these facts can be suggested. In fibroblasts, low levels of the affected protein cause excess copper to enter. Excess accumulation is known to reflect impaired efflux (29). Since the excess copper that enters can neither be utilized efficiently nor secreted, MT is induced by the elevated Cu-levels. In contrast to fibroblasts or kidney cells, hepatocytes have alternative mechanisms for secreting copper e.g. Cu-secretion into bile or efflux is perhaps mediated by the 38 kd-cytosolic Cu-binding fraction that is higher in hepatocytes. According to this hypothesis, the liver is low in copper due to low activity of the affected protein in delivering copper to cellular sites where it can be bound irreversibly. It should be noted that decreased Cu-accumulation by hepatocytes has been detected with isolated cells in culture and therefore cannot be attributed to a secondary effect of other cell types having elevated copper (13). Obviously, this is a speculative model which must be assessed by experiment.

REFERENCES

1. Ettinger MJ (1984). In Lontie R (Ed): "Copper Proteins II." Cleveland: CRC, p. 175.
2. Owen CA Jr. (1981). "Wilson's Disease", Park Ridge,

NJ, Noyes.
3. Danks DM (1977). Inorg. Persp. Biol. Med. 1: 73.
4. Ettinger MJ (1987). Life Chem. Rep. 5: 169.
5. Kuivaniemi H, Peltonen L, Kivirikko KI (1985). Am. J. Hum. Genet. 37: 798.
6. Rauch HJ (1980). Gen. 94: 84.
7. Rauch HJ (1983). Heredity 74: 141.
8. Sass-Kortsak A (1965). Adv. Clin. Chem. 8: 1.
9. Schmitt RC, Darwish HM, Cheney JC, Ettinger MJ (1983). Am. J. Physiol. 244: G183.
10. Darwish HM, Schmitt RC, Cheney JC, Ettinger MJ (1984). Am. J. Physiol. 246: G48.
11. Ettinger MJ, Darwish HM, Schmitt RC (1986). Fed. Proc. 45: 2800.
12. Darwish HM, Cheney JC, Schmitt RC, Ettinger MJ (1984). Am. J. Physiol. 246: G72.
13. Darwish HM, Hoke JE, Ettinger MJ (1983). J. Biol. Chem. 258: 13621.
14. Herd SM, Camakaris J, Christofferson R, Wookey P, Danks DM (1987). Biochem. J. 247: 341.
15. Hartter DE, Barnea A (1988). J. Biol.Chem. 263: 799.
16. Bearn AG, Kunkel HG (1954). Proc. Soc. Exp. Med. 85: 44.
17. Marceau N, Aspin N, Sass-Kortsak A (1970). Am. J. Physiol. 218: 377.
18. Weiss KC, Linder MC (1985). Am. J. Physiol. 249: E77.
19. Gordon DT, Leinart AS, Cousins RJ (1987). Am. J. Physiol. 252: E327.
20. Sarkar B, Kruck TPA (1966). In Peisach J, Aisen P, Blumberg WE (Eds): "The Biochemistry of Copper." Academic Press, p. 183.
21. Freeman HC, Martin R-P (1969). J. Biol. Chem. 244: 4823.
22. Lau S-J, Sarkar B (1971) J. Biol. Chem. 246: 5938.
23. Harris DIM, Sass-Kortsak A (1967). J. Clin. Invest. 46: 659.
24. Silverberg M, Neumann PZ, Rotenberg AD (1968). Birth Defects Orig. Artic. Ser. 4: 8.
25. Lau S-J, Sarkar B (1975). Can. J. Chem. 53: 710.
26. Hunt DM (1974). Nature (Lond.) 249: 852.
27. Schmidt CJ, Hamer DH, McBride OW (1984). Science 224: 1104.
28. Sone T, Yamaoka K, Minami Y, Tsundo H (1987). J. Biol. Chem. 262: 5878.
29. Beratis NG, Price P, LaBadie G, Hirschhorn K

(1978). Pediatr. Res. 12: 699.

30. Leone A, Pavlakis GN, Hamer DH (1985). Cell 40: 301.
31. Royce PM, Camakaris J, Mann JR, Danks DM (1982). Biochem. J. 202: 369.

Metal Ion Homeostasis: Molecular Biology and Chemistry, pages 459–468

BACTERIAL AND FUNGAL HIGH-AFFINITY TRANSPORT SYSTEMS SPECIFIC FOR MANGANESE[1]

Georg Auling

Institute of Microbiology, University of Hanover,

D-3000 Hannover 1, Schneiderberg 50, FRG

ABSTRACT Manganese deficiency is a prerequisite for both the nucleotide fermentation with *Brevibacterium ammoniagenes* and the production of citric acid with *Aspergillus niger*. In *B. ammoniagenes* manganese is essential for the proper functioning of the ribonucleotide reduction, the central step in DNA precursor biosynthesis. Although specific transport systems for manganese exist in both microorganisms nothing is known about the carriers involved.

INTRODUCTION

In prokaryots manganese transport systems have been described for *Escherichia coli*, a few *Bacillus* species, *Rhodobacter capsulatus* (*Rhodopseudomonas capsulata*), and *Lactobacillus plantarum* (1). However, there is either no growth requirement demonstrated for the metal or the molecular basis for the manganese requirement in these organisms is not entirely clear.

Coryneform bacteria and certain fungi require manganese for growth (2,3). This requirement is exploited in industrial processes for microbial production of nucleotides (as flavour-enhancing substances) and citric acid (4,5). However limited research has been done on the question why only manganese de-

---

[1]This work was supported in part by grants from the Deutsche Forschungsgemeinschaft, the Joh. A. Benckiser GmbH, and the BMFT (Schwerpunkt:Bioprozeßtechnik)

ficient cells are productive. Inhibition of DNA formation and cell division are the most obvious effects of manganese deficiency in B. ammoniagenes. We could trace back these effects to malfunctioning of the ribonucleotide reduction both in-vivo (6) and in-vitro (7). Inhibition of this key reaction for DNA precursor biosynthesis by addition of the ribonucleotide reductase inhibitor hydroxyurea (HU) allows nucleotide production in the presence of surplus manganese (8). Manganese dependence of ribonucleotide reduction was shown to exist in other genera of Gram-positive bacteria as well (9). The manganese character of the B. ammoniagenes enzyme was recently confirmed after purification to homogeneity (10).

In order to examine a possible involvement of manganese control in DNA formation of A. niger the methods which have proven successfully for the elucidation of the primary target of manganese deficiency in B. ammoniagenes are currently applied to this filamentous fungus.

As manganese can be taken up either as a low-affinity substrate by the magnesium transport system or via a specific manganese transport system with high-affinity we recently concentrated on uptake studies at submicromolar concentrations of ^{54}Mn in both industrial organisms, B. ammoniagenes and A. niger.

RESULTS AND DISCUSSION

Manganese Control of DNA Formation in A. niger

When A. niger was grown in the mineral medium of Shu and Johnson (11) with 14 % sucrose purified by passage through a cation exchange column the short, thick hyphae indicative of manganese deficiency were observed. Addition of 0.2 uM manganese were sufficient for balanced growth recognizable by the normal filamentous morphology (Fig.1). Determination of the nucleic acid synthesis rates by in-vivo labeling with radioactive adenine using the Schmidt and Thannhauser method (12) indicated a reduced DNA formation of these mycelia (more than 50 % reduction at 24 h after inoculation with spores). Addition of 0.2 μM manganese to these deficient mycelia restored their DNA formation activity even above the level of the manganese sufficient control. Synthesis of RNA was not affected in the same mycelia by varying the concentrations of the manganese ions (13).

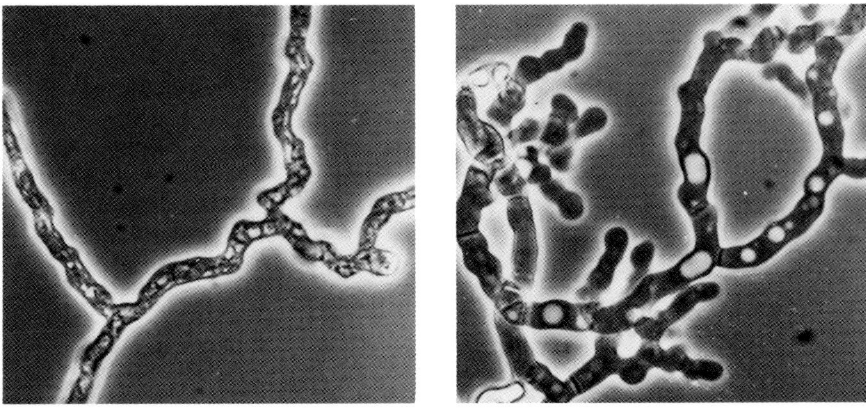

Fig. 1 Differential morphology of *A. niger* ATCC 11414 in the presence of 0.2 µM manganese (left) or in the absence of manganese (right) after 24 h growth in the mineral medium of Shu and Johnson (11), final magnification : 1200-fold.

An increased protein degradation due to a possible instability of ribosomes, giving rise to the high intracellular ammonium ion concentrations (14) necessary to counteract citrate inhibition of phosphofructokinase is responsible for citric acid production according to Kubicek and Röhr (15). Polyamines are generally assumed to stabilize ribosomes (16). Upon quantitative determination of polyamines by high performance liquid chromatography (17) during growth of *A. niger* in the presence or absence of 2 µM manganese no qualitative or quantitative differences were found. However, the sensitivity of DNA formation in *A. niger* towards HU indicated the presence of the Mn- or Fe-type of ribonucelotide reduction in this organism (18).

As the Mn-type of ribonucleotide reductase may be present in filamentous fungi as well and as the only fungal ribonucleotide reductase assayed so far (19) is not a B_{12}-enzyme (H.Hogenkamp, pers. comm.) as previously reported we are currently engaged to establish an in-vitro assay for ribonucleotide reduction in *A. niger*.

High-Affinity Manganese Uptake by *B. ammoniagenes*

The basic characterization of the manganese transport system in *B. ammoniagenes* was carried out in diluted tryptone broth (20) to be able to compare our kinetic data with those obtained for other microbial manganese transport systems in the same medium (Table 1). Upon addition of 0.4 to 3.7 kBq/ml ^{54}Mn to 2.1 x 10^8 mid-log cells/ml a 4000-fold higher intracellular concentration of the cation was observable after one hour of uptake compared to the initial concentration in the external environment.Manganese accumulation by *B. ammoniagenes* is an energy dependent process as it was suppressible by low temperature and inhibitors of energy metabolism. An inhibition of 96 % was observed by addition of 0.5 mM carbonyl cyanide *m*-chlorophenylhydrazone (CCCP) and of 50 % by addition of 1 mM 2,4-dinitrophenol. The inhibition by proton conductors indicates that manganese uptake in *B. ammoniagenes* may be driven by the electrochemical potential of protons as it is assumed for the manganese transport systems of *E. coli* and *L. plantarum* (1). Studies on the pH dependence of the manganese transport in *B. ammoniagenes* at 27°C (temperature optimum) revealed an unexpected high uptake activity between pH 8 and 10. Upon filtration of samples in controls without cells, no precipitation of the isotope was observed under these alkaline conditions within the duration of the uptake assays. As growth experiments confirmed that *B. ammoniagenes* is at least an alkalitolerant, if not an alkalophile bacterium we conclude that the high metabolic activity of the cells at alkaline pH enables the high uptake of manganese under these conditions. Accumulation of manganese obeys saturation kinetics. The kinetic data clearly characterize the manganese transport system of *B. ammoniagenes* as a high-affinity uptake (Table 1.).

The internal radioactivity of *B. ammoniagenes* cells preloaded with ^{54}Mn was exchangeable by addition of 10 uM nonradioactive manganese. Energy dependence of this efflux is assumed as the exchange was distinctly reduced by addition of the proton-conducting ionophore CCCP (0.5 mM). Rapid release of ^{54}Mn from preloaded *B. ammoniagenes* cells upon treatment with 1 % toluene is another argument that the bulk of the recently translocated manganese is not tightly bound intracellularily. What remains to be studied is the possible sequestration of manganese in low-molecular weight compounds, either precipitates of anorganic salts or metal binding peptides and/or small proteins.

The high-affinity manganese transport system of *B. ammoniagenes* is also highly specific for this cation, as it was not inhibited even by a 20 000-fold excess of Mg^{2+} or Ca^{2+}. When added at 1 or 2 µM Co^{2+}, Fe^{2+}, Zn^{2+}, and Cd^{2+}, but Ni^{2+} or Cu^{2+} inhibited uptake. As inhibition by Zn^{2+} and Cd^{2+} was

TABLE 1 MICROBIAL HIGH-AFFINITY MANGANESE TRANSPORT SYSTEMS

Organism		K_m (μM)	V_{max} ($\mu mol.min^{-1}.g^{-1}d.w.$)	inhibitors	K_i (μM)
Escherichia coli	(1970)	0.2	0.02	Co^{2+}, Fe^{2+}	20, 50
Bacillus subtilis	(1973)	1.0	0.01		
Rhodobacter capsulatus	(1978)	0.5	0.02	Co^{2+}, Fe^{2+}	100, 100
Staphylococcus aureus	(1982)	2.0	0.06	Cd^{2+}	1
Lactobacillus plantarum	(1984)	0.2	10.0	Cd^{2+}	0.9
Brevibacterium ammoniagenes	(1987)	1.0	0.1	Cd^{2+}, Zn^{2+}	0.2, 1.2
Candida utilis	(1986)	0.016	16.4	Zn^{2+}	8.0
Aspergillus niger	(1987)	3.0	0.01	Cu^{2+}, Cd^{2+}, Zn^{2+}	?

[a] cf references 1, 20, 23, and 24.

most severe it was studied in further detail and found to be competitive. For prokaryots a competitive inhibition of manganese uptake by zinc has not been reported previously. As no bacterial zinc transport system has been described so far we tried to detect active Zn^{2+} uptake via the manganese transport system of B. ammoniagenes. However,in uptake assays with ^{65}Zn energy dependent accumulation of Zn^{2+} could not be detected.

Further progress in the elucidation of the manganese transport system of B. ammoniagenes is expected from the study of mutants. Two types of mutants which were obtained after mutagenesis with nitrosoguanidine by selecting for resistance against 10 uM Cd^{2+} or 1mM Mn^{2+} on solid mineral medium are at present under investigation.

Screening for Manganese-Carriers in B. ammoniagenes

Energy-linked transporters which exhibit both high affinity and high specificity are generally detected only in cells grown at very low metal concentrations which presumably induce such pathways. The manganese deficient cells of B. ammoniagenes which are used for the production of nucleotides would be an ideal source of derepressed transporters, if such a regulation would exist for membrane bound manganese carriers in this organism. To follow this approach membranes were prepared by sucrose gradient centrifugation from cells grown in the standard fermentation medium (21). The changes in the protein composition of the membrane induced by manganese deficiency were visualized in 12 % gels according to Ames (22). Among the 14 out of 66 detectable polypeptide bands which changed in intensity when manganese was absent, only one was reduced and 13 were enhanced (Fig.2).

We assume that the bulk of these manganese regulated membrane proteins is not involved in manganese transport, but is related to inhibition of cell division by manganese deficiency. Upon malfunctioning of the manganese dependent ribonucleotide reductase the delivery of DNA precursors is curtailed which leads to inhibition of DNA formation and cell division. To prove this idea membranes were prepared from manganese sufficient cells after inhibition of the ribonucleotide reduction by addition of the specific inhibitor HU (20 mM). Indeed, these membranes resembled those from manganese deficient cells with regard to their protein composition. Roughly the same changes were also induced by inhibition of DNA replication with mitomycin C (30 µM) in the presence of manganese (21). By comparison of the depletion-like membrane patterns from the inhibitor-treated manganese-sufficient cells with thoses from the manganese deficient cells three polypeptides could be identified which seemed to be exclusively derepressed by manganese depletion. In our opinion these proteins are very promising candidates in the search for carriers involved in the high-affinity manganese transport of B. ammoniagenes.

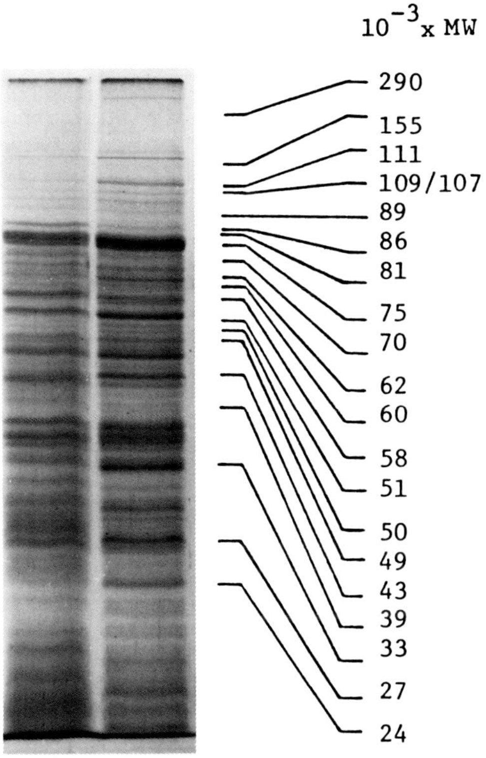

Fig. 2 Sodium dodecylsulfate polyacrylamide gel electrophoresis of membranes prepared from B. ammoniagenes after 48 h of growth in standard fermentation medium (21) in the absence (right) or presence of 450 µg Mn^{2+}/l (left).

High-Affinity Manganese Uptake by A. niger

Although a strict requirement for manganese has been reported as early as 1966 for filamentous fungi like Chaetomium globosum (3) and although it is well known since many years that the large biotechnological process of citric acid production with *A. niger* proceeds only in the absence of manganese (15) no research has been directed to the question how this essential trace element is taken up by these fungi. Obviously due to the difficulties encountered in transport assays with the mycelial growth, reports on uptake of divalent cations by filamentous fungi are rare covering preferentially the transport of iron or zinc. Among the energy dependent manganese transport systems described for yeasts only that of *Candida utilis* (23) displays high-affinity characteristics. However, as manganese uptake in the well-studied yeast *Saccharomyces cerevisiae* has been assayed only at the millimolar concentration range so far future research might detect in this yeast a high-affinity manganese transport system as well.

We were able to demonstrate a high-affinity uptake of manganese by *A. niger* with young mycelia which were only moderately branched and could be reproducibly handled in transport assays (24). As manganese deficiency is already fully expressed 24 hours after germination of inoculum spores, recognizable from the observation of inhibited DNA formation (13), reasonable uptake rates were measured with mycelia growing in citric acid production medium (Table 1). The kinetic parameters obtained are at present reevaluated because transport assays with the same mycelia in "Good"-buffers revealed a pH-optimum of 5.6. The suppressibility of manganese uptake by low temperature, anaerobiosis, and energy poisoning with inhibitors or by omission of the sugar substrate indicate energy dependence of this transport. With regard to is cation specificity the manganese uptake by *A. niger* reminds of the highly specific bacterial manganese transport systems (Table 1) because manganese uptake was unaffected by a 10^5 molar excess of Mg^{2+}or Ca^{2+}. The distinct inhibition of manganese uptake by A. niger upon addition of 10 μM Zn^{2+}, Cu^{2+}, and Cd^{2+} is currently investigated in further detail. The mechanism emerging seems to be a competitive one with K_i-values in the range of the K_m for Mn^{2+}.

To complete our knowledge of the manganese uptake by *A. niger*, the first high-affinity manganese transport system decribed so far for a filamentous fungus, further research has to be directed to the following questions:

1. How is manganese distributed in the different compartments of this eukaryot?
2. Is there an energy dependent efflux of different specificity for Cd^{2+} versus Mn^{2+}, Cu^{2+}, and Zn^{2+}?
3. Can manganese limitation induce derepression of membrane proteins as observed in *B. ammoniagenes*?

ACKNOWLEDGMENTS

The author appreciates the engagement of his coworkers Jan Schmid and Stefan Hockertz in the study of bacterial and fungal manganese transport and thanks W. Ahrens, H. Diekmann and H. Follmann for encouragement and continuing interest.

REFERENCES

1. Rosen BP, Silver S. (1987) "Ion transport in prokaryots" New York: Academic Press.
2. Oka T, Udagawa T, Kinoshita S (1968) Unbalanced growth and death due to depletion of Mn^{2+} in Brevibacterium ammoniagenes. J Bacteriol 96:1760.
3. Barnett HL, Lilly VG (1966) Manganese requirements and deficiency symptoms of some fungi. Mycologia 58:585.
4. Ogata K, Kinoshita S, Tsunoda T, Aida K (1976) "Microbial production of nucleic acid-related substances" Tokyo: Kodansha, New York: Wiley.
5. Milsom PE, Meers JL (1985) Citric Acid. In Moo-Young M (ed):"Comprehensive Biotechnology", Oxford: Pergamon Press, Vol. 3, p 665.
6. Auling G, Thaler M, Diekmann H (1980) Parameters of unbalanced growth and reversible inhibition of deoxyribonucleic acid synthesis in Brevibacterium ammoniagenes ATCC 6872 induced by depletion of Mn^{2+} . Inhibitor studies on on the reversibility of deoxyribonucleic acid synthesis. Arch Microbiol 127: 105.
7. Schimpff-Weiland G, Follmann H, Auling G (1981) A manganese-activated ribonucleotide reductase found in gram-positive bacteria. Biochem Biophys Res Commun 102: 1276.
8. Auling G (1983) The effect of manganese limitation on DNA precursor biosynthesis during nucleotide fermentation with Brevibacterium ammoniagenes and Micrococcus luteus. Eur J Appl Microbiol Biotechnol 18: 229.
9. Plönzig J, Auling G (1986) Manganese deficiency impairs ribonucleotidereduction but not DNA replication in Arthrobacter species. Arch Microbiol 146: 396.
10. Willing A, Follmann H, Auling G (1988) Ribonucleotide reductase of Brevibacterium ammoniagenes is a manganese enzyme. Eur J Biochem 170: 603.
11. Shu P, Johnson MJ (1948) Citric acid production by submerged fermentation with Aspergillus niger. Ind Engineering Chem Int Ed 40: 1202.

12. Schmidt G, Thannhauser SJ (1945) A method for the determination of deoxyribonucleic acid, ribonucleic acid and phosphoproteins in animal tissues. J Biol Chem 161: 83.
13. Hockertz S, Plönzig J, Auling G (1987) Impairment of DNA formation is an early event in Aspergillus niger under manganese starvation. Appl Microbiol Biotechnol 25: 590.
14. Ma H, Kubicek CP, Röhr M (1985) Metabolic effects of manganese deficiency in Aspergillus niger: evidence for increased protein degradation. Arch Microbiol 141: 266.
15. Kubicek CP, Röhr M (1986) Citric acid fermentation. CRC Critical Rev Biotechnol 3: 331.
16. Tabor CW, Tabor H (1985) Polyamines in microorganisms. Microbiol Rev 49: 81.
17. Busse J, Auling G (1988) Polyamine pattern as a chemotaxonomic marker within the Proteobacteria. Syst Appl Microbiol 10: (in press).
18. Lammers M, Follmann H (1983) The ribonucleotide reductases A unique group of metalloenzymes essential for cell proliferation. Struct Bonding 54: 27.
19. Stutzenberger F (1974) Ribonucleotide reductase of Pithomyces chartarum: requirement for B_{12} coenzyme. J Gen Microbiol 81: 501.
20. Schmid J, Auling G (1987) Manganese transport in Brevibacterium ammoniagenes ATCC 6872. J Bacteriol 169: 3385.
21. Schmid J (1985) Manganmangel-induzierte Veränderungen der Cytoplasmamembran und Mangantransport von Brevibacterium ammoniagenes. Ph. D. Thesis University of Hannover, FRG.
22. Ames GFL (1974) Resolution of bacterial proteins by polyacrylamide gel electrophoresis on slabs. J Biol Chem 249: 634.
23. Parkin MJ, Ross IS (1986) The specific uptake of manganese in the yeast Candida utilis. J Gen Microbiol 132: 2155.
24. Hockertz S, Schmid J, Auling G (1987) A specific transport system for manganese in the filamentous fungus Aspergillus niger. J Gen Microbiol 133: 3513.

Metal Ion Homeostasis: Molecular Biology and Chemistry, pages 469–477

COPPER TRANSPORT IN *ESCHERICHIA COLI*[1]

D. Rouch, J. Camakaris, and B.T.O. Lee

Department of Genetics, The University of Melbourne,
Parkville, Victoria 3052, Australia

ABSTRACT An experimental analysis of copper transport in *Escherichia coli* together with a consideration of copper as both a toxic agent and a biological requirement have been used to construct a model for copper metabolism in *E. coli*. The model consists of 5 major components (1) Copper uptake, (2) Cytoplasmic transport, storage and utilization of copper, (3) Copper efflux, (4) Cellular damage by copper and its limitations, and (5) Regulation of internal copper levels. The properties of mutants altered in each component of the model can be predicted. The analysis of the mutants already isolated, namely, copper sensitive and/or copper dependent, indicates six mutant types, which we have named *cut* (copper transport) A-F. These mutants are defective in various copper transport functions including uptake, efflux and storage and have been assigned as follows: *cut*A non-specific copper uptake; *cut*B copper-specific uptake; *cut*C copper efflux; *cut*D copper efflux; *cut*E intracellular copper storage; *cut*F intracellular copper carrier.

INTRODUCTION

In order to study heavy metal transport it is necessary to make a clear distinction between two types of heavy metal effects. (1) Heavy metals that are toxic *per se*, and (2) those heavy metals that have a physiological role but which are toxic at high concentrations. Enterobacterial transport

---

[1]This work was supported by the Australian Research Grants Scheme and funds made available by the Murdoch Institute of the Royal Childrens Hospital, Parkville.

systems for many non-toxic biological cations including potassium (1), sodium (2), magnesium (3)(4), calcium (5)(6), manganese (7) and iron (8) have been relatively well described. In contrast, the transport of copper has been little studied even though the metal has a significant physiological role (9). Copper is toxic at higher than physiological levels so that cells must possess intracellular facilities to control this toxicity while supplying the metal to copper-dependent metalloenzymes. In this paper we present a model of copper utilization in *Escherichia coli* and its use in the isolation of copper mutants in the organism.

RESULTS AND DISCUSSION

The Model of Copper Transport in *E. coli*

The basic feature of this model (Figure 1) is that cells may acquire and utilize copper for functions dependent on the metal without allowing the expression of toxic effects. This can only occur if the movement of copper within a cell is carefully controlled. The model consists of five major elements: (i) uptake, (ii) cytoplasmic transport storage and utilization, (iii) efflux, (iv) cellular damage and its limitation, and (v) regulation. From this cellular model of copper metabolism it is possible to predict the types of copper transport mutants that may be isolated and to predict their properties. The comparison of the actual properties of copper transport mutants with their predicted properties can be used to test the model and to refine it.

Isolation of Copper Transport Mutants and Transductional Analysis

As one could predict from the model, the two major classes of mutants would be copper-sensitive and copper-sensitive/copper-dependent. Both conditional and unconditional copper-sensitive mutants were derived from ED8739 (used as the wild-type) (10) by NTG mutagenesis. Potential temperature sensitive/copper sensitive mutants were those which showed growth on Luria agar plus 4 mM $CuSO_4$ at 30°C but not at 42°C. Of six sensitive mutants (three conditional

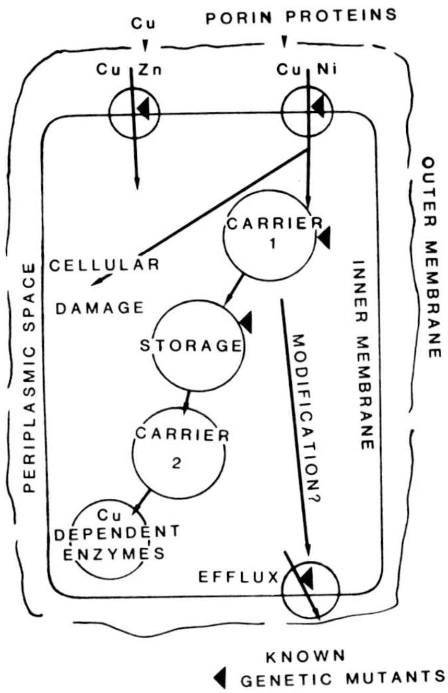

FIGURE 1. Cu utilization in *E. coli*

and three unconditional) chosen for analysis all but one, GME115, produced spontaneous revertants. This indicates that the copper tolerance phenotypes of GME111, GME112, GME135, GME137 and GME138 are a result of point mutations. The mutants were further characterized by qualitative transduction analysis (Table 1).

The genes defined by the mutants are designated *cut*A-F (copper transport). All transduction between different mutants produced wild-type recombinants at similar numbers to positive controls. This indicates that the six mutants carry mutations in different genes which therefore affect different components of cellular copper metabolism.

TABLE 1
QUALITATIVE TRANSDUCTION ANALYSIS OF COPPER SENSITIVE MUTANTS

Recipient (strain)	Donor (phage) cutC	cutB	cutA	cutE	cutF	cutD	ED8739 (wild-type)
GME111 *cutC*	-	+	+	+	+	+	+
GME112 *cutB*		-	+	+	+	+	+
GME115 *cutA*			-	+	+	+	+
GME135 *cutE*				-	+	+	+
GME137 *cutF*					-	+	+
GME138 *cutD*						-	+

\- No wild-type transductants obtained
\+ Wild-type transductants obtained

The Transport Properties of Copper-Sensitive Mutants

The transport activities affected in these mutants was examined using the following parameters (Table 2), copper tolerance, copper accumulation (using ^{64}Cu), kinetic analysis of copper uptake and a qualitative analysis of copper efflux. Valinomycin was used to block copper efflux, so that, the difference between copper accumulation with and without valinomycin is a measure of the efflux capability of the mutant strain.

Copper-Sensitive Mutants and Their Relationships to the Model

Copper uptake. Since copper is a trace requirement for growth, copper uptake functions are essential. Channels and/or transport systems must exist in the membranes of the *E. coli* cell envelope which offer localized low resistance pathways for the passage of physiological ions across the impermeable membrane lipids. Copper cations should preferentially cross the outer membrane *via* the *OmpF*, *OmpC* and protein K porins due to their selectivity for cations (11). The importance of the *OmpF* and *OmpC* porins in copper uptake is demonstrated in *E. coli* K12. Double *OmpC*, *OmpF* mutants

TABLE 2

THE TRANSPORT PROPERTIES OF COPPER SENSITIVE MUTANTS

Strain	Cu tolerance	^{64}Cu accumulation	Effect of valinomycin (% change*)	Uptake Kinetics Apparent Km		Inferred Activity of Efflux	
				Apparent Km	Apparent Vmax	Efflux	Uptake
GME111 cutC	Cu^{s}	+	-3	N	N	-	N
GME112 cutB	Cu^{sDEP}	-	+57	N	+	N	-
GME115 cutA	Cu^{s}	+	+85	N	+	N	+
GME135 cutE	Cu^{sDEP}	N to +	ND	N	N	N	N
GME137 cutF	Cu^{sDEP}	+	ND	N	N	-	N
GME138 cutD	Cu^{s}	+	ND	N	+	-	+
ED8739	wild-type	N	+46	N	N	N	N

+ increased; - decreased; N normal; ND not determined; Cu^{s} copper sensitive; Cu^{sDEP} copper sensitive and copper dependent; * the difference between copper accumulation of ^{64}Cu with and without valinomycin as a measure of efflux capacity.

show reduced copper accumulation. These mutants are not only resistant to copper but dependent on elevated levels of copper for optimal growth (12). In our studies two of the mutants isolated fit the parameters of copper uptake, GME115 (_cutA_) and GME112 (_cutB_).

cutA (GME115) has an increased apparent Vmax for copper uptake thus an increased uptake capability. This strain accumulates copper in the presence of valinomycin (a Cu efflux blocker) so it is not an efflux mutant - it is also zinc tolerant (zinc being an efficient competitive inhibitor of copper uptake), which therefore suggests an alteration in a Cu/Zn system. Reduced copper tolerance presumably occurs due to saturation of intracellular copper handling because of increased accumulation.

cutB (GME112) has an impaired uptake, normal efflux and normal zinc tolerance. It is more tolerant to nickel than ED8739 (which has a similar atomic radius to copper). It is also copper dependent (0.1 mM for growth). Thus we suggest that it is impaired in a second uptake system.

Cytoplasmic copper transport storage and utilization. A premise of the copper transport model is that uncontrolled access to copper by cellular components is prevented by careful control of intracellular copper movement. Storage facilities are necessary to both provide copper for copper-dependent enzymes and to act as a buffer against intracellular changes in levels of the metal. The putative storage mutant _cutE_ (GME135) is both copper sensitive and dependent, and thus has properties consistent with this proposal. Cytoplasmic copper carrier proteins also will ensure that controlled copper transport between other components of the copper utilization system can take place. Two types of carrier are postulated, those which transfer copper from the membrane to the storage protein and those which transfer from the membrane uptake facilities both to storage and to the efflux system. Two mutants, GME138 (_cutD_) and GME137 (_cutF_) are proposed to be of this type.

cutD (GME138) - this mutant is copper sensitive but not copper dependent at 42°C. Growth of this mutant was only possible in media depleted of copper. It does not show significant net efflux at 42°C but does at 30°C, suggesting that it is an efflux-deficient mutant. It shows an increased uptake capability due to increase in Vmax. This suggests that uptake and efflux are linked.

cutE (GME135) - this mutant is temperature sensitive, copper sensitive (max. 0.4 mM Cu++) and copper dependent. It has normal copper uptake and efflux facilities. We therefore propose that it is deficient in intracellular copper storage. A storage deficiency would lead to the existing free copper binding to cellular components and, thus, sensitivity. The dependence would occur because of decreased efficiency of the pathway that donates copper to copper dependent functions.

cutF (GME137) - this mutant also is temperature sensitive/copper sensitive (max. 0.5 mM Cu++). This mutant shows greatly increased copper dependence. It has normal uptake but decreased efflux. Therefore, despite increased copper accumulation it is copper dependent. We therefore postulate a second carrier involved in both transfer of copper to copper dependent enzymes and to the efflux system.

Copper efflux. The transfer of excess intracellular Cu to membrane efflux system for export from the cell would be useful in controlling intracellular Cu levels. The removal of excess Cu from the cell by an energy-dependent efflux system is potentially a more efficient and flexible way of dealing with excess metal than a storage facility alone. The efficiency of transport systems is in part due to the rapid substrate turnover possible, which by definition is not a property of storage proteins.

The inhibition of Cu efflux by the metabolic inhibitor valinomycin demonstrates that *E. coli* does have an energy-dependent efflux system for copper. The regulation of such an efflux system may have an important role in the control of cellular Cu levels; that is, in the maintenance of copper homeostasis.

The copper efflux pathway may contain a modification step. This would involve chemical modification of copper prior to extrusion from the cell so it would no longer be recognised by the copper uptake systems. Preventing re-entry of metal processed by cellular systems would maximise the cell's copper tolerance. Such a modification process then could be inducible; that is, expressed when extracellular copper levels are high. A mutant GME111 (*cutC*) is thought to be of this type.

cutC (GME111) shows normal uptake but increased total accumulation. It has a reduced response to the efflux inhibitor valinomycin. The copper sensitive phenotype is consistent as saturation of the internal copper management systems will result from increased copper levels. This mutant is sensitive to copper and zinc in normal medium, which suggests that the wild-type efflux system exports zinc as well as copper.

Copper damage to the cell and its limitation. The

cytoplasm of the cell is likely to be more susceptible to copper toxicity than the cell envelope, so that the prevention of copper induced damage requires copper management systems to control intracellular copper movements. If copper escapes control, systems may exist which can partially nullify its toxicity. Mutants of this type may have a decreased capacity to scavenge radicals caused by copper toxicity, or since DNA is a target for copper damage they may be of the DNA repair deficient phenotype. We have not isolated any mutants of this type.

Regulation of Cu metabolism. The facilities for the control of intracellular Cu movements may have a limited capacity to maintain this control when intracellular Cu levels are elevated. Also, a minimum intracellular Cu level is necessary to maintain Cu dependent functions. Cells therefore require control of intracellular Cu levels.

If intracellular Cu levels are elevated, resulting from elevated extracellular Cu levels, only the Cu storage facility may be available to bind excess free Cu. Bacterial cells may, however have limited storage capacity to buffer against high external Cu levels since excess storage capacity may tend to strip Cu from Cu metalloenzymes. Thus cells require control of intracellular Cu levels to provide Cu homeostasis. Under homeostatic control internal Cu is maintained at levels sufficient to fulfil metabolic requirements but not sufficient to cause toxicity. A mutant of this class has been identified by its interaction with a copper resistance plasmid regulatory region in the plasmid pRJ1004 (13), but it has not been characterized.

The cellular model of copper transport proposed is consistent with the types of copper transport mutants that have been isolated so far.

REFERENCES

1. Rhoads DB, Laimins LA, Epstein W (1978). Functional organization of the kdp genes of Escherichia coli K12. J Bacteriol 135:445.
2. Borbolla MG, Rosen BP (1984). Energetics of sodium efflux from Escherichia coli. Arch Biochem Biophys 229:98.
3. Nelson DL, Kennedy EP (1972). Transport of magnesium by a repressible and non-repressible system in Escherichia coli. Proc Natl Acad Sci (USA) 69:1091.
4. Park MH, Wong BB, Lusk JE (1976). Mutants in three

genes affecting transport of magnesium in Escherichia coli; genetics and physiology. J Bacteriol 126:1096.

5. Silver S, Toth K, Scribner H (1975). Facilitated transport of calcium by cells and subcellular membranes of Bacillus subtilis and Escherichia coli. J Bacteriol 122:880.
6. Tsujibo H, Rosen BP (1983). Energetics of calcium efflux from cells of Escherichia coli. J Bacteriol 154:854.
7. Silver S, Johnseine P, King K (1970). Manganese active transport in Escherichia coli. J Bacteriol 104:1299.
8. Neilands JB (1982). Microbial envelope proteins selected to iron. Ann Rev Microbiol 36:285.
9. Trevors JT (1987). Copper resistance in bacteria. Microbiological Sciences 4:29.
10. Borck K, Beggs JD, Brammar WJ, Hopkins AS, Murray NE (1976). The construction in vitro of transducing derivatives of phage lambda. Mol Gen Genet 146:199.
11. Benz R, Schmid A, Hancock REW (1985). Ion selectivity of gram negative bacterial porins. J Bacteriol 162:722.
12. Lutkenhaus JF (1977). Role of a major outer membrane protein in Escherichia coli. J Bacteriol 131:631.
13. Rouch D, Camakaris J, Lee BTO, Luke RKJ (1985). Inducible plasmid-mediated copper resistance in Escherichia coli. J Gen Microbiol 131:939.

Metal Ion Homeostasis: Molecular Biology and Chemistry, pages 479–486

STUDIES OF THE HLA CLASS I REGION IN HEREDITARY HEMOCHROMATOSIS

Joanna L. Hansen and James P. Kushner
Department of Medicine, University of Utah
Salt Lake City, Utah 84132

ABSTRACT Iron homeostasis is perturbed in hereditary hemochromatosis, an autosomal recessive trait characterized by excess iron absorption. The gene responsible for this disease maps to the class I region of the HLA locus on the short arm of chromosome six. This region of the HLA locus has been examined in several pedigrees with hemochromatosis by hybridization analysis of Southern blots with three probes originating in the HLA class I region. Polymorphisms were detected for five different restriction endonucleases. The inheritance of these polymorphisms is compared with the genotype assigned by HLA haplotyping and clinical analysis of iron stores. The catalogue of restriction fragment length polymorphisms appears to be as useful as HLA typing for identifying homozygotes within a pedigree. Characterization of polymorphisms in the HLA region should facilitate the isolation of the gene responsible for hereditary hemochromatosis.

INTRODUCTION

Hereditary hemochromatosis is an inherited, autosomal recessive trait characterized by excess intestinal absorption of dietary iron (1). Since there

---

This work was supported by University of Utah BRS Grant #5-34205 and NIH Grant 5R01DK20630 is no normal excretory pathway for iron, the excess iron accumulates in parenchymal tissues. Clinical symptoms due to organ dysfunction typically arise in the fourth or fifth decade. Affected homozygotes can be detected by measuring the percent of transferrin in serum that is saturated with iron (transferrin saturation, TS). Typically, affected male homozygotes have TS > 62% while normal individuals cluster about 25% TS (2). Although the %TS is elevated in affected homozygotes (providing a convenient screening test) the expression and regulation of transferrin receptors appears normal (3).

Because the biochemical abnormality responsible for hemochromatosis is unknown, we have chosen to use molecular genetics to attempt to isolate the responsible gene without any knowledge of its precise function. Previous investigations demonstrated that the gene for hereditary hemochromatosis (h) was closely linked to the class I region of the HLA locus on chromosome six (1). Rare recombination events in pedigrees with hemochromatosis support the contention that the gene maps to the region between HLA-B and HLA-A (see figure 1) (4). As a first step towards isolating the hemochromatosis gene, we are characterizing polymorphisms that reside in this region. Three cloned DNA probes (generously donated by Drs. B. Koller and H. Orr) which map between HLA-B and HLA-A were used to look for restriction fragment length polymorphisms (RFLPs) in pedigrees with hemochromatosis. The inheritance of these RFLPs is compared with the genotypes of pedigree members assigned by HLA haplotyping and assessment of iron stores.

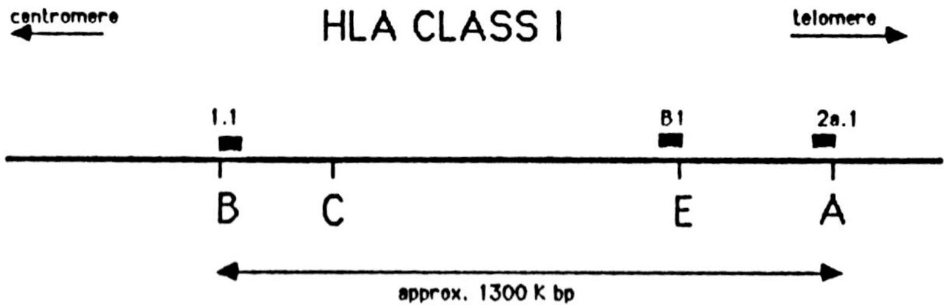

Figure 1. The Class I Region of the HLA Locus on Chromosome six.

The class I genes B, C, E, and A span about 1.3 million bp. The DNA probes used map to the areas indicated above. 1.1 is derived from an HLA-B cDNA while 2a.1 and B1 were isolated from genomic clones. The hemochromatosis gene maps between HLA-B and HLA-A.

METHODS

Genomic DNA was isolated from peripheral blood leukocytes or Epstein Barr virus transformed lymphocytes and digested to completion with the indicated restriction enzymes, using conditions recommended by the supplier (New England Biolabs). Southern blots were prepared by the method of Reed and Mann (7), using Gelman Biotrace membranes. Hybridizations were performed in 50% formamide, 5 X SSC, 10 mM $NaPO_4$, pH 6.7, 10 µg/ml carrier DNA, 10% Dextran sulphate, 1 X Denhardts with 8 X 10^5 cpm/ml of radiolabeled probe (8) at 42 C for 36 hr followed by stringent washes in 0.2 X SSC, 0.1% SDS at 65 C for 1 hr and autoradiography.

RESULTS AND DISCUSSION

The hybridization probes used in this study were isolated from either cDNA or genomic clones (5,6) and mapped to the region between HLA-B and HLA-A by analysis of chromosome deletion mutants (fig. 1). Initially each of these three probes were hybridized with DNA from normal individuals, digested with 20 different restriction enzymes to ascertain RFLPs. Seven polymorphisms were followed: three for probe 2a.1 (Taq I, Msp I and Hind III), one for probe B1 (Pst I) and three for probe 1.1 (Bgl II, Msp I and Taq I). Although most of the polymorphisms consisted of two alleles (individuals were either 1,1 or 1,2 or 2,2, where the numbers indicate the presence of a DNA fragment of specific size), other polymorphisms consisted of up to five alleles (with individuals inheriting any two of those five).

Seven pedigrees with hemochromatosis were chosen to characterize the inheritance of RFLPs in this region of chromosome six. Some of the probands were identified from a large scale population screening study (9) and were detected prior to any organ dysfunction, while others were first seen after the onset of clinical symptoms. Each individual in these pedigrees was assessed for iron stores (% TS and serum ferritin concentration) and HLA haplotype. In addition, some members underwent liver biopsy to assess liver iron stores. Homozygote status was assigned on the basis of repeatedly elevated transferrin saturation (>62%), elevated liver iron stores, clinical symptoms, and/or HLA identity with the proband.

Data for two pedigrees are presented in figures 2 and 3. In figure 2 it is apparent by comparing the list

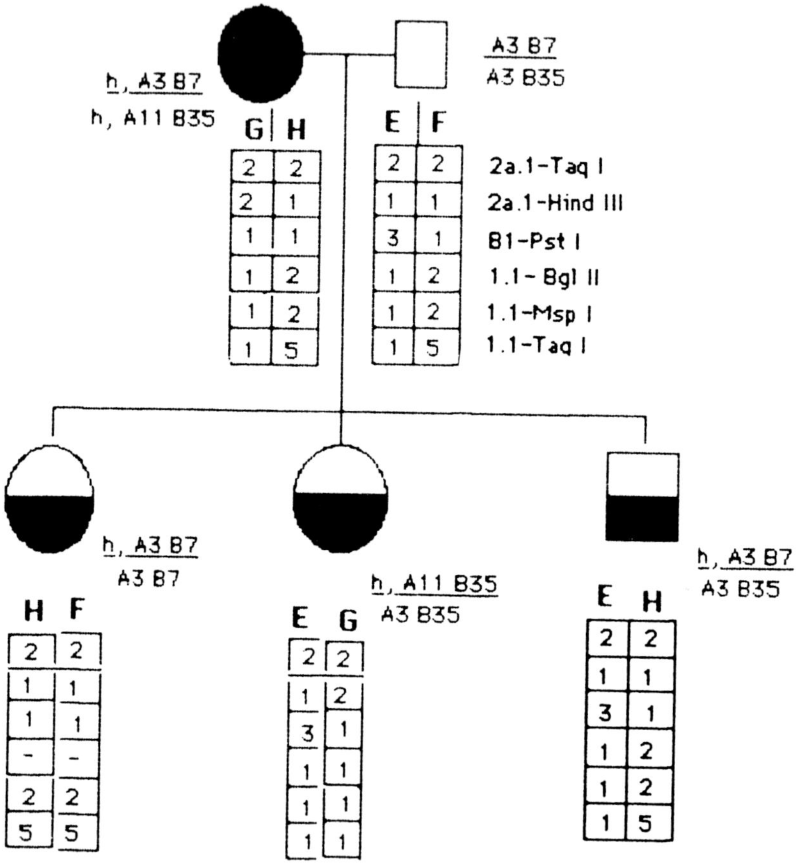

Figure 2. RFLPs and HLA haplotypes of first hemochromatosis pedigree. The presence of the h gene was deduced as defined in the text. RFLP alleles are numbered and listed in the order shown for the probe and restriction enzyme used. Each chromosome six is indicated by a capital letter: H and F are A3B7, G is A11B35, and E is A3B35. □ = nn (normal): ⬓ = hn ■ = hh (homozygous for h gene).

of six polymorphisms that one RFLP segregates with the h gene found on chromosome G- that being allele "2" detected by probe 2a.1 with Hind III digests. This observation holds up only for this pedigree, however, as there was no segregation of 2a.1-Hind III - 2 with h genes in the other 6 pedigrees analyzed (data not shown). In the pedigree shown in fig. 2 both parents carry an A3B7 chromosome. The mother has hemochromatosis and thus her A3B7 chromosome (H) must carry the h gene (or mutation) while the father appears normal by all criteria so his A3B7 chromosome (F) lacks the h gene. Neither the catalogue of polymorphisms nor HLA haplotyping is able to distinguish between these two A3B7 chromosomes.

In figure 3 three generations of a second pedigree are shown. In this pedigree one h gene segregates with the "4" allele detected by probe 1.1 on MspI digests (chromosome I). The other h gene is carried on chromosome K and appears to segregate with the "2" allele detected by probe B1 on Pst I digests. Thus, these alleles could act as DNA markers for the h gene within this pedigree. They are not indicators of the h gene independent of pedigree information, however, as both the 1.1.-MspI-4 and B1-PstI-2 alleles were found in similar frequency in 30 unrelated normal individuals and 30 homozygotes for hemochromatosis (data not shown). One additional point is that the A3B7 chromosome in this pedigree (chromosome L) has the same RFLP catalogue as chromosome H and F in the first pedigree (figure 2).

Analysis of the other five pedigrees (data not shown) reveals similar findings. Within some pedigrees, certain RFLPs do segregate with the h gene and thus serve as linked markers for the gene (or mutation). These same RFLPs are not specific for the h gene however, as they are found at similar frequencies in both unrelated normal individuals and homozygotes for hemochromatosis.

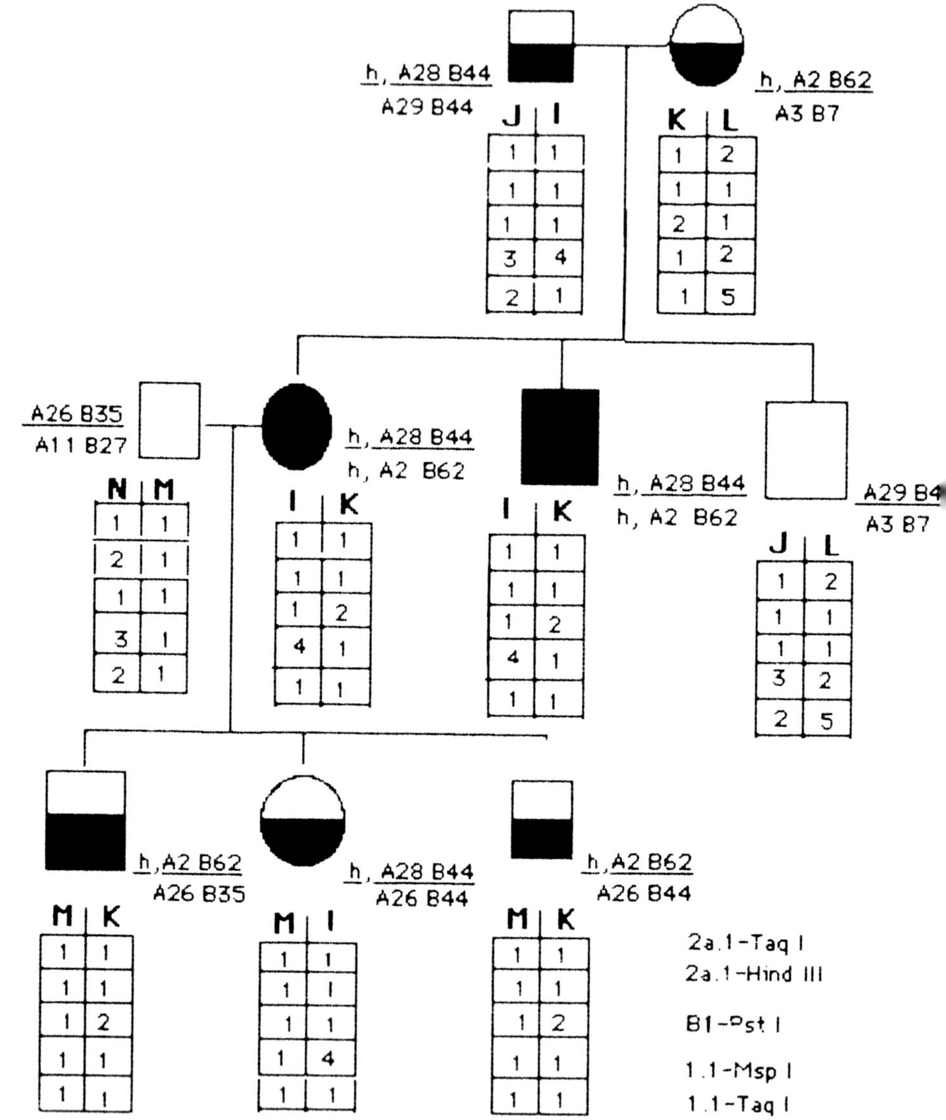

Figure 3. RFLPs and HLA haplotypes of second hemochromatosis pedigree. All parameters are as indicated in the legend to figure 2. Chromosome J is A29B44, I is A28B44, K is A2B62, Lis A3B7, N is A11 B27, and M is A 26B35.

CONCLUSIONS

Seven different RFLPs were followed in seven pedigrees with hemochromatosis. In all instances, the catalogue of RFLPs was as good as HLA haplotyping in identifying homozygotes within a pedigree. Within some pedigrees, one or more RFLPs are linked to the h gene. In the absence of pedigree studies however, these RFLPs are not informative for the presence of the h gene as they were observed in similar frequencies in both normal and affected individuals. Further work with additional hybridization probes will be necessary to identify an RFLP specific for the gene or mutation responsible for hemochromatosis.

ACKNOWLEDGMENTS

We wish to acknowledge the technical assistance of Nancy Law, Kathy Metcalf and Evette Radisky.

REFERENCES

1. Bothwell TH, Charlton RW, Motulsky AG (1983). Idiopathic Hemochromatosis. In Stanbury, JB, Wyngaarden, JB, Fredrickson, DS (eds): "The Metabolic Basis of Inherited Diseases". MacGraw-Hill, New York, p.1269.
2. Dadone MM, Kushner JP, Edwards CQ, Bishop DT, Skolnick MH (1982). Hereditary Hemochromatosis: Analysis of laboratory expression of the disease by genotpye in 18 pedigrees. Am. J. Clin. Path. 78:196.
3. Ward JH, Kushner JP, Ray FA, Kaplan J (1984). Transferrin receptor function in hereditary hemochromatosis. J. Lab. Clin. Med. 103:246.
4. Edwards CQ, Griffen LM, Dadone MM, Skolnick MH, Kushner JP (1986). Mapping the locus for hereditary hemochromatosis: localization between HLA-B and HLA-A. Am J. Hum. Genet. 38:805.
5. Koller BH, Sidwell B., DeMars R, Orr HT (1984). Isolation of HLA locus-specific DNA probes from the 3'-untranslated region. Proc. Natl. Acad. Sci, USA 81:5175.

6. Orr HT, Koller BH, Geraghty D, Shimizu Y, DeMars R (1987). Immunol. Rev. in press.
7. Reed KC, Mann DA (1985). Rapid transfer of DNA from agarose gels to nylon membranes. Nucl. Acids Res. 13:7207.
8. Feinberg AP, Vogelstein B (1983). A technique for radiolabeling DNA restriction endonuclease fragments to high specific activity. Analyt. Biochem. 132:6.
9. Edwards CQ, Griffen LM, Goldgar D, Drummond C, Skolnick MH, Kushner JP (1988). Detection of hemochromatosis in 11,065 healthy blood donors. New Engl. J. Med. in press.

Index